Carbon-Based Nanomaterials 2.0

Carbon-Based Nanomaterials 2.0

Editor

Ana María Díez-Pascual

MDPI • Basel • Beijing • Wuhan • Barcelona • Belgrade • Manchester • Tokyo • Cluj • Tianjin

Editor
Ana María Díez-Pascual
Química Analítica, Química
Física e Ingeniería Química
Universidad de Alcalá
Alcalá de Henares, Madrid
Spain

Editorial Office
MDPI
St. Alban-Anlage 66
4052 Basel, Switzerland

This is a reprint of articles from the Special Issue published online in the open access journal *International Journal of Molecular Sciences* (ISSN 1422-0067) (available at: www.mdpi.com/journal/ijms/special_issues/carbon_nano_2).

For citation purposes, cite each article independently as indicated on the article page online and as indicated below:

LastName, A.A.; LastName, B.B.; LastName, C.C. Article Title. *Journal Name* **Year**, *Volume Number*, Page Range.

ISBN 978-3-0365-6549-1 (Hbk)
ISBN 978-3-0365-6548-4 (PDF)

© 2023 by the authors. Articles in this book are Open Access and distributed under the Creative Commons Attribution (CC BY) license, which allows users to download, copy and build upon published articles, as long as the author and publisher are properly credited, which ensures maximum dissemination and a wider impact of our publications.

The book as a whole is distributed by MDPI under the terms and conditions of the Creative Commons license CC BY-NC-ND.

Contents

About the Editor vii

Preface to "Carbon-Based Nanomaterials 2.0" ix

Ana María Díez-Pascual
Carbon-Based Nanomaterials
Reprinted from: *Int. J. Mol. Sci.* **2021**, *22*, 7726, doi:10.3390/ijms22147726 1

Md. Motiar Rahman, Mst Gulshan Ara, Mohammad Abdul Alim, Md. Sahab Uddin, Agnieszka Najda and Ghadeer M. Albadrani et al.
Mesoporous Carbon: A Versatile Material for Scientific Applications
Reprinted from: *Int. J. Mol. Sci.* **2021**, *22*, 4498, doi:10.3390/ijms22094498 7

Banendu Sunder Dash, Gils Jose, Yu-Jen Lu and Jyh-Ping Chen
Functionalized Reduced Graphene Oxide as a Versatile Tool for Cancer Therapy
Reprinted from: *Int. J. Mol. Sci.* **2021**, *22*, 2989, doi:10.3390/ijms22062989 29

Lucja Dybowska-Sarapuk, Weronika Sosnowicz, Jakub Krzeminski, Anna Grzeczkowicz, Ludomira H. Granicka and Andrzej Kotela et al.
Printed Graphene Layer as a Base for Cell Electrostimulation—Preliminary Results
Reprinted from: *Int. J. Mol. Sci.* **2020**, *21*, 7865, doi:10.3390/ijms21217865 51

Carlos Sainz-Urruela, Soledad Vera-López, María Paz San Andrés and Ana M. Díez-Pascual
Graphene-Based Sensors for the Detection of Bioactive Compounds: A Review
Reprinted from: *Int. J. Mol. Sci.* **2021**, *22*, 3316, doi:10.3390/ijms22073316 65

Anastasios Gotzias, Elena Tocci and Andreas Sapalidis
On the Consistency of the Exfoliation Free Energy of Graphenes by Molecular Simulations
Reprinted from: *Int. J. Mol. Sci.* **2021**, *22*, 8291, doi:10.3390/ijms22158291 107

María Paz San Andrés, Marina Baños-Cabrera, Lucía Gutiérrez-Fernández, Ana María Díez-Pascual and Soledad Vera-López
Fluorescence Study of Riboflavin Interactions with Graphene Dispersed in Bioactive Tannic Acid
Reprinted from: *Int. J. Mol. Sci.* **2021**, *22*, 5270, doi:10.3390/ijms22105270 119

Łukasz Wasyluk, Vitalii Boiko, Marta Markowska, Mariusz Hasiak, Maria Luisa Saladino and Dariusz Hreniak et al.
Graphene Coating Obtained in a Cold-Wall CVD Process on the Co-Cr Alloy (L-605) for Medical Applications
Reprinted from: *Int. J. Mol. Sci.* **2021**, *22*, 2917, doi:10.3390/ijms22062917 139

Mónica Cicuéndez, Laura Casarrubios, Nathalie Barroca, Daniela Silva, María José Feito and Rosalía Diez-Orejas et al.
Benefits in the Macrophage Response Due to Graphene Oxide Reduction by Thermal Treatment
Reprinted from: *Int. J. Mol. Sci.* **2021**, *22*, 6701, doi:10.3390/ijms22136701 161

Ana R. Silva, Ana J. Cavaleiro, O. Salomé G. P. Soares, Cátia S.N. Braga, Andreia F. Salvador and M. Fernando R. Pereira et al.
Detoxification of Ciprofloxacin in an Anaerobic Bioprocess Supplemented with Magnetic Carbon Nanotubes: Contribution of Adsorption and Biodegradation Mechanisms
Reprinted from: *Int. J. Mol. Sci.* **2021**, *22*, 2932, doi:10.3390/ijms22062932 177

Jaroslaw Szczepaniak, Joanna Jagiello, Mateusz Wierzbicki, Dorota Nowak, Anna Sobczyk-Guzenda and Malwina Sosnowska et al.
Reduced Graphene Oxides Modulate the Expression of Cell Receptors and Voltage-Dependent Ion Channel Genes of Glioblastoma Multiforme
Reprinted from: *Int. J. Mol. Sci.* **2021**, *22*, 515, doi:10.3390/ijms22020515 . . . . **197**

Sebastian Muraru and Mariana Ionita
Towards Performant Design of Carbon-Based Nanomotors for Hydrogen Separation through Molecular Dynamics Simulations
Reprinted from: *Int. J. Mol. Sci.* **2020**, *21*, 9588, doi:10.3390/ijms21249588 . . . . **215**

Dmytro Nozdrenko, Olga Abramchuk, Svitlana Prylutska, Oksana Vygovska, Vasil Soroca and Kateryna Bogutska et al.
Analysis of Biomechanical Parameters of Muscle Soleus Contraction and Blood Biochemical Parameters in Rat with Chronic Glyphosate Intoxication and Therapeutic Use of C_{60} Fullerene
Reprinted from: *Int. J. Mol. Sci.* **2021**, *22*, 4977, doi:10.3390/ijms22094977 . . . . **227**

About the Editor

Ana María Díez-Pascual

Ana María Díez-Pascual graduated in Chemistry in 2001 from Complutense University (Madrid, Spain), where she also carried out her Ph.D. studies (2002–2005) . In 2005, she worked at the Max Planck Institute of Colloids and Interfaces (Germany). From 2006 to 2008, she was a Postdoctoral Researcher at the Physical Chemistry Institute of the RWTH Aachen University (Germany). Then she moved to the Institute of Polymer Science and Technology (Madrid, Spain) . Currently, she is a Permanent Professor at Alcala University (Madrid), where she focuses on the development of polymer/nanofiller systems for biomedical applications. She has participated in 25 research projects (15 international and 10 national, being principal investigator in 6 of them). She has published 112 SCI articles and has an h-index of 41 and more than 3500 total citations. She has published 22 book chapters and 2 monographs; she is the first author of an international patent. She has contributed to 65 international conferences. She was awarded the TR35 2012 Prize by the Massachusetts Institute of Technology (MIT) for her innovative work in the field of nanotechnology.

Preface to "Carbon-Based Nanomaterials 2.0"

This reprint, a collection of 9 original contributions and 3 reviews, provides a selection of the most recent advances in the preparation, characterization, and applications of carbon-based nanomaterials, such as carbon nanotubes, fullerenes, and graphene and its derivatives. The research on this type of nanomaterials has experienced a sharp exponential growth over the last years. The infinite possibilities to modify and tailor carbon nanomaterials, is associated with their small size, approaching the size of many fundamental biomolecules. Their large specific surface area, high electrical and thermal conductivity, unique optical properties, and superior mechanical properties, have paved the way for a broad range of applications ranging from flexible electronics to biomedicine, bioimaging, and sensing. This reprint gathers contributions from renowned researchers in the field, making it a reference for the scientific community working on the fundamental and applied research of carbon nanomaterials.

Ana María Díez-Pascual
Editor

Editorial

Carbon-Based Nanomaterials

Ana María Díez-Pascual

Universidad de Alcalá, Facultad de Ciencias, Departamento de Química Analítica, Química Física e Ingeniería Química, Ctra. Madrid-Barcelona, Km. 33.6, 28805 Alcalá de Henares, Madrid, España (Spain); am.diez@uah.es; Tel.: +34-918-856-430

Citation: Díez-Pascual, A.M. Carbon-Based Nanomaterials. *Int. J. Mol. Sci.* 2021, 22, 7726. https://doi.org/10.3390/ijms22147726

Received: 6 July 2021
Accepted: 7 July 2021
Published: 20 July 2021

Publisher's Note: MDPI stays neutral with regard to jurisdictional claims in published maps and institutional affiliations.

Copyright: © 2021 by the author. Licensee MDPI, Basel, Switzerland. This article is an open access article distributed under the terms and conditions of the Creative Commons Attribution (CC BY) license (https://creativecommons.org/licenses/by/4.0/).

Research on carbon-based nanomaterials, such as carbon nanotubes, graphene and its derivatives, nanodiamonds, fullerenes, and other nanosized carbon allotropes, has experienced sharp exponential growth over recent years. The infinite possibilities to modify and tailor carbon nanomaterials are associated with their small size, approaching the size of many fundamental biomolecules, their large specific surface area, high electrical and thermal conductivity, unique optical properties, and superior mechanical properties, which have paved the way for a broad range of applications. In particular, fullerene derivatives have been applied to solar energy scavenging, graphene has been widely used in flexible electronics, carbon nanotubes have been tailored to have molecular recognition capability, graphene quantum dots have been extensively used for bio-imaging and sensing owing to their photoluminescence properties, and nanodiamonds have been demonstrated to be useful in super-resolution imaging and nanoscale temperature sensing.

This Special Issue "Carbon-Based Nanomaterials" (https://www.mdpi.com/journal/ijms/special_issues/carbon-ijms (accessed on 30 September 2020)) and Special Issue "Carbon-Based Nanomaterials 2.0" (https://www.mdpi.com/journal/ijms/special_issues/carbon_nano_2 (accessed on 30 September 2020)) with a collection of 13 original contributions and 5 literature reviews, provides selected examples on surface modifications of carbon nanomaterials to tailor their physicochemical properties as well as their applications in a variety of fields, such as electronics, energy storage, biomedicine, and sensing.

The last two decades have witnessed a lot of research addressing the possible biomedical applications of carbon nanotubes (CNTs) such as drug delivery, tissue engineering, diagnostics, and biosensing [1]. In particular, CNTs can act as contrast agents in different imaging methods [2]. Upon functionalization and conjugation with various biomarkers, they can indicate the presence and localization of targeted cells with good spatial resolution. CNTs are suitable candidates to be employed in solving the theragnostic challenges associated with neurological diseases such as ischemic stroke. In this regard, Komane et al. [3] recently synthesized vertically aligned multiwalled carbon nanotubes (VA-MWCNTs), which were purified, carboxylated, acylated, and PEGylated, for use in targeting studies. The functionalized VA-MWCNTs were found to be nontoxic towards PC-12 neuronal cells, a type of catecholamine cell that synthesizes, stores, and releases norepinephrine and dopamine.

Nerve regeneration via cell electrostimulation will turn out to be essential in regenerative medicine, particularly in body reconstruction, artificial organs, or nerve prostheses. Indirect electrical cell stimulation needs a nontoxic, highly electrically conductive substrate material allowing for an accurate and effective cell electrostimulation. This can be effectively achieved using graphene nanoplatelets (GNPs); however, their strong agglomerating tendency hinders the quality of the manufactured coatings. Therefore, the choice of an appropriate amount of surfactant to achieve both a high conductivity and quality of the coating is crucial. In this regard, Dybowska-Sarapuk et al. [4] developed graphene inks with different surfactant contents in the range of 2–20 wt% and demonstrated an observable effect of electrostimulation on the behavior of the neuronal stem cells embedded in the graphene layer. The use of cellular electrostimulation may provide a solution to the currently irredeemable neurological disorders, e.g., possibilities for the restoration of the spinal cord and nerve connections.

On the other hand, biocompatible and water-soluble C_{60} fullerenes can inactivate free radicals, including methyl radicals, superoxide anion radicals, and hydroxyl radicals, protecting cell membranes from oxidation. They are powerful scavengers of free radicals during the development of ischemia and fatigue processes in skeletal muscle [5]. The usage of safe doses of C_{60} fullerene at the initiation of various pathologies leads to significant positive therapeutic effects, in particular, during acute liver injury, colorectal cancer, obesity, acute cholangitis, and hemiparkinsonism and can be effective nanotherapeutics in the treatment of certain herbicide poisoning such as glyphosate [6].

Another interesting biomedical application of carbon nanomaterials is in vitro bio-imaging. In this regard, Parasuraman et al. [7] synthesized "dahlia-like" hydrophilic fluorescent carbon nanohorns (CNH) via a simple hydrothermal chemical oxidation method from Nafion-encapsulated carbon nanorice particles at a mild temperature of 100 °C. The CNHs obtained could be used as bio-imaging probes because of the presence of structural defects such as 5-hydroxymethylfurfural or other aromatic moieties generated during the carbon nanorice oxidation. The synthesis method developed in this study will pave a new way for the application of CNHs as bio-imaging agents and drug carriers.

Graphene and its derivatives are very promising nanomaterials for the preparation of scaffolds for tissue repair. The response of immune cells to these graphene-based materials appears to be critical in promoting regeneration; thus, the study of this response is essential before they are used to prepare any type of scaffold. Another relevant factor is the variability of graphene-based materials, including the surface chemistry, namely, the type and quantity of oxygen functional groups. Thus, Cicuéndez et al. [8] investigated the response of RAW-264.7 macrophages, monocyte-like cells, originating from an Abelson leukemia virus-transformed cell line derived from BALB/c mice, to graphene oxide (GO) and two types of reduced GO, rGO15 and rGO30, obtained after vacuum-assisted thermal treatment of 15 and 30 min, respectively. The results demonstrate that the GO reduction led to a decrease in both oxidative stress and proinflammatory cytokine secretion, considerably enhancing its biocompatibility and potential for the preparation of novel 3D scaffolds able to trigger the immune response for tissue regeneration.

Another approach is to take advantage of the chemically inert properties of graphene for coating medical devices. For effective coating, it is necessary to prepare the surface of the substrate to be coated and to optimize the chemical vapor deposition process, allowing better attachment, e.g., by using the defective interface side of the graphene layer for bonding. In this regard, Wasyluk et al. [9] synthesized a graphene coating on a Co-Cr alloy by a cold wall chemical vapor deposition (CW-CVD) method, with good mechanical properties (namely, hardness and elastic modulus). The results of the hemocompatibility test indicate that the developed coating does not have a pro-coagulant effect, thus corroborating its potential for medical applications, particularly in the field of cardiovascular diseases.

Graphene and its derivatives can also be used in anticancer therapy. For instance, they exhibit antitumor effects on glioblastoma multiforme cells in vitro. In this regard, the antitumor activity of rGO with different contents of oxygen-containing functional groups and graphene has been compared [10]. Cell membrane damage, changes in the cell membrane potential, the gene expression of voltage-dependent ion channels, and extracellular receptors were analyzed. A reduction in the potential of the U87 glioma cell membrane was observed after treatment with rGO/ammonium thiosulphate and rGO/thiourea dioxide flakes. Treatment with graphene or rGO led to reduced endoglin expression, stimulated cell adhesion, and, hence, reduced the ability of cancer cell migration.

rGO has also emerged as a good candidate for cancer photothermal therapy due to its huge specific surface area for drug loading, high biocompatibility, targeted delivery, outstanding photothermal conversion in the near-infrared range, and functional groups for functionalization with molecules such as photosensitizers, siRNA, and ligands [11]. Multifunctional rGO-based nanosystems can be designed, which possess promising temperature/pH-dependent drug/gene delivery abilities for multimodal cancer therapy.

On the other hand, pharmaceuticals are recently being considered emergent micropollutants due to their potential environmental, ecotoxicological, and sociological risks. The anaerobic removal of pharmaceuticals assisted by nanomaterials constitutes a promising strategy that is worthy of investigation. In this regard, Silva et al. [12] evaluated the potential of commercial and custom-made CNTs as redox mediators in the anaerobic removal of ciprofloxacin, a fluoroquinolone antibiotic widely used for the treatment of bacterial infections in humans and animals. Functionalized CNTs with different surface chemical groups (acidic and basic), and magnetic CNTs incorporating iron were prepared and used in biological experiments, since they are easier to recover and may be reused, which is important for applied biological treatment processes. The potential contribution of adsorption and biodegradation processes was assessed. The amount of carbon nanomaterial needed was found to be very low, only 0.1 g L^{-1}, which minimizes the quantity of the carbon nanomaterial that can be released to the medium.

Clean energy technologies constitute another hot topic of research in the field of carbon nanomaterials. Hydrogen fuel, a valuable alternative to fossil fuels, shows drawbacks such as the poisoning by impurities of the metal catalyst controlling the reaction involved in its production. Thus, separating H_2 from other gases such as CO, CO_2, CH_4, N_2, and H_2O is essential. Muraru and Ionita [13] presented a rotating partially double-walled CNT membrane designed for hydrogen separation and assessed its performance using molecular dynamics simulations by imposing three discrete angular velocities. They demonstrated that the angular velocity plays a noteworthy role in hydrogen separation through a rotating nanotube.

Owing to both the extremely high in-plane/shell strength and modulus and extremely low inter-plane/shell friction, CNTs and graphene are widespread in developing nanodevices, e.g., oscillators, nanomotors, nanobearing, and nanoresonators. Generally, MWCNTs are used in experiments or simulations for investigating the inter-tube friction. In particular, it was found that an abrupt jump in the output torque moment from a rotation transmission nanosystem made from CNTs occurred when reducing the system temperature. When the rotor was subjected to an external resistant torque moment, it could not rotate opposite to the motor, even if it deformed severely. Combining molecular dynamics simulations with the bi-sectioning algorithm, the critical value of the resistant torque moment was determined [14].

Graphene and its derivatives are also promising materials for the development of a new generation of gas separation membranes, mainly because of their thinness, excellent membrane formation capabilities, permeability, and selectivity. In particular, GO-based membranes can allow extremely high fluxes because of their fineness and layered structure. In addition, their high selectivity is due to the molecular sieving or diffusion effect resulting from their narrow pore size distribution and unique surface chemistry. In this regard, the different mechanisms of gas transport and gas separation by GO membranes, as well as the methods for GO membrane preparation and characterization, have been reviewed [15].

Fullerenols, nanosized water-soluble polyhydroxylated derivatives of fullerenes, are bioactive compounds that can be used for drug development. Kovel et al. [16] investigated the antioxidant activity and toxicity of a series of fullerenols with a different number of oxygen substituents. All fullerenols caused toxic effects at high concentrations (>0.01 g L^{-1}), whereas their antioxidant activity was demonstrated at low and ultralow concentrations (<0.001 g L^{-1}). The toxic and antioxidant characteristics of the fullerenols were found to be dependent on the number of oxygen substituents: the fewer the oxygen substituents, the lower the toxicity, and the better the antioxidant activity. The differences in fullerenol properties were ascribed to their catalytic activity due to the reversible electron acceptance, radical trapping, and balance of reactive oxygen species in aqueous solutions. The results provide the basis for the selection of carbon nanomaterials with appropriate toxic and antioxidant characteristics.

On the other hand, the use of carbon nanomaterials as antibacterial agents has been widely investigated. It constitutes a very promising way to fight microorganisms due to their high specific surface area and intrinsic or chemically incorporated antibacterial action. In particular, graphene and its derivatives, GO and rGO, are highly suitable candidates for restricting microbial infections. However, the mechanisms of antimicrobial action, their cytotoxicity, and other issues remain unclear. Selected examples on the preparation of antimicrobial nanocomposites incorporating inorganic nanoparticles and graphene or its derivatives have been provided [17]. The antibacterial property of composites of rGO with Ag and ZnO nanoparticles synthesized using a microwave-assisted approach has been investigated in detail [18]. A strong antibacterial activity against *E. coli* and *S. aureus* was found since they decreased the reductase activity and affected the membrane integrity in the bacteria. These nanocomposites can be applied not only as antibacterial agents but also in a variety of biomedical materials such as sensors, photothermal therapy, drug delivery, and catalysis.

Over recent years, diverse nanomaterials have been investigated to design highly selective and sensitive sensors, reaching nano/picomolar concentrations of biomolecules, which is crucial for medical sciences and the healthcare industry in order to assess physiological and metabolic parameters. In particular, G and its derivatives are becoming crucial in the field of optical and electrochemical sensors. G-based nanomaterials can be combined with inorganic nanoparticles, including metals and metal oxides, quantum dots, organic polymers, and biomolecules, to yield a wide range of nanocomposites with enhanced sensitivity for sensor applications. Recent research on G-based nanocomposites for the detection of bioactive compounds, including melatonin, gallic acid, tannic acid, resveratrol, oleuropein, hydroxytyrosol, tocopherol, ascorbic acid, and curcumin, has been reviewed [19], and the sensitivity and selectivity of G-based electrochemical and fluorescent sensors have been analyzed. Some of these bioactive compounds, such as tannic acid, can be used as effective dispersing agents for graphene in aqueous solutions [20], and tannic acid's interaction with a water-soluble vitamin, riboflavin, has been studied under different experimental conditions. Tannic acid induces quenching of riboflavin fluorescence, and the effect is stronger with an increasing tannic acid concentration, due to π–π interactions through the aromatic rings, and hydrogen bonding interactions between the hydroxyl moieties of both compounds.

Finally, the recent advances in the synthesis of mesoporous carbon, with a high surface area, large pore volume, and good thermostability, make it a promising material with a wide range of applications. It can act as a catalytic support, can be used in energy storage devices, controls the body's oral drug delivery system, and adsorbs poisonous metals from water and various other molecules from aqueous solutions. The potential applications of mesoporous carbon in many scientific disciplines have been reviewed [21]. Particular emphasis has been placed on some areas, especially surface modification, oxidation, and functionalization time for certain applications. The role of the reaction pH, reaction time, time of carbon surface oxidation, functionalization parameters, solvent uses during adsorption, and optimal time in adsorption has been discussed. Moreover, the outlook for further improvement in mesoporous carbon has been demonstrated.

Conflicts of Interest: The author declares no conflict of interest.

References

1. Simon, J.; Flahaut, E.; Golzio, M. Overview of Carbon Nanotubes for Biomedical Applications. *Materials* **2019**, *12*, 624. [CrossRef] [PubMed]
2. Gong, H.; Peng, R.; Liu, Z. Carbon Nanotubes for Biomedical Imaging: The Recent Advances. *Adv. Drug Deliv. Rev.* **2013**, *65*, 1951–1963. [CrossRef] [PubMed]
3. Komane, P.P.; Kumar, P.; Choonara, Y.E.; Pillay, V. Functionalized, Vertically Super-Aligned Multiwalled Carbon Nanotubes for Potential Biomedical Applications. *Int. J. Mol. Sci.* **2020**, *21*, 2276. [CrossRef]
4. Dybowska-Sarapuk, L.; Sosnowicz, W.; Krzeminski, J.; Grzeczkowicz, A.; Granicka, L.H.; Kotela, A.; Jakubowska, M. Printed Graphene Layer as a Base for Cell Electrostimulation—Preliminary Results. *Int. J. Mol. Sci.* **2020**, *21*, 7865. [CrossRef] [PubMed]

5. Nozdrenko, D.M.; Zavodovsky, D.O.; Matvienko, T.Y.; Zay, S.Y.; Bogutska, K.I.; Prylutskyy, Y.I.; Ritter, U.; Scharff, P. C_{60} fullerene as promising therapeutic agent for the prevention and correction of functioning skeletal muscle at ischemic injury. *Nanoscale Res. Lett.* **2017**, *12*, 1–9. [CrossRef]
6. Nozdrenko, D.; Abramchuk, O.; Prylutska, S.; Vygovska, O.; Soroca, V.; Bogutska, K.; Khrapatyi, S.; Prylutskyy, Y.; Scharff, P.; Ritter, U. Analysis of Biomechanical Parameters of Muscle Soleus Contraction and Blood Biochemical Parameters in Rat with Chronic Glyphosate Intoxication and Therapeutic Use of C60 Fullerene. *Int. J. Mol. Sci.* **2021**, *22*, 4977. [CrossRef]
7. Parasuraman, P.S.; Parasuraman, V.R.; Anbazhagan, R.; Tsai, H.-C.; Lai, J.-Y. Synthesis of "Dahlia-Like" Hydrophilic Fluorescent Carbon Nanohorn as a Bio-Imaging PROBE. *Int. J. Mol. Sci.* **2019**, *20*, 2977. [CrossRef]
8. Cicuéndez, M.; Casarrubios, L.; Barroca, N.; Silva, D.; Feito, M.J.; Diez-Orejas, R.; Marques, P.A.A.P.; Portolés, M.T. Benefits in the Macrophage Response Due to Graphene Oxide Reduction by Thermal Treatment. *Int. J. Mol. Sci.* **2021**, *22*, 6701. [CrossRef]
9. Wasyluk, Ł.; Boiko, V.; Markowska, M.; Hasiak, M.; Saladino, M.L.; Hreniak, D.; Amati, M.; Gregoratti, L.; Zeller, P.; Biały, D.; et al. Graphene Coating Obtained in a Cold-Wall CVD Process on the Co-Cr Alloy (L-605) for Medical Applications. *Int. J. Mol. Sci.* **2021**, *22*, 2917. [CrossRef]
10. Szczepaniak, J.; Jagiello, J.; Wierzbicki, M.; Nowak, D.; Sobczyk-Guzenda, A.; Sosnowska, M.; Jaworski, S.; Daniluk, K.; Szmidt, M.; Witkowska-Pilaszewicz, O.; et al. Reduced Graphene Oxides Modulate the Expression of Cell Receptors and Voltage-Dependent Ion Channel Genes of Glioblastoma Multiforme. *Int. J. Mol. Sci.* **2021**, *22*, 515. [CrossRef]
11. Dash, B.S.; Jose, G.; Lu, Y.-J.; Chen, J.-P. Functionalized Reduced Graphene Oxide as a Versatile Tool for Cancer Therapy. *Int. J. Mol. Sci.* **2021**, *22*, 2989. [CrossRef] [PubMed]
12. Silva, A.R.; Cavaleiro, A.J.; Soares, O.S.G.P.; Braga, C.S.N.; Salvador, A.F.; Pereira, M.F.R.; Alves, M.M.; Pereira, L. Detoxification of Ciprofloxacin in an Anaerobic Bioprocess Supplemented with Magnetic Carbon Nanotubes: Contribution of Adsorption and Biodegradation Mechanisms. *Int. J. Mol. Sci.* **2021**, *22*, 2932. [CrossRef] [PubMed]
13. Muraru, S.; Ionita, M. Towards Performant Design of Carbon-Based Nanomotors for Hydrogen Separation through Molecular Dynamics Simulations. *Int. J. Mol. Sci.* **2020**, *21*, 9588. [CrossRef]
14. Wu, P.; Shi, J.; Wang, J.; Shen, J.; Cai, K. Critical Output Torque of a GHz CNT-Based Rotation Transmission System Via Axial Interface Friction at Low Temperature. *Int. J. Mol. Sci.* **2019**, *20*, 3851. [CrossRef]
15. Alen, S.K.; Nam, S.; Dastgheib, S.A. Recent Advances in Graphene Oxide Membranes for Gas Separation Applications. *Int. J. Mol. Sci.* **2019**, *20*, 5609. [CrossRef] [PubMed]
16. Kovel, E.S.; Sachkova, A.S.; Vnukova, N.G.; Churilov, G.N.; Knyazeva, E.M.; Kudryasheva, N.S. Antioxidant Activity and Toxicity of Fullerenols via Bioluminescence Signaling: Role of Oxygen Substituents. *Int. J. Mol. Sci.* **2019**, *20*, 2324. [CrossRef]
17. Díez-Pascual, A.M. Antibacterial Action of Nanoparticle Loaded Nanocomposites Based on Graphene and Its Derivatives: A Mini-Review. *Int. J. Mol. Sci.* **2020**, *21*, 3563. [CrossRef]
18. Hsueh, Y.-H.; Hsieh, C.-T.; Chiu, S.-T.; Tsai, P.-H.; Liu, C.-Y.; Ke, W.-J. Antibacterial Property of Composites of Reduced Graphene Oxide with Nano-Silver and Zinc Oxide Nanoparticles Synthesized Using a Microwave-Assisted Approach. *Int. J. Mol. Sci.* **2019**, *20*, 5394. [CrossRef]
19. Sainz-Urruela, C.; Vera-López, S.; San Andrés, M.P.; Díez-Pascual, A.M. Graphene-Based Sensors for the Detection of Bioactive Compounds: A Review. *Int. J. Mol. Sci.* **2021**, *22*, 3316. [CrossRef]
20. San Andrés, M.P.; Baños-Cabrera, M.; Gutiérrez-Fernández, L.; Díez-Pascual, A.M.; Vera-López, S. Fluorescence Study of Riboflavin Interactions with Graphene Dispersed in Bioactive Tannic Acid. *Int. J. Mol. Sci.* **2021**, *22*, 5270. [CrossRef]
21. Rahman, M.M.; Ara, M.G.; Alim, M.A.; Uddin, M.S.; Najda, A.; Albadrani, G.M.; Sayed, A.A.; Mousa, S.A.; Abdel-Daim, M.M. Mesoporous Carbon: A Versatile Material for Scientific Applications. *Int. J. Mol. Sci.* **2021**, *22*, 4498. [CrossRef]

Review

Mesoporous Carbon: A Versatile Material for Scientific Applications

Md. Motiar Rahman [1,2,3,*], Mst Gulshan Ara [2,3], Mohammad Abdul Alim [4,5], Md. Sahab Uddin [6,7], Agnieszka Najda [8], Ghadeer M. Albadrani [9], Amany A. Sayed [10], Shaker A. Mousa [11] and Mohamed M. Abdel-Daim [12]

1. Shenzhen Institute of Advanced Technology (SIAT) of the Chinese Academy of Sciences (CAS), Shenzhen 518055, China
2. Nanotechnology and Catalysis Research Center (NanoCat), University of Malaya, Kuala Lumpur 50603, Malaysia; aramstgulshan63@gmail.com
3. Department of Biochemistry and Molecular Biology, University of Rajshahi, Rajshahi 6205, Bangladesh
4. Department of Chemistry, Bangabandhu Sheikh Mujibur Rahman Science and Technology University, Gopalganj 8100, Bangladesh; alimbsmrstu@gmail.com
5. Graduate School of Innovative Life Science, University of Toyama, Gofuku 3190, Toyama 930-8555, Japan
6. Department of Pharmacy, Southeast University, Dhaka 1213, Bangladesh; msu-neuropharma@hotmail.com
7. Pharmakon Neuroscience Research Network, Dhaka 1207, Bangladesh
8. Laboratory of Quality of Vegetables and Medicinal Plants, Department of Vegetable Crops and Medicinal Plants, University of Life Sciences in Lublin, 15 Akademicka Street, 20-950 Lublin, Poland; agnieszka.najda@up.lublin.pl
9. Department of Biology, College of Science, Princess Nourah bint Abdulrahman University, Riyadh 11474, Saudi Arabia; gmalbadrani@pnu.edu.sa
10. Zoology Department, Faculty of Science, Cairo University, Giza 12613, Egypt; amanyasayed@sci.cu.edu.eg
11. Pharmaceutical Research Institute, Albany College of Pharmacy and Health Sciences, Rensselaer, NY 12144, USA; shaker.mousa@acphs.edu
12. Pharmacology Department, Faculty of Veterinary Medicine, Suez Canal University, Ismailia 41522, Egypt; abdeldaim.m@vet.suez.edu.eg
* Correspondence: md_motiar@siat.ac.cn

Abstract: Mesoporous carbon is a promising material having multiple applications. It can act as a catalytic support and can be used in energy storage devices. Moreover, mesoporous carbon controls body's oral drug delivery system and adsorb poisonous metal from water and various other molecules from an aqueous solution. The accuracy and improved activity of the carbon materials depend on some parameters. The recent breakthrough in the synthesis of mesoporous carbon, with high surface area, large pore-volume, and good thermostability, improves its activity manifold in performing functions. Considering the promising application of mesoporous carbon, it should be broadly illustrated in the literature. This review summarizes the potential application of mesoporous carbon in many scientific disciplines. Moreover, the outlook for further improvement of mesoporous carbon has been demonstrated in detail. Hopefully, it would act as a reference guidebook for researchers about the putative application of mesoporous carbon in multidimensional fields.

Keywords: mesoporous carbon; surface modification; catalytic support; adsorbent; drug delivery; capacitor

1. Introduction

According to IUPAC, the term "mesoporous carbon" refers to solid-based material, which might have either disordered or ordered networks with narrow or broad pores distribution in the range from 2 to 50 nm [1]. Historically, Ryoo et al. [2] reported the first structural composition of highly ordered mesoporous carbon material using mesoporous silica template, and later, much consideration was given to its synthetic methods and applications [3–5]. Therefore, rapid development was started for the synthesis of mesoporous carbon with the use of carbon precursors and self-assembly of copolymer molecular arrays [6–9]. Recently, modification, as well as the introduction of physicochemical properties on carbon, has been applied by incorporating inorganic constituents either anchored onto the mesoporous walls or decorated within the channels. Thus, the emergence of so-called modified mesoporous carbon has mostly driven the advancement of carbon-based electrodes or catalysts for its putative applications [10–13]. On the other hand, the substantial development of clean energy technology has motivated carbon-based researchers to exploit this porous material in the formation of energy storage devices [14–18].

Porous carbon materials, including carbon molecular sieves and activated carbon, have been synthesized on a large scale from fruit shells, wood, coal, or polymers by simple pyrolysis and physical or chemical treatment at elevated temperatures [19–23]. Moreover, these carbons have relatively broad pore-size distributions in both mesopore and micropore ranges and have been synthesized from defects resulting from heteroatoms elimination during carbonization. However, the carbon materials produced in these methods provide poor conductivity, structural integrity, and mass transport, owing to the presence of heteroatoms, lack of structural control, and restricted flow pathways. These limitations have been resolved by the introduction of mesopores for potential applications [24,25]. Mesoporous carbons with good physical properties are being synthesized on a large scale by carbonization from organic sources, such as sucrose, under acid treatment using silica templates. In this protocol, a transient silica-carbon complex is formed, and finally, following the alkali treatment, the silica is removed, leaving mesoporous carbon with analogous properties of silica materials (Figure 5) [24,25]. The mesoporous carbon synthesis, in this way, possesses good structural features and meets high physiological demands in modern scientific research [17,21–23].

Carbon materials with well-ordered mesoporous structures are ubiquitous and indispensable in modern scientific research, which attracted huge attention for their promising applications in various fields, including adsorption and separation of large biomolecules, electrical double-layer capacitors, catalytic support, wastewater treatment, and air purification (Figure 1) [26–32]. The emergence of inorganic porous materials, such as silica, zeolite, zeotype materials, carbon, metal oxides, and composites, has paved the way for the progression of drug delivery systems and other applications due to their high adsorption efficiency and ideal biocompatibility [33–39]. Compared with conventional porous materials, an inorganic carrier with small particle sizes, well-control shape, high stability, surface functionalization properties, low density, and wide availability offers substantial benefits in terms of application [40,41]. Recent advances in nanotechnology have provided a new dimension in carbon structure through continuous development of conventional methods as well as the introduction of new synthesis techniques [42]. Considering the well-controlled structure and promising application of mesoporous carbon in many scientific disciplines, it should be broadly compiled in the scientific literature from different points of view. To this line, the article aimed to provide a brief but elaborate overview of the applications of mesoporous carbon in many scientific research fields with special emphasis on catalytic reaction, adsorption, and drug delivery. Moreover, the outlook for further improvements and plausible challenges of mesoporous carbon has been demonstrated in detail. The authors, therefore, hope that this article would serve as a reference guidebook for future researchers regarding the potential application of mesoporous carbon in diverse research fields

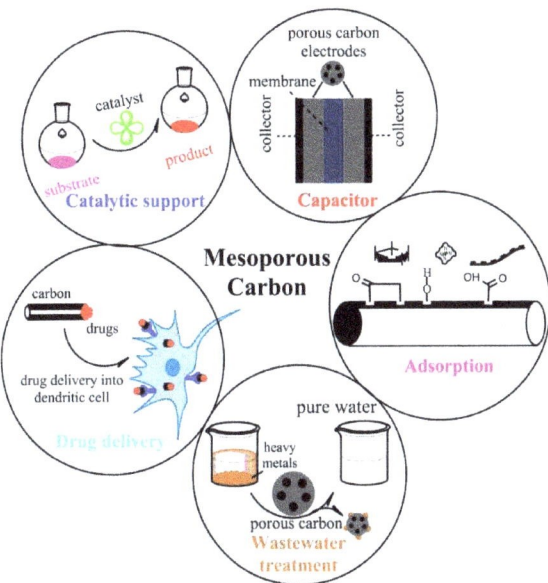

Figure 1. Potential applications of mesoporous carbon in modern scientific research.

2. Versatility of Mesoporous Carbon

Porous carbon nanomaterials are of great interest in current research due to their physicochemical properties and high surface area. However, traditional methods produce indiscriminate porous carbon materials with little to no control over their mesostructures and pore-size distributions, etc. The latest advances in the synthesis of other porous nanomaterials have led to developing alternative syntheses of mesoporous carbon with exceptionally ordered mesostructures and high surface areas, which play a pivotal role in separating media, cutting-edge electronic materials, and catalytic support (Figure 1, Table 1) [20,43]. Here, we outlined versatile applications of mesoporous carbon with the current scenarios in the respective fields.

Table 1. Application of mesoporous carbon in various scientific disciplines.

Disciplines	Form of Mesoporous Carbon	Preparation Method	Application	Ref.
Catalytic supports	Au/C	Deposition–precipitation, cationic adsorption	Glucose oxidation	[44]
	Au/C	Incipient wetness impregnation	Glucose oxidation	[45]
	Au–Pd/C	Impregnation	Glyoxal and glucose oxidation	[46]
	Au/C	Immobilization	Glucose and Alcohol oxidation	[47]
	OMCs/tungsten carbide composites	Soft template	Methanol electrooxidation	[48]
	Pt-Ru/C	Co-precipitation	Methanol electrooxidation	[49]
	Pt/C and PtCO$_3$O$_4$/C	Microwave and Impregnation	Methanol oxidation	[50]

Table 1. *Cont.*

Disciplines	Form of Mesoporous Carbon	Preparation Method	Application	Ref.
	Au/MnO$_x$/C	Electrodeposition	CO oxidation	[51]
	Au/TiO2/C	Sonochemical and Microwave	CO oxidation	[52]
	Porous carbon supported gold catalysts	Antigalvanic reduction	Oxygen reduction reaction	[53]
	Mesoporous carbon supported gold	Hydrothermal synthesis	Reduction of nitroarenes	[54]
	Ordered mesoporous carbon supported gold	Wet chemical	Oxygen reduction	[55]
	Au/FeO$_x$	Deposition	CO oxidation	[56]
Adsorbents	CMK-3–100 (3 nm) and CMK-3-150 (6.5 nm)	Template synthesis	Lysozyme adsorption	[57]
	CMK-1 and CMK-3	Template synthesis	Vitamin E adsorption	[58]
	Activated carbon with mesopores	Commercial	Sugars adsorption	[59]
	Activated carbon with mesopores	Commercial	Sugars adsorption	[60]
	Mesoporous carbon	Template synthesis	Antibiotics adsorption	[61]
	Glucose-based mesoporous carbon	Template synthesis	Antibiotics adsorption	[62]
	Activated carbon with mesopores	Commercial	Sugars adsorption	[63]
Wastewater treatment	Magnetic mesoporous carbon	Wet impregnation	Removal of methyl orange and methyl blue	[64]
	Iron containing mesoporous carbon	Template synthesis	Removal of Arsenic (As)	[65]
	Magnetically graphitic mesoporous carbon	Template synthesis	Removal of Chinese medical waste	[66]
	Magnetically encapsulated mesoporous carbon	Template synthesis	Removal of methylene blue, Congo red	[67]
	Ordered mesoporous carbon	Template synthesis	Removal of malachite green	[68]
	Boron-doped mesoporous carbon nitride	Template synthesis	Removal of malachite green	[69]
	Mesoporous carbon nanofibers	Hydrothermal	Removal of methylene blue, methyl orange	[70]
	S-doped magnetic mesoporous carbon	Template synthesis	Removal of methyl orange	[71]
	Magnetic mesoporous carbon nanospheres	Template synthesis	Removal of hexavalent chromium	[72]
	Polyacrylic acid modified magnetic mesoporous carbon	Template synthesis, co-impregnation	Removal of Cd(II)	[73]
	Ordered mesoporous carbon electrodes	Template synthesis	Copper (II)	[74]

Table 1. Cont.

Disciplines	Form of Mesoporous Carbon	Preparation Method	Application	Ref.
	Boron doped ordered mesoporous carbon	Template synthesis	Pb(II)	[75]
	Functionalized mesoporous carbon	Chemical modification	Pb(II)	[76]
	Phosphate modified ordered mesoporous carbon	Template synthesis	Pb(II)	[77]
	Ordered mesoporous carbon	Template synthesis	AV90 dye	[78]
	Modified mesoporous carbon	Template synthesis	Bisphenol-A	[79]
	Octyl modified ordered mesoporous carbon	Template synthesis	Phenol	[80]
	Functional mesoporous carbon	Hydrothermal carbonization	Bisphenol and diuron	[81]
	Mesoporous carbon microsphere	Template synthesis	Removal of hexavalent chromium	[82]
Drug delivery	Mesoporous carbon	Template synthesis	Celecoxib	[83]
	Mesoporous carbon nanoparticles	Template synthesis	Ruthenium dye	[84]
	ZnO gated mesoporous carbon nanoparticles	Template synthesis	Mitoxantrone	[85]
	CMK-1 type mesoporous carbon nanoparticle	Template synthesis	Fura-2	[86]
	Folate functionalized mesoporous carbon	Template synthesis	Doxorubicin	[87]
	Hyaluronic acid modified mesoporous carbon nanoparticles	Template synthesis	Doxorubicin	[88]
	Mesoporous carbon nanospheres	Hydrothermal synthesis	Doxorubicin	[89]
Capacitors	Mesoporous carbon	Carbonization	Electric double layer capacitors	[90]
	Mesoporous carbon	Defluorination	Electric double layer capacitors	[91]

2.1. Catalytic Supports

Currently, mesoporous carbon materials have drawn significant attention in the field of catalysis due to their promising structural features, such as unique optoelectronic and physio-chemical properties, uniform and tunable pore size, high surface area (up to 2500 m^2/g), ordered pore architecture, good electrical conductivity and large pore volume, thermal stability, high recyclability, and biocompatibility [25,92,93]. These features offer high dispersion of metal nanoparticles, and therefore, increase the transport of ions, or electrons, and molecules, via the nanochannels/nanopores, and enhance product yields of a catalytic reaction with the shortest duration and minimum cost.

Delidovich et al. [44] comparatively analyzed the aerobic oxidation of glucose over Au/Al$_2$O$_3$ and Au/C catalysts at various glucose to Au molar ratios. The analysis showed that the Au/C catalysts were lower active than alumina-supported catalysts at high glucose to Au ratios. At these ratios, Au/Al$_2$O$_3$ (with various metal dispersion) exhibited high TOF (turnover frequency), which is a feature of the alumina-supported catalysts with metal particles size 1–5 nm. At these conditions, more than 90% of Au/Al$_2$O$_3$ catalysts showed a homogeneous gold distribution into the solid supports (Figure 2a). For Au/C catalysts with a non-homogeneous gold distribution, the superficial reaction rate was disturbed by internal diffusion, although the interface of gas–liquid–solid oxygen transfer influenced the overall reaction kinetics as well. In the aqueous phase, the reaction rate was controlled by oxygen dissolution at low glucose to Au ratios. In this condition, the carbon-supported catalysts were more active compared to the alumina-supported gold catalysts because the hydrophobic carbon supports highly adhere to the gas–liquid interface, and thus, facilitate the oxygen mass transfer towards catalytic sites (Figure 2b).

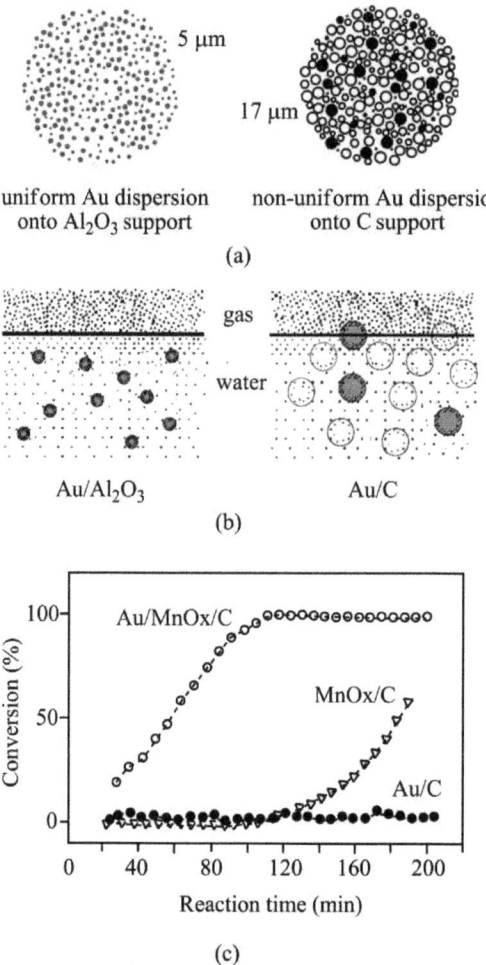

Figure 2. Application of carbon supports in various oxidation reactions. Comparative analyses on Au/Al$_2$O$_3$ and Au/C catalysts (**a**,**b**). Uniform gold distribution onto alumina supports at high glucose to Au ratios (**a**). Hydrophobic carbon supports greatly adhere to gas–liquid interface assisting oxygen mass transfer towards catalytic sites (**b**), adapted with permission from [44]. Effects of carbon functionalization in carbon-monoxide oxidation (**c**), adapted with permission from [51].

Ma et al. [94] have reported that although Au/C catalysts showed potential activity for specific liquid-phase oxidation; however, they are poorly active for gas-phase CO oxidation. This might be owing to several factors: first, carbon supports are acidic with low isoelectric points; second, during catalyst syntheses, carbon materials (acting as reducing agents) might interreact with Au^{3+} to form large/bulky gold nanoparticles [95]. In order to activate catalysts, carbon supports were modified in aqueous KMnO$_4$ [51]. In aqueous solution, mesoporous carbon acts as a sacrificial reductant undergoing a self-limiting reaction for the decoration of MnO_x on its surface, and therefore, facilitates gold loading onto MnO_x/C support through deposition-precipitation [96]. This modified ($Au/MnO_x/C$) catalyst showed higher activity for CO oxidation (Figure 2c) [51]. The surface modification of carbon support has also been performed using various metal oxides (ZnO, FeO_x, TiO_2, etc.) for supported gold catalysts preparation [52,56,97].

Ketchie et al. [95] comparatively studied carbon-supported gold (5–42 nm) and unsupported gold powder in the aqueous-phase oxidation of glycerol and carbon monoxide (CO). For aqueous-phase oxidation of CO at 27 °C and pH = 14, supported-gold nanoparticles (5 nm) showed a TOF of 5 s^{-1}, whereas large (42 nm) and bulky gold particles showed a TOF of only 0.5 and 0.4 s^{-1}, respectively. Moreover, the observed rate of peroxide production during CO oxidation was higher over small gold particles. Small-sized gold particles showed similar effects in aqueous phase glycerol oxidation (60 °C and pH 13.8), in which 5 nm gold nanoparticles showed TOF 17 s^{-1}, and larger and bulky golds showed an order of magnitude lower activity. Surprisingly, larger-sized gold (>20 nm) nanoparticles exhibited higher selectivity to glyceric acid formation. The poor selectivity of small Au particles has been attributed to the high rate of H_2O_2 formation during glycerol oxidation, since peroxide production promotes C-C bond cleavage. Prati et al. [98] prepared gold nanoparticles on various support materials by different methods for the liquid-phase oxidation of diols. Under normal conditions, the activity of Au/C catalysts showed different conversion and selectivity for ethane-1, 2 diol oxidations to glycolate.

2.2. Adsorbent

Nowadays, the carbons with well-ordered mesopores provide crucial applications in diverse fields, and one such example is the adsorption and separation of biomolecules, including enzymes, sugars, vitamins, proteins, amino acids, and useful antibiotics [42,99–102]. The carbon materials have good physicochemical properties with tunable pore diameter, high pore volume, and high surface area; however, the hydrophobicity and inertness of mesoporous carbons might not be favorable for this application. Therefore, surface modification of the carbon materials is required for their development and to change the hydrophobic and hydrophilic properties of the carbon surface [103]. Functional groups, such as the carboxyl group, are generally introduced into carbon surfaces upon oxidation using various oxidizing agents, for example, ammonium persulfate solution (APS), concentrated nitric acid, ozone, and sulfuric acids, and thus, enhances the wettability of polar solvents and activates the carbon surface to covalently immobilize DNA, nanocrystals, proteins, metal-containing complexes, etc. The extent of biomolecules adsorption on the carbon surface predominantly depends on the strength and density of the functional agent of the solid surface. Contemporary studies have shown that porous carbon treatment with APS provides more carboxyl groups onto the carbon surface compared to the treatment with H_2SO_4 or HNO_3 because treatments with these strong acids offer large quantities of oxygen interfaces that affect the overall structure as well as textural properties [57,104].

The Vinu group [57] comparatively analyzed the ability of lysozyme adsorption using various APS-treated functionalized carbons namely CMK-3–0 h, CMK-3–12 h, CMK-3–24 h, and CMK-3–48 h (depending on the APS concentrations and oxidation times). However, these materials did not show promising activity for the adsorption of large biomolecules, owing to their poor textural features, disordered porous structure, and small pore size. The carboxyl group functionalization onto APS-treated-mesoporous carbon provides higher adsorption ability because the adsorbent molecules easily access the interior portion of mesoporous carbon materials and the oxidation of small carbon rods that are usually organized either vertically or horizontally inside the functionalized mesoporous channels. It has also been noticed that the anchoring capacity of carboxyl groups placed at the entrance and inside the mesopores may hinder the release of protein biomolecules from porous channels of carboxy functionalized mesoporous carbon, which results in elevated protein adsorption [57,105].

It has also been found that the lysozyme adsorption isotherms of carboxy group functionalization on APS-treated mesoporous carbon exhibited a sharp initial rise, advocating an extraordinary affinity between the carbon adsorbent and lysozyme. Among the carboxyl-functionalized-APS-treated materials, surprisingly, CMK-3–12 h and CMK-3–24 h exhibited higher adsorption abilities than untreated mesoporous material and CMK-3–48 h (Figure 3a). Moreover, the amount of protein adsorption depends on solution pH, which increases with increasing pH (Figure 3b) [57].

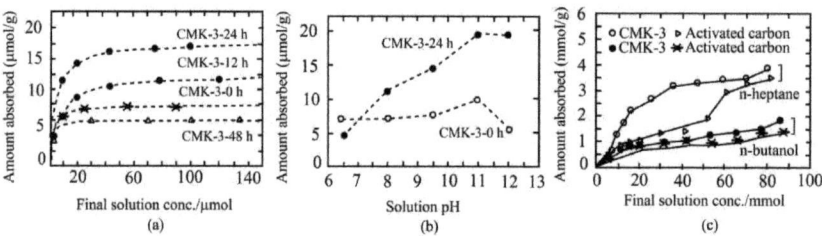

Figure 3. Adsorption of lysozyme and vitamin E over mesoporous carbon surfaces. Lysozyme adsorption dependency on the APS concentration and oxidation times, and pH of the solution (**a**,**b**). Comparative analyses of lysozyme adsorption on various forms of activated carbons (**a**) and dependency of pH on the adsorption properties of lysozyme enzyme (**b**), adapted with permission from [57]. Adsorption ability of vitamin E onto carbon surfaces from n-heptane and n-butanol at 293 K (**c**), adapted with permission from [58].

Hartmann et al. [58] demonstrated the adsorption ability of vitamin E onto different mesoporous carbon (CMK-1 and CMK-3) materials from two different solvents, n-butanol and n-heptane. The adsorption efficacy of these mesoporous materials was also compared with conventional adsorbents, including activated carbon and microporous carbon. The extent of vitamin E adsorption was found to largely rely on solvent types, as well as surface area and mesopore volume of the adsorbent. Moreover, it has been demonstrated that nonpolar n-heptane is preferred for higher vitamin E adsorption onto solid surfaces than polar solvent n-butanol (Figure 3c). This reduced adsorption ability of carbon materials in n-butanol apparently due to the increased production of n-butanol solvent clusters that may interact with ether and hydroxyl groups of the vitamin E. On the other hand, the interaction between the hydrophobic tail of vitamin E and the surface of hydrophobic mesoporous carbon was not influenced by the nonpolar n-heptane. In this report, CMK-3 (5.94 mmol/g) showed higher efficiency than CMK-1 (5.01 mmol/g) and activated carbon (4.10 mmol/g) in terms of vitamin E adsorption. XRD and N_2 adsorption analyses showed that the vitamin E was closely packed inside mesopores of CMK-1 and CMK-3. This group also suggested that nonpolar solvent *n*-heptane is a promising candidate for higher loading of adsorbents with vitamin E. However, although further analysis is desired, when carbons would be loaded in medical applications with vitamin E, ethanol and water might be useful as solvents.

Abe et al. [59] studied the adsorption ability of monosaccharides and disaccharides onto activated carbon from the complex aqueous solution. The adsorption features were demonstrated by the physical characteristics of adsorbates, e.g., molecular refraction and/or parachor. Moon and Cho [63] demonstrated that the activated carbon materials can act as the effective adsorbent for the separation and purification of maltopentaose from the complex mixture of other maltooligosaccharides, including maltotriose, maltose, and maltotetraose. Three distinct carbon materials were used for the purification of maltopentaose followed by enzymatic degradation. They employed activated carbon into the column and found 61% selectivity of maltopentaose by increasing the adsorption cycle. Lee et al. [60] analyzed the adsorption kinetics of monosaccharides (galactose, glucose, and fructose), disaccharides (sucrose, maltose, and lactose), and maltooligosaccharides (maltotriose, maltotetraose, and maltopentaose) onto 35 nm of mesoporous carbon. Among the nine saccharides used in this experiment, the adsorption of maltose was superior compared to other adsorbates at 100 °C. It was also found that the adsorption level of disaccharides, compared to monosaccharides, was extremely dependent on the chemical structure.

Antibiotics have extensive use in biomedical applications for disease inhibition and treatment, in aquaculture and in farming for growth promotion [106]. Owing to their unhealthy management, antibiotics are posing a potential hazard to ecosystems and public health [107]. Currently, a number of antibiotics are repeatedly measured in surface water, municipal wastewater, drinking water, and groundwater [108,109]. Among the exclusion methods for antibiotics from wastewater, adsorption is regarded as one of the most encouraging techniques due to its extraordinary removal efficacy, ecofriendly properties, and ease of adsorbent synthesis [110]. Activated carbon is regarded as a promising candidate for this application.

Ji et al. [61] prepared ordered micro-and mesoporous carbon materials to comparatively analyze the adsorption properties towards three antibiotics, namely tylosin, sulfamethoxazole, and tetracycline. Nonporous graphite, two commercial microporous activated carbons, and single-walled carbon nanotubes were also included for comparative analyses. The adsorption efficiency of smaller-sized sulfamethoxazole was higher onto activated carbons compared to other carbonaceous materials, which may be the result of the pore-filling effect. On the other hand, because of the size-exclusion effect, the adsorption of bulky tylosin and tetracycline was significantly lower over activated carbons, particularly for the more microporous absorbent, compared to the synthesized carbons. Following the normalization of the surface area of the adsorbent materials, adsorption of tylosin and tetracycline over prepared carbons was comparable with nonporous graphite, demonstrating the accessibility of the adsorbent surface area in adsorption. Compared to other porous adsorbents, the template synthesized carbons exhibited faster adsorption of tylosin and tetracycline, which may be due to their ordered-shaped, accessible as well as interconnected 3D pore structure. These results suggest that meso-and microporous carbons synthesized (template synthesis) in this study were promising candidates for the separation of antibiotics from an aqueous mixture.

Wang et al. [62] investigated the adsorption ability of three commonly used antibiotics, including sulfadiazine (SDZ), ciprofloxacin (CIP), and tetracycline (TC), from the aqueous phase over glucose-based mesoporous carbon (GMC). The highest adsorption efficiency of GMC for SDZ, CIP, and TC were 246.73, 369.85, and 297.91 mg/g, respectively. They also showed that the efficiency of antibiotics adsorption depends significantly on pH and the ideal pH values for the release of SDZ, CIP, and TC were measured as 4, 6, and 6, respectively. Antibiotic adsorption was not reported to be affected by ionic conditions, including NaCl, $MgCl_2$, and $CaCl_2$; however, the adsorption increased with an increasing temperature between 288 and 308 K of target antibiotics. The thermodynamic analyses showed that the adsorption process of selected antibiotics onto GMC was regarded spontaneous as well as exothermic in nature. The research proposed that several mechanisms, e.g., π-π interactions or hydrogen bonding, electrostatic interactions, and a hydrophobic effect, might be involved in the adsorption of antibiotics (Figure 4). Apart from these, several other reports on antibiotic adsorption have been illustrated in the literature [111–115].

2.3. Waste-Water Treatment

As per the Global Risk Report of 2015, the water catastrophe is positioned as the highest long-durable risk [116]. Approximately 783 million people have no access to safe and clean water, and 1 in 9 (11%) of the world's population do not have access to drinking water [117]. Since population, as well as industry, are growing constantly, various pollutants, including endocrine disrupters, nitrosoamines, and heavy metals, are released into the water. These impurities have an adverse effect on environmental flora and fauna [118–122]. Currently, the conventional approaches of wastewater treatment, including sedimentation, coagulation, membrane process, decontamination, disinfection, ion exchange, and filtration, have been reported; however, these methods are intensive chemically and operationally [123–133]. Moreover, various water disinfectants have been applied for the chemical treatment of water, which may generate side products, for example, hydrochloric acid, chlorine, alum, ammonia, ozone, permanganate, and ferric salts, that are dangerous to freshwater resources [119]. Hence, there is an urgent need for efficient, large-scale, and low-cost systems for developing water quality. Many adsorbents have also been tested for metal removal, such as granular ferric hydroxide (GFH) [134,135], activated alumina [136,137], iron oxide-coated sand (IOCS) [138–141], and ferric (hydro)oxides [142,143]. Nowadays, ordered mesoporous carbons have drawn much attraction for this purpose due to their promising physical properties [144].

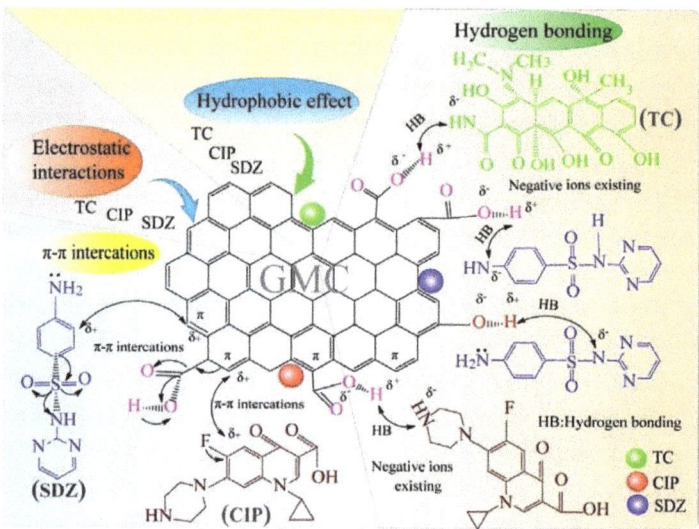

Figure 4. The mechanisms of the target antibiotics adsorption on GMC surface from aqueous solutions. Adapted with permission from [62].

Gu et al. [65] impregnated iron onto mesoporous carbon, which diffuses ferrous (II) iron into the interior of porous carbon, so that the iron can disperse more consistently in the mesoporous carbon for improved arsenic adsorption (As^{III} and As^{V}) from the drinking water (Figure 5). This mesoporous material showed a maximum of 5.15 mg and 5.96 mg As/g adsorption for arsenate and arsenite, respectively. Zhou et al. [66] synthesized iron-containing graphitic mesoporous carbon by carbonization using silica template and sucrose via impregnation. The synthesized complexes were shown high and rapid adsorption efficiency of pigments from *Carthamus tinctorius* flowers through solid-liquid magnetic separation. In a copious study, Zhang et al. [67] encapsulated mesoporous carbon with Fe_3O_4 for wastewater treatment. TEM analysis showed the complexes possess a rattle-like shape and Fe_3O_4 nanoparticles have reached the internal site of the mesoporous carbon. The porous wall of carbon materials provides enough spaces that enhance adsorption abilities as well as rates of pollutants adsorption from the aqueous solution. The synthesized composites show superparamagnetic properties with a saturation magnetization of 5.5 emu g^{-1} (electromagnetic unit per gram) at room temperature, which is the precondition for the high-magnetic separation of pollutants in wastewater treatment. The prepared samples showed higher organic pollutants adsorption when compared with commercially available activated carbon, and their highest adsorption ability for phenol, Congo red, and methylene blue reached 108.38, 1656.9, and 608.04, mg g^{-1}, respectively.

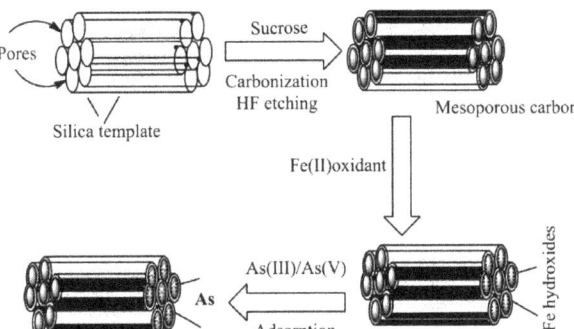

Figure 5. A graphical model for the preparation of carbon-based adsorbents and their application in arsenic adsorption. Silica has been used as the template for the synthesis of mesoporous carbon by carbonization followed by iron coating for the removal of metal ions from drinking water.

Recently, novel magnetic mesoporous activated carbon (MMAC) was synthesized from activated carbon, isolated from rice husk via $ZnCl_2$ chemical activation [64]. Mesostructural properties were induced through silica formation and the magnetic component was integrated by magnetite via wetness impregnation. This silica increases the porosity and gives a 3D network to stimulate the molecule adsorption. The adsorption ability of methyl orange and methyl blue dyes on the activated carbon, mesoporous activated carbon, and magnetic mesoporous carbon complex was studied. The effect of various parameters, for example, dosing of adsorbents, initial dye concentration, and pH of the dye solution, on the adsorption efficacy were analyzed. The MMAC adsorbed 98.5% methyl orange and 82% methyl blue dye in 30 min. The adsorbent material was recycled four times for methyl orange and methyl blue removal without a substantial loss in the adsorption ability. Comparative analyses showed that MMAC was more efficient in adsorbing wastewater than mesoporous activated carbon and activated carbon because of its high pore volume and high surface area. The magnetic properties of MMAC separate the adsorbent material after the adsorption process, showing that it is a promising material for water purification. Various other data on wastewater treatment using mesoporous carbon materials have also been reported in the recently published articles for the adsorption of malachite green dyes [68,69], methylene blue and methyl orange [70,71], hexavalent chromium [72–82], Cd(II) [73], copper [74], lead(II) [75–77], AV90 dye [78], pesticides [145], phenolic compound [79–81], diuron (herbicides), etc. [81].

2.4. Drug Delivery

Among various drug delivery routes, oral drug administration is still considered the preferred route because of patient compliance and simplicity [146–148]. The oral sustained-release formulation is the most common process and has drawn huge attraction in the disciplines of novel drug delivery. However, the oral delivery method is highly challenging because of its poor bioavailability as a result of biopharmaceutical (e.g., poor solubility and permeability and/or instability in the gastrointestinal environment) and pharmacokinetic (rapid clearance) challenges in the delivery process [149]. Until now, there is a surprisingly significant number of active drug constituents that show poor dissolution rates. For example, 30–40% of the leading 200 oral drug stuffs from the USA, Japan, Spain, and the UK are weakly dissolved in water [150]. Therefore, this is an emergency issue to explore novel drug formulation methods that might increase the water solubility of such drugs, predominantly the Biopharmaceutics Classification System (BCS) Class 2 compounds (poor solubility and high permeability) [151].

To this line, many delivery techniques, such as silica-lipid hybrid microcapsules [152], micelles [153], nanodispersions [154], silica nanoparticles [155–157], solid-lipid nanoparticles [158], self-microemulsification [159], and mesoporous carbon [43,160,161], have been established to upgrade drug bioavailability via improving the solubility and dissolution rate. There are several reports of drug delivery using mesoporous carbon with novel characteristics, e.g., higher drug loading, high adsorption efficiency, larger specific surface area and pore volume, and chemically inert [43,162]. In a recent study, mesoporous carbon with pore size (9.74 nm) and pore volume (1.53 cm^3/g) was prepared for the delivery of celecoxib [83]. The drug was loaded into porous channels of carbon material. The performance was tested in terms of celecoxib release. The results showed that the rate of celecoxib release from mesoporous carbon was significantly higher than that of pure celecoxib administration. Mesoporous carbon increased the inhibitory function of celecoxib on the migration and invasion of human breast cancer cell lines (MDAMB-231 cells) that might be owing to the increased dissolution efficiency. It has been expected that the mesoporous carbon may have potential in the field of anticancer metastasis [83]. Wang et al. [162] selected a model drug (ibuprofen) and loaded it onto ordered mesoporous carbons to study the effect of carbon materials on drug release. They applied a two-step release process and observed a first rapid release, and subsequently, a slower release, and the delivery of ibuprofen was correlated with time. Moreover, the pore size was regarded as a vital factor for ibuprofen delivery, which stimulated positively with an increasing pore size. Miguel et al. [84] functionalized mesoporous carbon materials with a pH-sensitive self-immolative polymer and examined them as drug nanocarriers. A vial release assay of a Ru dye at pH 5 and 7.4 indicates the pH-sensitivity of the hybrid methods, exhibiting that only small quantities of the payload are released at neutral pH, while at acidic pH, self-immolation occurs and a substantial extent of the cargo is liberated (Figure 6). Cytotoxicity assay in human osteosarcoma cells confirms that the complex nanocarriers have shown no cytotoxicity by themselves but stimulate noteworthy cell growth inhibition while loading with doxorubicin, a chemotherapeutic drug. Under physiological pH, in vivo report showed no significant cargo release over 96 h. However, temporary exposure to acidic pH discharges an experimental fluorescent cargo for over 72 h.

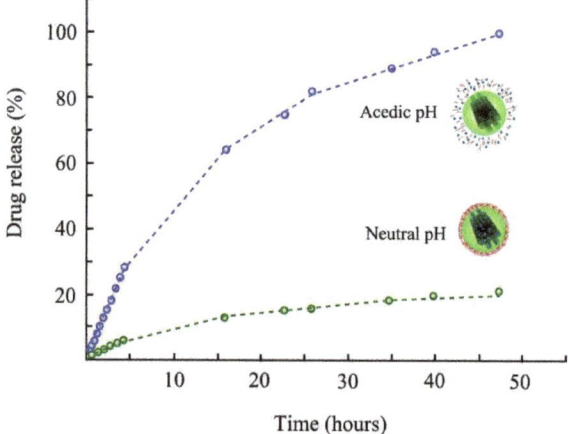

Figure 6. Schematic diagram of the pH-responsive mesoporous carbons. At physiological pH, the self-immolative coating remains collapsed on the surface. Conversely, at acidic pH, the polymers undergo self-immolation, leading to cargo release. Adapted with permission from [84].

2.5. Capacitors

Capacitors are energy storage devices which deliver high power density as well as incredible energy. Recently, these features have offered great advantages as electric vehicles, electronic components, and backup power systems due to their energy storage benefits. These devices are based on an energy storage mechanism resulting from the synthesis of an electric double layer at the interface between an electrolyte solution and an electronically conductive material [24,163–165]. A classic supercapacitor consists of two carbon electrodes interposed by a porous matrix (separator) (Figure 1). The energy accumulated in carbon-based supercapacitors is a function of several parameters, such as pore structure and specific surface area of electrodes, voltage stability, and ionic conductivity of electrolyte. Moreover, the usual materials employed in traditional electrodes should have a high surface area, e.g., carbon aerogels [166,167], carbon fibers [168], and activated carbons [169–171]. The electrolytes commonly utilized in supercapacitors are an aqueous solution of H_2SO_4 or KOH [172] or non-aqueous solutions of tetraethyl-ammonium-tetrafluoroborate (ET_4NBF_4) in organic solvents [169,173]. The addition of solid polymer electrolytes in supercapacitors, as a liquid electrolytes substitute, may offer several benefits, including flexible structure, compact geometry, and easier packaging. Additionally, since this device is free from outflows of dangerous and corrosive liquids, it offers safety profiles. If mesoporous carbon with a well-ordered pore structure and a high specific surface area are used, the well-engineered solid-state supercapacitors would provide higher energy as well as power density compared to liquid electrolyte-based supercapacitors. Moreover, this capacitor would provide an advantage in terms of ease of design and realization of lightweight and flexible devices.

Recently, nanostructured mesoporous carbon materials achieved much interest as negative electrodes of rechargeable lithium batteries [174,175] and supercapacitor electrodes [171]. Tamai et al. [90] revealed that the introduction of mesoporous carbon in the electrodes improved the specific capacitance of carbon-based supercapacitors. Yamada et al. [91] comparatively studied various carbons of similar specific surface areas with various pore size distributions and found that highly mesoporous materials exhibited better capacitance ability. Yoon et al. [176] demonstrated that higher fractionated mesopores carbons delivered high power densities and exhibited a smaller time constant compared to microporous carbons. However, these experiments were based on liquid electrolyte supercapacitors [107] and the microstructural features of such carbon may affect the generation of effective polymer electrolyte-based supercapacitors.

Lufrano et al. [173] synthesized mesoporous carbons using SBA-15 silica templates and sucrose as a carbon source. These mesoporous carbon materials were used to develop composite electrodes by a casting method for solid-state supercapacitors. The electrochemical features of mesoporous carbons in supercapacitors were analyzed by electrochemical impedance spectroscopy and cyclic voltammetry. The CMK-3A mesoporous carbon showed a value of specific capacitance of 132 Fg^{-1} (for a single electrode), which was 68% higher than that of reference carbon material. The double-layer capacitor performance of the mesoporous carbon was 127% (12.05 $\mu F\,cm^{-2}$ vs. 5.3 $\mu F\,cm^{-2}$) when compared with the reference carbon. These findings were explained based on the high pseudocapacitance and other physicochemical properties of the mesoporous carbon, including accessibility of pores to the electrolyte, pore structure, and surface wettability by surface functional groups.

3. Future Challenges and Opportunities

Mesoporous carbon is regarded as the next generation inorganic material for many applications, including biomedical science, owing to its promising combinatorial properties, such as carbonaceous composition, distinct mesoporous structure, and high biocompatible nature [31]. However, in some applications, such as controlled drug release, mesoporous carbon materials are still at the developing stage. Additional improvements are necessary to boost the clinical application of this material: (a) scalable synthetic protocols with optimized structural as well as compositional parameters; to date, a very limited number of standard and controllable methodologies for the synthesis of mesoporous carbon have been reported, particularly for pore size-tunable and hydrophilic spherical material. (b) Surface modification of mesoporous carbon still remains challenging. The current oxidization scheme of mesoporous carbon might provide the carrier with many organic groups for additional modification; however, the strong oxidization partly affects the carbonaceous framework as well as the mesostructure of mesoporous material, leading to structural distortion and declined photothermal-conversion capability. More significantly, surface modification is crucial for well-ordered catalysts preparation, to supply drug carriers with smart drug delivery, and controlled and stimuli-responsive drug release, which has been hardly applied to mesoporous carbon, although extensively used in mesoporous silica. Compared to other materials, including graphene and mesoporous silica, mesoporous carbon has been rarely applied in the biomedical field and diagnostic imaging, although it has been expected to show good performance due to its unique structural composition and physicochemical features. For considering the use of mesoporous carbon in clinical translation in the upcoming future, it is urgently necessary to evaluate its biosafety, which intensely depends on the methodology of material synthesis. In addition, biosafety evaluation should be focused on some crucial parameters, including biodegradation, excretion, biodistribution, and other disease-specific toxicities, such as embryonic toxicity, reproductive toxicity, neurotoxicity, etc. An in vivo quantitative assay of mesoporous carbons might be challenging owing to their carbonaceous composition that may be affected by carbon present in the living system. In this case, radiolabeling of mesoporous carbon should be promising for biosafety evaluation.

Although the functional modification on mesoporous carbon surface gives excellent activity in various fields, it is still necessary to invent a cost-effective approach for large-scale applications with high stability as well as reliability. In addition to surface modification, the future challenge is to generate contamination-free functionalized material, since this material might be mixed up with the nanoporous material during synthesis. Functional materials with a high surface area were reported to be excellent materials for adsorption. Future challenges might arise to use the mesoporous material with both absorption and built-in electrochemical properties that may stimulate carbon material to be employed as electrocatalyst. Uncontrollable doping and irreversible agglomeration at the molecular level during high-temperature pyrolysis in their synthetic procedure are more challenging and need to be solved for better application; for example, when used as catalytic supports or as adsorbents. Furthermore, the lower synthesis of mesoporous carbon materials by the current methodology remains a problem from an industrial point of view, which should be taken into consideration during material synthesis.

4. Conclusions

In this review, the authors provide a general overview of the recent literature regarding various applications of mesoporous carbon. The limitation of the synthesis procedure of mesoporous carbon for potential applications is presented. Particular emphasis is provided to some areas, especially in surface modification, oxidation as well as functionalization time for suitable application. The role of reaction pH, reaction time, time of carbon-surface oxidation, functionalization parameters, solvent uses during adsorption, and optimal time in adsorption has been demonstrated for better understanding. Future research should focus more on the synthesis procedures, surface modification, and novel modification parameters for the promising application of mesoporous carbon.

Author Contributions: M.M.R. designed the study, wrote the manuscript, and drew the figures and the table. M.G.A. provided necessary suggestions and edited the article. M.A.A., M.S.U., A.N., G.M.A., A.A.S., S.A.M. and M.M.A.-D. revised and edited the manuscript. All authors have read and agreed to the published version of the manuscript.

Funding: This work was funded by the Deanship of Scientific Research at Princess Nourah bint Abdulrahman University through the Fast-Track Research Funding Program.

Institutional Review Board Statement: Not applicable.

Informed Consent Statement: Not applicable.

Data Availability Statement: Not applicable.

Acknowledgments: This work was funded by the Deanship of Scientific Research at Princess Nourah bint Abdulrahman University through the Fast-Track Research Funding Program.

Conflicts of Interest: The authors declare no conflict of interest.

References

1. Rouquerol, J.; Avnir, D.; Fairbridge, C.W.; Everett, D.H.; Haynes, J.M.; Pernicone, N.; Ramsay, J.D.F.; Sing, K.S.W.; Unger, K.K. Recommendations for the characterization of porous solids. *Pure Appl. Chem.* **1994**, *68*, 1739–1758. [CrossRef]
2. Ryoo, R.; Joo, S.H.; Jun, S. Synthesis of highly ordered carbon molecular sieves via template-mediated structural transformation. *J. Phys. Chem. B* **1999**, *103*, 7743–7746. [CrossRef]
3. Lu, A.H.; Schmidt, W.; Spliethoff, B.; Schüth, F. Synthesis of Ordered Mesoporous Carbon with Bimodal Pore System and High Pore Volume. *Adv. Mater.* **2003**, *15*, 1602–1606. [CrossRef]
4. Kim, T.W.; Park, I.S.; Ryoo, R. A synthetic route to ordered mesoporous carbon materials with graphitic pore walls. *Angew. Chem. Int. Ed.* **2003**, *115*, 4511–4515. [CrossRef]
5. Lu, A.H.; Smått, J.H.; Lindén, M.; Schüth, F. Synthesis of carbon monoliths with a multi-modal pore system by a one step impregnation technique. *New Carbon Mater.* **2003**, *18*, 181–185.
6. Liang, C.; Hong, K.; Guiochon, G.A.; Mays, J.W.; Dai, S. Synthesis of a large-scale highly ordered porous carbon film by self-assembly of block copolymers. *Angew. Chem. Int. Ed.* **2004**, *43*, 5785–5789. [CrossRef]
7. Zhang, F.; Meng, Y.; Gu, D.; Yan, Y.; Yu, C.; Tu, B.; Zhao, D. A facile aqueous route to synthesize highly ordered mesoporous polymers and carbon frameworks with Ia3d bicontinuous cubic structure. *J. Am. Chem. Soc.* **2005**, *127*, 13508–13509. [CrossRef]
8. Meng, Y.; Gu, D.; Zhang, F.; Shi, Y.; Yang, H.; Li, Z.; Yu, C.; Tu, B.; Zhao, D. Ordered mesoporous polymers and homologous carbon frameworks: Amphiphilic surfactant templating and direct transformation. *Angew. Chem. Int. Ed.* **2005**, *117*, 7215–7221. [CrossRef]
9. Liang, C.; Dai, S. Synthesis of mesoporous carbon materials via enhanced hydrogen-bonding interaction. *J. Am. Chem. Soc.* **2006**, *128*, 5316–5317. [CrossRef] [PubMed]
10. Huang, Y.; Miao, Y.E.; Tjiu, W.W.; Liu, T. High-performance flexible supercapacitors based on mesoporous carbon nanofibers/Co_3O_4/MnO_2 hybrid electrodes. *RSC Adv.* **2015**, *5*, 18952–18959. [CrossRef]
11. Xu, J.; Wu, F.; Wu, H.T.; Xue, B.; Li, Y.X.; Cao, Y. Three-dimensional ordered mesoporous carbon nitride with large mesopores: Synthesis and application towards base catalysis. *Microporous Mesoporous Mater.* **2014**, *198*, 223–229. [CrossRef]
12. Li, F.; Chan, K.Y.; Yung, H.; Yang, C.; Ting, S.W. Uniform dispersion of 1:1 PtRu nanoparticles in ordered mesoporous carbon for improved methanol oxidation. *Phys. Chem. Chem. Phys.* **2013**, *15*, 13570–13577. [CrossRef] [PubMed]
13. Thieme, S.; Brückner, J.; Bauer, I.; Oschatz, M.; Borchardt, L.; Althues, H.; Kaskel, S. High capacity micro-mesoporous carbon-sulfur nanocomposite cathodes with enhanced cycling stability prepared by a solvent-free procedure. *J. Mater. Chem. A* **2013**, *1*, 9225–9234. [CrossRef]
14. Miao, L.; Song, Z.; Zhu, D.; Li, L.; Gan, L.; Liu, M. Recent advances in carbon-based supercapacitors. *Mater. Adv.* **2020**, *1*, 945–966. [CrossRef]
15. Xu, M.; Rong, Y.; Ku, Z.; Mei, A.; Liu, T.; Zhang, L.; Li, X.; Han, H. Highly ordered mesoporous carbon for mesoscopic $CH_3NH_3PbI_3$/TiO_2 heterojunction solar cell. *J. Mater. Chem. A* **2014**, *2*, 8607–8611. [CrossRef]
16. Trifonov, A.; Herkendell, K.; Tel-Vered, R.; Yehezkeli, O.; Woerner, M.; Willner, I. Enzyme-capped relay-functionalized mesoporous carbon nanoparticles: Effective bioelectrocatalytic matrices for sensing and biofuel cell applications. *ACS Nano* **2013**, *7*, 11358–11368. [CrossRef]
17. Fang, Y.; Lv, Y.; Gong, F.; Wu, Z.; Li, X.; Zhu, H.; Zhou, L.; Yao, C.; Zhang, F.; Zheng, G.; et al. Interface tension-induced synthesis of monodispersed mesoporous carbon hemispheres. *J. Am. Chem. Soc.* **2015**, *137*, 2808–2811. [CrossRef] [PubMed]
18. Chang, P.; Huang, C.; Doong, R. Ordered mesoporous carbon–TiO_2 materials for improved electrochemical performance of lithium ion battery. *Carbon N. Y.* **2012**, *50*, 4259–4268. [CrossRef]
19. Gaffney, T.R. Porous solids for air separation. *Curr. Opin. Solid State Mater. Sci.* **1996**, *1*, 69–75. [CrossRef]
20. Liang, C.; Li, Z.; Dai, S. Mesoporous carbon materials: Synthesis and modification. *Angew. Chem. Int. Ed.* **2008**, *47*, 3696–3717. [CrossRef]
21. Ma, T.Y.; Liu, L.; Yuan, Z.Y. Direct synthesis of ordered mesoporous carbons. *Chem. Soc. Rev.* **2013**, *42*, 3977–4003. [CrossRef]
22. Yang, R.T. *Adsorbents: Fundamentals and Applications*; John Wiley & Sons, Inc.: Hoboken, NJ, USA, 2003; p. 410.
23. Adsorbents: Fundamentals and applications. *Focus Catal.* **2004**, *6*, 2004.
24. Zhai, Y.; Dou, Y.; Zhao, D.; Fulvio, P.F.; Mayes, R.T.; Dai, S. Carbon Materials for Chemical Capacitive Energy Storage. *Adv. Mater.* **2011**, *23*, 4828–4850. [CrossRef] [PubMed]
25. Liu, B.; Liu, L.; Yu, Y.; Zhang, Y.; Chen, A. Synthesis of mesoporous carbon with tunable pore size for supercapacitors. *New J. Chem.* **2020**, *44*, 1036–1044. [CrossRef]
26. Böhme, K.; Einicke, W.D.; Klepel, O. Templated synthesis of mesoporous carbon from sucrose-the way from the silica pore filling to the carbon material. *Carbon N. Y.* **2005**, *43*, 1918–1925. [CrossRef]
27. Lee, D.W.; Yu, C.Y.; Lee, K.H. Facile synthesis of mesoporous carbon and silica from a silica nanosphere-sucrose nanocomposite. *J. Mater. Chem.* **2009**, *19*, 299–304. [CrossRef]

28. Hu, Z.; Srinivasan, M. Mesoporous high-surface-area activated carbon. *Microporous Mesoporous Mater.* **2001**, *43*, 267–275. [CrossRef]
29. Lee, J.; Kim, J.; Hyeon, T. Recent progress in the synthesis of porous carbon materials. *Adv. Mater.* **2006**, *18*, 2073–2094. [CrossRef]
30. Biener, J.; Stadermann, M.; Suss, M.; Worsley, M.A.; Biener, M.M.; Rose, K.A.; Baumann, T.F. Advanced carbon aerogels for energy applications. *Energy Environ. Sci.* **2011**, *4*, 656–667. [CrossRef]
31. Marsh, H.; Rodríguez-Reinoso, F. Characterization of Activated Carbon. *Act. Carbon* **2006**, 143–242.
32. Eftekhari, A.; Fan, Z. Ordered mesoporous carbon and its applications for electrochemical energy storage and conversion. *Mater. Chem. Front.* **2017**, *1*, 1001–1027. [CrossRef]
33. Chen, Y.; Shi, J. Mesoporous carbon biomaterials. *Sci. China Mater.* **2015**, *58*, 241–257. [CrossRef]
34. Zhao, P.; Wang, L.; Sun, C.; Jiang, T.; Zhang, J.; Zhang, Q.; Sun, J.; Deng, Y.; Wang, S. Uniform mesoporous carbon as a carrier for poorly water-soluble drug and its cytotoxicity study. *Eur. J. Pharm. Biopharm.* **2012**, *80*, 535–543. [CrossRef]
35. Zheng, H.; Gao, F.; Valtchev, V. Nanosized inorganic porous materials: Fabrication, modification and application. *J. Mater. Chem. A* **2016**, *4*, 16756–16770. [CrossRef]
36. Butt, A.R.; Ejaz, S.; Baron, J.C.; Ikram, M.; Ali, S. CaO nanoparticles as a potential drug delivery agent for biomedical applications. *Dig. J. Nanomater. Biostructures* **2015**, *10*, 799–809.
37. Huo, Q. Synthetic Chemistry of the Inorganic Ordered Porous Materials. In *Modern Inorganic Synthetic Chemistry*; Elsevier: Amsterdam, The Netherlands, 2011; pp. 339–373. ISBN 9780444535993.
38. Qiao, Z.A.; Huo, Q.S. Synthetic Chemistry of the Inorganic Ordered Porous Materials. In *Modern Inorganic Synthetic Chemistry*, 2nd ed.; Elsevier: Amsterdam, The Netherlands, 2017; pp. 389–428. ISBN 9780444635914.
39. Doane, T.L.; Burda, C. The unique role of nanoparticles in nanomedicine: Imaging, drug delivery and therapy. *Chem. Soc. Rev.* **2012**, *41*, 2885–2911. [CrossRef]
40. Wei, A.; Mehtala, J.G.; Patri, A.K. Challenges and opportunities in the advancement of nanomedicines. *J. Control. Release* **2012**, *164*, 236–246. [CrossRef]
41. Taratula, O.; Kuzmov, A.; Shah, M.; Garbuzenko, O.B.; Minko, T. Nanostructured lipid carriers as multifunctional nanomedicine platform for pulmonary co-delivery of anticancer drugs and siRNA. *J. Control. Release* **2013**, *171*, 349–357. [CrossRef]
42. Son, S.J.; Bai, X.; Lee, S. Inorganic hollow nanoparticles and nanotubes in nanomedicine. Part 2: Imaging, diagnostic, and therapeutic applications. *Drug Discov. Today* **2007**, *12*, 657–663.
43. Son, S.J.; Bai, X.; Lee, S.B. Inorganic hollow nanoparticles and nanotubes in nanomedicine. Part 1. Drug/gene delivery applications. *Drug Discov. Today* **2007**, *12*, 650–656.
44. Xin, W.; Song, Y. Mesoporous carbons: Recent advances in synthesis and typical applications. *RSC Adv.* **2015**, *5*, 83239–83285. [CrossRef]
45. Zhao, Q.; Lin, Y.; Han, N.; Li, X.; Geng, H.; Wang, X.; Cui, Y.; Wang, S. Mesoporous carbon nanomaterials in drug delivery and biomedical application. *Drug Deliv.* **2017**, *24*, 94–107. [CrossRef] [PubMed]
46. Delidovich, I.V.; Moroz, B.L.; Taran, O.P.; Gromov, N.V.; Pyrjaev, P.A.; Prosvirin, I.P.; Bukhtiyarov, V.I.; Parmon, V.N. Aerobic selective oxidation of glucose to gluconate catalyzed by Au/Al$_2$O$_3$ and Au/C: Impact of the mass-transfer processes on the overall kinetics. *Chem. Eng. J.* **2013**, *223*, 921–931. [CrossRef]
47. Zhang, M.; Zhu, X.; Liang, X.; Wang, Z. Preparation of highly efficient Au/C catalysts for glucose oxidation via novel plasma reduction. *Catal. Commun.* **2012**, *25*, 92–95. [CrossRef]
48. Hermans, S.; Deffernez, A.; Devillers, M. Au-Pd/C catalysts for glyoxal and glucose selective oxidations. *Appl. Catal. A Gen.* **2011**, *395*, 19–27. [CrossRef]
49. Prati, L.; Porta, F. Oxidation of alcohols and sugars using Au/C catalysts: Part 1. Alcohols. *Appl. Catal. A Gen.* **2005**, *291*, 199–203. [CrossRef]
50. Wang, Y.; He, C.; Brouzgou, A.; Liang, Y.; Fu, R.; Wu, D.; Tsiakaras, P.; Song, S. A facile soft-template synthesis of ordered mesoporous carbon/tungsten carbide composites with high surface area for methanol electrooxidation. *J. Power Sources* **2012**, *200*, 8–13. [CrossRef]
51. Wang, K.W.; Huang, S.Y.; Yeh, C.T. Promotion of carbon-supported platinum-ruthenium catalyst for electrodecomposition of methanol. *J. Phys. Chem. C* **2007**, *111*, 5096–5100. [CrossRef]
52. Amin, R.S.; Elzatahry, A.A.; El-Khatib, K.M.; Elsayed Youssef, M. Nanocatalysts prepared by microwave and impregnation methods for fuel cell application. *Int. J. Electrochem. Sci.* **2011**, *6*, 4572–4580.
53. Ma, Z.; Liang, C.; Overbury, S.H.; Dai, S. Gold nanoparticles on electroless-deposition-derived MnO$_x$/C: Synthesis, characterization, and catalytic CO oxidation. *J. Catal.* **2007**, *252*, 119–126. [CrossRef]
54. George, P.P.; Gedanken, A.; Perkas, N.; Zhong, Z. Selective oxidation of CO in the presence of air over gold-based catalysts Au/TiO$_2$/C (sonochemistry) and Au/TiO$_2$/C (microwave). *Ultrason. Sonochem.* **2008**, *15*, 539–547. [CrossRef]
55. Wang, L.; Tang, Z.; Yan, W.; Yang, H.; Wang, Q.; Chen, S. Porous Carbon-Supported Gold Nanoparticles for Oxygen Reduction Reaction: Effects of Nanoparticle Size. *ACS Appl. Mater. Interfaces* **2016**, *8*, 20635–20641. [CrossRef]
56. Chen, S.; Fu, H.; Zhang, L.; Wan, Y. Nanospherical mesoporous carbon-supported gold as an efficient heterogeneous catalyst in the elimination of mass transport limitations. *Appl. Catal. B Environ.* **2019**, *248*, 22–30. [CrossRef]
57. Wang, L.; Tang, Z.; Liu, X.; Niu, W.; Zhou, K.; Yang, H.; Zhou, W.; Li, L.; Chen, S. Ordered mesoporous carbons-supported gold nanoparticles as highly efficient electrocatalysts for oxygen reduction reaction. *RSC Adv.* **2015**, *5*, 103421–103427. [CrossRef]

58. Bulushev, D.A.; Kiwi-Minsker, L.; Yuranov, I.; Suvorova, E.I.; Buffat, P.A.; Renken, A. Structured Au/FeO$_x$/C catalysts for low-temperature CO oxidation. *J. Catal.* **2002**, *210*, 149–159. [CrossRef]
59. Vinu, A.; Hossian, K.Z.; Srinivasu, P.; Miyahara, M.; Anandan, S.; Gokulakrishnan, N.; Mori, T.; Ariga, K.; Balasubramanian, V.V. Carboxy-mesoporous carbon and its excellent adsorption capability for proteins. *J. Mater. Chem.* **2007**, *17*, 1819–1825. [CrossRef]
60. Hartmann, M.; Vinu, A.; Chandrasekar, G. Adsorption of vitamin E on mesoporous carbon molecular sieves. *Chem. Mater.* **2005**, *17*, 829–833. [CrossRef]
61. Abe, I.; Hayashi, K.; Kitagawa, M. Adsorption of saccharides from aqueous solution onto activated carbon. *Carbon N. Y.* **1983**, *21*, 189–192. [CrossRef]
62. Lee, J.W.; Kwon, T.O.; Moon, I.S. Adsorption of monosaccharides, disaccharides, and maltooligosaccharides on activated carbon for separation of maltopentaose. *Carbon N. Y.* **2004**, *42*, 371–380. [CrossRef]
63. Ji, L.; Liu, F.; Xu, Z.; Zheng, S.; Zhu, D. Adsorption of pharmaceutical antibiotics on template-synthesized ordered micro- and mesoporous carbons. *Environ. Sci. Technol.* **2010**, *44*, 3116–3122. [CrossRef]
64. Wang, B.; Xu, X.; Tang, H.; Mao, Y.; Chen, H.; Ji, F. Highly efficient adsorption of three antibiotics from aqueous solutions using T glucose-based mesoporous carbon. *Appl. Surf. Sci.* **2020**, *528*, 147048. [CrossRef]
65. Moon, I.S.; Cho, G. Production of maltooligosaccharides from starch and separation of maltopentaose by adsorption of them on activated carbon (I). *Biotechnol. Bioprocess. Eng.* **1997**, *2*, 19–22. [CrossRef]
66. Azam, K.; Raza, R.; Shezad, N.; Shabir, M.; Yang, W.; Ahmad, N.; Shafiq, I.; Akhter, P.; Razzaq, A.; Hussain, M. Development of recoverable magnetic mesoporous carbon adsorbent for removal of methyl blue and methyl orange from wastewater. *J. Environ. Chem. Eng.* **2020**, *8*, 104220. [CrossRef]
67. Gu, Z.; Deng, B. Use of iron-containing mesoporous carbon (IMC) for arsenic removal from drinking water. *Environ. Eng. Sci.* **2007**, *24*, 113–121. [CrossRef]
68. Hong, Z.Q.; Li, J.X.; Zhang, F.; Zhou, L.H. Synthesis of magnetically graphitic mesoporous carbon from hard templates and its application in the adsorption treatment of traditional Chinese medicine wastewater. *Wuli Huaxue Xuebao/Acta Phys. Chim. Sin.* **2013**, *29*, 590–596.
69. Zhang, Y.; Xu, S.; Luo, Y.; Pan, S.; Ding, H.; Li, G. Synthesis of mesoporous carbon capsules encapsulated with magnetite nanoparticles and their application in wastewater treatment. *J. Mater. Chem.* **2011**, *21*, 3664–3671. [CrossRef]
70. Anbia, M.; Ghaffari, A. Removal of malachite green from dye wastewater using mesoporous carbon adsorbent. *J. Iran. Chem. Soc.* **2011**, *8*, S67–S76. [CrossRef]
71. Azimi, E.B.; Badiei, A.; Ghasemi, J.B. Efficient removal of malachite green from wastewater by using boron-doped mesoporous carbon nitride. *Appl. Surf. Sci.* **2019**, *469*, 236–245. [CrossRef]
72. Li, S.; Jia, Z.; Li, Z.; Li, Y.; Zhu, R. Synthesis and characterization of mesoporous carbon nanofibers and its adsorption for dye in wastewater. *Adv. Powder Technol.* **2016**, *27*, 591–598. [CrossRef]
73. Xu, J.; Zhai, S.; Zhu, B.; Liu, J.; Lu, A.; Jiang, H. S-Doped Magnetic Mesoporous Carbon for Efficient Adsorption of Methyl Orange from Aqueous Solution. *Clean SoilAirWater* **2021**, *49*, 2000285.
74. Wang, G.; Gao, G.; Yang, S.; Wang, Z.; Jin, P.; Wei, J. Magnetic mesoporous carbon nanospheres from renewable plant phenol for efficient hexavalent chromium removal. *Microporous Mesoporous Mater.* **2021**, *310*, 110623. [CrossRef]
75. Zeng, G.; Liu, Y.; Tang, L.; Yang, G.; Pang, Y.; Zhang, Y.; Zhou, Y.; Li, Z.; Li, M.; Lai, M.; et al. Enhancement of Cd(II) adsorption by polyacrylic acid modified magnetic mesoporous carbon. *Chem. Eng. J.* **2015**, *259*, 153–160. [CrossRef]
76. Huang, C.C.; He, J.C. Electrosorptive removal of copper ions from wastewater by using ordered mesoporous carbon electrodes. *Chem. Eng. J.* **2013**, *221*, 469–475. [CrossRef]
77. Liu, Y.; Xiong, Y.; Xu, P.; Pang, Y.; Du, C. Enhancement of Pb (II) adsorption by boron doped ordered mesoporous carbon: Isotherm and kinetics modeling. *Sci. Total Environ.* **2020**, *708*, 134918. [CrossRef] [PubMed]
78. Lian, Q.; Yao, L.; Uddin Ahmad, Z.; Gang, D.D.; Konggidinata, M.I.; Gallo, A.A.; Zappi, M.E. Enhanced Pb(II) adsorption onto functionalized ordered mesoporous carbon (OMC) from aqueous solutions: The important role of surface property and adsorption mechanism. *Environ. Sci. Pollut. Res.* **2020**, *20*, 23616–23630. [CrossRef] [PubMed]
79. Lian, Q.; Yao, L.; Ahmad, Z.U.; Konggidinata, M.I.; Zappi, M.E.; Gang, D.D. Modeling mass transfer for adsorptive removal of Pb(II) onto phosphate modified ordered mesoporous carbon (OMC). *J. Contam. Hydrol.* **2020**, *228*, 103562. [CrossRef] [PubMed]
80. Koyuncu, D.D.E.; Okur, M. Removal of AV 90 dye using ordered mesoporous carbon materials prepared via nanocasting of KIT-6: Adsorption isotherms, kinetics and thermodynamic analysis. *Sep. Purif. Technol.* **2021**, *257*, 117657. [CrossRef]
81. Jeong, Y.; Cui, M.; Choi, J.; Lee, Y.; Kim, J.; Son, Y.; Khim, J. Development of modified mesoporous carbon (CMK-3) for improved adsorption of bisphenol-A. *Chemosphere* **2020**, *238*, 124559. [CrossRef] [PubMed]
82. He, J.; Ma, K.; Jin, J.; Dong, Z.; Wang, J.; Li, R. Preparation and characterization of octyl-modified ordered mesoporous carbon CMK-3 for phenol adsorption. *Microporous Mesoporous Mater.* **2009**, *121*, 173–177. [CrossRef]
83. Zbair, M.; Bottlinger, M.; Ainassaari, K.; Ojala, S.; Stein, O.; Keiski, R.L.; Bensitel, M.; Brahmi, R. Hydrothermal Carbonization of Argan Nut Shell: Functional Mesoporous Carbon with Excellent Performance in the Adsorption of Bisphenol A and Diuron. *Waste Biomass Valorization* **2020**, *11*, 1565–1584. [CrossRef]
84. Zhou, J.; Wang, Y.; Wang, J.; Qiao, W.; Long, D.; Ling, L. Effective removal of hexavalent chromium from aqueous solutions by adsorption on mesoporous carbon microspheres. *J. Colloid Interface Sci.* **2016**, *462*, 200–207. [CrossRef]

85. Zhu, W.; Zhao, Q.; Zheng, X.; Zhang, Z.; Jiang, T.; Li, Y.; Wang, S. Mesoporous carbon as a carrier for celecoxib: The improved inhibition effect on MDA-MB-231 cells migration and invasion. *Asian J. Pharm. Sci.* **2014**, *9*, 82–91. [CrossRef]
86. Gisbert-Garzarán, M.; Berkmann, J.C.; Giasafaki, D.; Lozano, D.; Spyrou, K.; Manzano, M.; Steriotis, T.; Duda, G.N.; Schmidt-Bleek, K.; Charalambopoulou, G.; et al. Engineered pH-Responsive Mesoporous Carbon Nanoparticles for Drug Delivery. *ACS Appl. Mater. Interfaces* **2020**, *12*, 14946–14957.
87. Huang, X.; Wu, S.; Du, X. Gated mesoporous carbon nanoparticles as drug delivery system for stimuli-responsive controlled release. *Carbon N. Y.* **2016**, *101*, 135–142. [CrossRef]
88. Kim, T.W.; Chung, P.W.; Slowing, I.I.; Tsunoda, M.; Yeung, E.S.; Lin, V.S.Y. Structurally Ordered Mesoporous Carbon Nanoparticles as Transmembrane Delivery Vehicle in Human Cancer Cells. *Nano Lett.* **2008**, *8*, 3724–3727. [CrossRef]
89. Zhu, J.; Liao, L.; Zhu, L.; Kong, J.; Liu, B. Folate functionalized mesoporous carbon nanospheres as nanocarrier for targetted delivery and controlled release of doxorubicin to HeLa cells. *Acta Chim. Sin.* **2013**, *71*, 69–74. [CrossRef]
90. Wan, L.; Jiao, J.; Cui, Y.; Guo, J.; Han, N.; Di, D.; Chang, D.; Wang, P.; Jiang, T.; Wang, S. Hyaluronic acid modified mesoporous carbon nanoparticles for targeted drug delivery to CD44-overexpressing cancer cells. *Nanotechnology* **2016**, *27*, 135102. [CrossRef] [PubMed]
91. Zhu, J.; Liao, L.; Bian, X.; Kong, J.; Yang, P.; Liu, B. PH-controlled delivery of doxorubicin to cancer cells, based on small mesoporous carbon nanospheres. *Small* **2012**, *8*, 2715–2720. [CrossRef] [PubMed]
92. Tamai, H.; Kouzu, M.; Morita, M.; Yasuda, H. Highly mesoporous carbon electrodes for electric double-layer capacitors. *Electrochem. Solid-State Lett.* **2003**, *6*, A214–A217.
93. Yamada, Y.; Tanaike, O.; Liang, T.T.; Hatori, H.; Shiraishi, S.; Oya, A. Electric double layer capacitance performance of porous carbons prepared by defluorination of polytetrafluoroethylene with potassium. *Electrochem. Solid-State Lett.* **2002**, *5*, A283.
94. Ghimbeu, C.M.; Vidal, L.; Delmotte, L.; Le Meins, J.M.; Vix-Guterl, C. Catalyst-free soft-template synthesis of ordered mesoporous carbon tailored using phloroglucinol/glyoxylic acid environmentally friendly precursors. *Green Chem.* **2014**, *16*, 3079–3088. [CrossRef]
95. Deng, Y.; Cai, Y.; Sun, Z.; Gu, D.; Wei, J.; Li, W.; Guo, X.; Yang, J.; Zhao, D. Controlled synthesis and functionalization of ordered large-pore mesoporous carbons. *Adv. Funct. Mater.* **2010**, *20*, 3658–3665. [CrossRef]
96. Ma, Z.; Dai, S. Development of novel supported gold catalysts: A materials perspective. *Nano Res.* **2011**, *4*, 3–32. [CrossRef]
97. Ketchie, W.C.; Fang, Y.L.; Wong, M.S.; Murayama, M.; Davis, R.J. Influence of gold particle size on the aqueous-phase oxidation of carbon monoxide and glycerol. *J. Catal.* **2007**, *1*, 94–101. [CrossRef]
98. Huang, X.; Yue, H.; Attia, A.; Yang, Y. Preparation and Properties of Manganese Oxide/Carbon Composites by Reduction of Potassium Permanganate with Acetylene Black. *J. Electrochem. Soc.* **2007**, *154*, A26. [CrossRef]
99. Khanderi, J.; Hoffmann, R.C.; Engstler, J.; Schneider, J.J.; Arras, J.; Claus, P.; Cherkashinin, G. Binary Au/MWCNT and ternary Au/ZnO/MWCNT nanocomposites: Synthesis, characterisation and catalytic performance. *Chem. A Eur. J.* **2010**, *16*, 2300–2308. [CrossRef]
100. Prati, L.; Martra, G. New gold catalysts for liquid phase oxidation. *Gold Bull.* **1999**, *32*, 96–101. [CrossRef]
101. Vinu, A.; Miyahara, M.; Ariga, K. Biomaterial immobilization in nanoporous carbon molecular sieves: Influence of solution pH, pore volume, and pore diameter. *J. Phys. Chem. B* **2005**, *109*, 6436–6441. [CrossRef] [PubMed]
102. Vinu, A.; Miyahara, M.; Mori, T.; Ariga, K. Carbon nanocage: A large-pore cage-type mesoporous carbon material as an adsorbent for biomolecules. *J. Porous Mater.* **2006**, *13*, 379–383. [CrossRef]
103. Vinu, A.; Streb, C.; Murugesan, V.; Hartmann, M. Adsorption of cytochrome c on new mesoporous carbon molecular sieves. *J. Phys. Chem. B* **2003**, *107*, 8297–8299. [CrossRef]
104. Kyotani, T. Porous Carbon. In *Carbon Alloys: Novel Concepts to Develop Carbon Science and Technology*; Elsevier: Amsterdam, The Netherlands, 2003; p. 584.
105. Shim, J.W.; Park, S.J.; Ryu, S.K. Effect of modification with HNO_3 and NaOH on metal adsorption by pitch-based activated carbon fibers. *Carbon N. Y.* **2001**, *39*, 1635–1642. [CrossRef]
106. Moreno-Castilla, C.; Carrasco-Marín, F.; Mueden, A. The creation of acid carbon surfaces by treatment with $(NH_4)_2S_2O_8$. *Carbon N. Y.* **1997**, *35*, 1619–1626. [CrossRef]
107. Vinu, A.; Miyahara, M.; Hossain, K.Z.; Takahashi, M.; Balasubramanian, V.V.; Mori, T.; Ariga, K. Lysozyme adsorption onto mesoporous materials: Effect of pore geometry and stability of adsorbents. *J. Nanosci. Nanotechnol.* **2007**, *7*, 828–832. [CrossRef]
108. Chen, B.; Lin, L.; Fang, L.; Yang, Y.; Chen, E.; Yuan, K.; Zou, S.; Wang, X.; Luan, T. Complex pollution of antibiotic resistance genes due to beta-lactam and aminoglycoside use in aquaculture farming. *Water Res.* **2018**, *134*, 200–208. [CrossRef] [PubMed]
109. Fang, X.; Wu, S.; Wu, Y.; Yang, W.; Li, Y.; He, J.; Hong, P.; Nie, M.; Xie, C.; Wu, Z.; et al. High-efficiency adsorption of norfloxacin using octahedral UIO-66-NH_2 nanomaterials: Dynamics, thermodynamics, and mechanisms. *Appl. Surf. Sci.* **2020**, *518*, 146226. [CrossRef]
110. Yang, J.F.; Ying, G.G.; Zhao, J.L.; Tao, R.; Su, H.C.; Chen, F. Simultaneous determination of four classes of antibiotics in sediments of the Pearl Rivers using RRLC-MS/MS. *Sci. Total Environ.* **2010**, *408*, 3424–3432. [CrossRef] [PubMed]
111. Michael, I.; Rizzo, L.; McArdell, C.S.; Manaia, C.M.; Merlin, C.; Schwartz, T.; Dagot, C.; Fatta-Kassinos, D. Urban wastewater treatment plants as hotspots for the release of antibiotics in the environment: A review. *Water Res.* **2013**, *47*, 957–995. [CrossRef]

112. Peng, X.; Hu, F.; Huang, J.; Wang, Y.; Dai, H.; Liu, Z. Preparation of a graphitic ordered mesoporous carbon and its application in sorption of ciprofloxacin: Kinetics, isotherm, adsorption mechanisms studies. *Microporous Mesoporous Mater.* **2016**, *228*, 196–206. [CrossRef]
113. Carrales-Alvarado, D.H.; Ocampo-Pérez, R.; Leyva-Ramos, R.; Rivera-Utrilla, J. Removal of the antibiotic metronidazole by adsorption on various carbon materials from aqueous phase. *J. Colloid Interface Sci.* **2014**, *436*, 276–285. [CrossRef]
114. Peng, X.; Hu, F.; Dai, H.; Xiong, Q.; Xu, C. Study of the adsorption mechanisms of ciprofloxacin antibiotics onto graphitic ordered mesoporous carbons. *J. Taiwan Inst. Chem. Eng.* **2016**, *65*, 472–481. [CrossRef]
115. Peng, X.; Hu, F.; Lam, F.L.Y.; Wang, Y.; Liu, Z.; Dai, H. Adsorption behavior and mechanisms of ciprofloxacin from aqueous solution by ordered mesoporous carbon and bamboo-based carbon. *J. Colloid Interface Sci.* **2015**, *460*, 349–360. [CrossRef] [PubMed]
116. Zou, J.; Zefeng, S.; Yuesuo, Y. Preparation of low-cost sludge-based mesoporous carbon and its adsorption of tetracycline antibiotics. *Water Sci. Technol.* **2019**, *79*, 676–687.
117. Hu, X.; Qi, J.; Lu, R.; Sun, X.; Shen, J.; Han, W.; Wang, L.; Li, J. Efficient removal of tylosin by nitrogen-doped mesoporous carbon nanospheres with tunable pore sizes. *Environ. Sci. Pollut. Res.* **2020**, *27*, 30844–30852. [CrossRef] [PubMed]
118. Kumar, S.; Ahlawat, W.; Bhanjana, G.; Heydarifard, S.; Nazhad, M.M.; Dilbaghi, N. Nanotechnology-based water treatment strategies. *J. Nanosci. Nanotechnol.* **2014**, *14*, 1838–1858. [CrossRef] [PubMed]
119. WHO and UNICEF Progress on sanitation and drinking-water. *World Heal. Organ. Unicef* **2014**, *4*, 1.
120. Joseph, L.; Jun, B.M.; Flora, J.R.V.; Park, C.M.; Yoon, Y. Removal of heavy metals from water sources in the developing world using low-cost materials: A review. *Chemosphere* **2019**, *229*, 142–159. [CrossRef]
121. Shannon, M.A.; Bohn, P.W.; Elimelech, M.; Georgiadis, J.G.; Marĩas, B.J.; Mayes, A.M. Science and technology for water purification in the coming decades. *Nature* **2008**, *452*, 301–310. [CrossRef]
122. Ali, I. New generation adsorbents for water treatment. *Chem. Rev.* **2012**, *112*, 5073–5091. [CrossRef] [PubMed]
123. Hamidi, A.; Parham, K.; Atikol, U.; Shahbaz, A.H. A parametric performance analysis of single and multi-effect distillation systems integrated with open-cycle absorption heat transformers. *Desalination* **2015**, *371*, 37–45. [CrossRef]
124. Chan, G.Y.S.; Chang, J.; Kurniawan, T.A.; Fu, C.X.; Jiang, H.; Je, Y. Removal of non-biodegradable compounds from stabilized leachate using VSEPRO membrane filtration. *Desalination* **2007**, *202*, 310–317. [CrossRef]
125. Choong, T.S.Y.; Chuah, T.G.; Robiah, Y.; Gregory Koay, F.L.; Azni, I. Arsenic toxicity, health hazards and removal techniques from water: An overview. *Desalination* **2007**, *217*, 139–166. [CrossRef]
126. Han, B.; Runnells, T.; Zimbron, J.; Wickramasinghe, R. Arsenic removal from drinking water by flocculation and microfiltration. *Desalination* **2002**, *145*, 293–298. [CrossRef]
127. Hering, J.G.; Chen, P.Y.; Wilkie, J.A.; Elimelech, M. Arsenic removal from drinking water during coagulation. *J. Environ. Eng.* **1997**, *123*, 800–807. [CrossRef]
128. McNeill, L.S.; Edwards, M. Predicting as removal during metal hydroxide precipitation. *J. / Am. Water Work. Assoc.* **1997**, *89*, 75–86.
129. Yoon, K.; Hsiao, B.S.; Chu, B. High flux nanofiltration membranes based on interfacially polymerized polyamide barrier layer on polyacrylonitrile nanofibrous scaffolds. *J. Memb. Sci.* **2009**, *326*, 484–492. [CrossRef]
130. Molinari, R.; Palmisano, L.; Drioli, E.; Schiavello, M. Studies on various reactor configurations for coupling photocatalysis and membrane processes in water purification. *J. Memb. Sci.* **2002**, *206*, 399–415. [CrossRef]
131. Molinari, R.; Mungari, M.; Drioli, E.; Di Paola, A.; Loddo, V.; Palmisano, L.; Schiavello, M. Study on a photocatalytic membrane reactor for water purification. *Catal. Today* **2000**, *55*, 71–78. [CrossRef]
132. Geise, G.M.; Lee, H.S.; Miller, D.J.; Freeman, B.D.; McGrath, J.E.; Paul, D.R. Water purification by membranes: The role of polymer science. *J. Polym. Sci. Part. B Polym. Phys.* **2010**, *48*, 1685–1718. [CrossRef]
133. Weidlich, C.; Mangold, K.M.; Jüttner, K. Conducting polymers as ion-exchangers for water purification. *Electrochim. Acta* **2001**, *47*, 741–745. [CrossRef]
134. Houri, B.; Legrouri, A.; Barroug, A.; Forano, C.; Besse, J.P. Use of the ion-exchange properties of layered double hydroxides for water purification. *Collect. Czechoslov. Chem. Commun.* **1998**, *63*, 732–740. [CrossRef]
135. Dąbrowski, A.; Hubicki, Z.; Podkościelny, P.; Robens, E. Selective removal of the heavy metal ions from waters and industrial wastewaters by ion-exchange method. *Chemosphere* **2004**, *56*, 91–106. [CrossRef] [PubMed]
136. Kersten, M.; Karabacheva, S.; Vlasova, N.; Branscheid, R.; Schurk, K.; Stanjek, H. Surface complexation modeling of arsenate adsorption by akagenéite (β-FeOOH)-dominant granular ferric hydroxide. *Colloids Surf. A Phys. Eng. Asp.* **2014**, *448*, 73–80. [CrossRef]
137. Sun, J.; Zhou, J.; Shang, C.; Kikkert, G.A. Removal of aqueous hydrogen sulfide by granular ferric hydroxide-Kinetics, capacity and reuse. *Chemosphere* **2014**, *117*, 324–329. [CrossRef]
138. Chatterjee, S.; De, S. Adsorptive removal of fluoride by activated alumina doped cellulose acetate phthalate (CAP) mixed matrix membrane. *Sep. Purif. Technol.* **2014**, *125*, 223–238. [CrossRef]
139. Sankararamakrishnan, N.; Jaiswal, M.; Verma, N. Composite nanofloral clusters of carbon nanotubes and activated alumina: An efficient sorbent for heavy metal removal. *Chem. Eng. J.* **2014**, *235*, 1–9. [CrossRef]
140. Yang, J.S.; Kwon, M.J.; Park, Y.T.; Choi, J. Adsorption of Arsenic from Aqueous Solutions by Iron Oxide Coated Sand Fabricated with Acid Mine Drainage. *Sep. Sci. Technol.* **2015**, *50*, 267–275. [CrossRef]

141. Gupta, V.K.; Saini, V.K.; Jain, N. Adsorption of As(III) from aqueous solutions by iron oxide-coated sand. *J. Colloid Interface Sci.* **2005**, *288*, 55–60. [CrossRef]
142. Otero-González, L.; Mikhalovsky, S.V.; Václavíková, M.; Trenikhin, M.V.; Cundy, A.B.; Savina, I.N. Novel nanostructured iron oxide cryogels for arsenic (As(III)) removal. *J. Hazard. Mater.* **2020**, *381*, 120996. [CrossRef]
143. Zelmanov, G.; Semiat, R. Boron removal from water and its recovery using iron (Fe+3) oxide/hydroxide-based nanoparticles (NanoFe) and NanoFe-impregnated granular activated carbon as adsorbent. *Desalination* **2014**, *333*, 107–117. [CrossRef]
144. Guo, X.; Chen, F. Removal of arsenic by bead cellulose loaded with iron oxyhydroxide from groundwater. *Environ. Sci. Technol.* **2005**, *39*, 6808–6818. [CrossRef] [PubMed]
145. Du, J.; Jing, C.; Duan, J.; Zhang, Y.; Hu, S. Removal of arsenate with hydrous ferric oxide coprecipitation: Effect of humic acid. *J. Environ. Sci.* **2014**, *26*, 240–247. [CrossRef]
146. Gu, Z.; Deng, B. Arsenic sorption and redox transformation on iron-impregnated ordered mesoporous carbon. *Appl. Organomet. Chem.* **2007**, *21*, 750–757. [CrossRef]
147. Du, J.; Liu, L.; Yu, Y.; Zhang, Y.; Chen, A. Mesoporous carbon materials with different morphology for pesticide adsorption. *Appl. Nanosci.* **2020**, *10*, 151–157. [CrossRef]
148. Sastry, S.V.; Nyshadham, J.R.; Fix, J.A. Recent technological advances in oral drug delivery—A review. *Pharm. Sci. Technol. Today* **2000**, *3*, 138–145. [CrossRef]
149. Zamani, F.; Jahanmard, F.; Ghasemkhah, F.; Amjad-Iranagh, S.; Bagherzadeh, R.; Amani-Tehran, M.; Latifi, M. Nanofibrous and nanoparticle materials as drug-delivery systems. In *Nanostructures for Drug Delivery*; Elsevier: Amsterdam, The Netherlands, 2017; pp. 239–270.
150. Jain, K.K. An overview of drug delivery systems. *Drug Deliv. Syst.* **2020**, *2059*, 1–54.
151. Pouton, C.W. Formulation of poorly water-soluble drugs for oral administration: Physicochemical and physiological issues and the lipid formulation classification system. *Eur. J. Pharm. Sci.* **2006**, *29*, 278–287. [CrossRef]
152. Takagi, T.; Ramachandran, C.; Bermejo, M.; Yamashita, S.; Yu, L.X.; Amidon, G.L. A provisional biopharmaceutical classification of the top 200 oral drug products in the United States, Great Britain, Spain, and Japan. *Mol. Pharm.* **2006**, *3*, 631–643. [CrossRef] [PubMed]
153. Amidon, G.L.; Lennernäs, H.; Shah, V.P.; Crison, J.R. A Theoretical Basis for a Biopharmaceutic Drug Classification: The Correlation of in Vitro Drug Product Dissolution and in Vivo Bioavailability. *Pharm. Res. Off. J. Am. Assoc. Pharm. Sci.* **1995**, *12*, 413–420.
154. Tan, A.; Simovic, S.; Davey, A.K.; Rades, T.; Prestidge, C.A. Silica-lipid hybrid (SLH) microcapsules: A novel oral delivery system for poorly soluble drugs. *J. Control. Release* **2009**, *134*, 62–70. [CrossRef] [PubMed]
155. Yu, B.; Tai, H.C.; Xue, W.; Lee, L.J.; Lee, R.J. Receptor-targeted nanocarriers for therapeutic delivery to cancer. *Mol. Membr. Biol.* **2010**, *27*, 286–298. [CrossRef]
156. Li, C.; Li, C.; Le, Y.; Chen, J.F. Formation of bicalutamide nanodispersion for dissolution rate enhancement. *Int. J. Pharm.* **2011**, *404*, 257–263. [CrossRef] [PubMed]
157. Zhang, Y.; Zhi, Z.; Jiang, T.; Zhang, J.; Wang, Z.; Wang, S. Spherical mesoporous silica nanoparticles for loading and release of the poorly water-soluble drug telmisartan. *J. Control. Release* **2010**, *145*, 257–263. [CrossRef]
158. Luo, W.; Xu, X.; Zhou, B.; He, P.; Li, Y.; Liu, C. Formation of enzymatic/redox-switching nanogates on mesoporous silica nanoparticles for anticancer drug delivery. *Mater. Sci. Eng. C* **2019**, *100*, 855–861. [CrossRef]
159. Gisbert-Garzarán, M.; Manzano, M.; Vallet-Regí, M. Mesoporous silica nanoparticles for the treatment of complex bone diseases: Bone cancer, bone infection and osteoporosis. *Pharmaceutics* **2020**, *12*, 83. [CrossRef]
160. Liu, Z.; Zhang, X.; Wu, H.; Li, J.; Shu, L.; Liu, R.; Li, L.; Li, N. Preparation and evaluation of solid lipid nanoparticles of baicalin for ocular drug delivery system in vitro and in vivo. *Drug Dev. Ind. Pharm.* **2011**, *37*, 475–481. [CrossRef] [PubMed]
161. Singh, A.K.; Chaurasiya, A.; Awasthi, A.; Mishra, G.; Asati, D.; Khar, R.K.; Mukherjee, R. Oral bioavailability enhancement of exemestane from self-microemulsifying drug delivery system (SMEDDS). *Aaps Pharmscitech* **2009**, *10*, 906–916. [CrossRef] [PubMed]
162. Feng, S.; Mao, Y.; Wang, X.; Zhou, M.; Lu, H.; Zhao, Q.; Wang, S. Triple stimuli-responsive ZnO quantum dots-conjugated hollow mesoporous carbon nanoplatform for NIR-induced dual model antitumor therapy. *J. Colloid Interface Sci.* **2020**, *559*, 51–64. [CrossRef] [PubMed]
163. Asgari, S.; Pourjavadi, A.; Hosseini, S.H.; Kadkhodazadeh, S. A pH-sensitive carrier based-on modified hollow mesoporous carbon nanospheres with calcium-latched gate for drug delivery. *Mater. Sci. Eng. C* **2020**, *109*, 110517. [CrossRef] [PubMed]
164. Wang, X.; Liu, P.; Tian, Y. Ordered mesoporous carbons for ibuprofen drug loading and release behavior. *Microporous Mesoporous Mater.* **2011**, *142*, 334–340. [CrossRef]
165. Kötz, R.; Carlen, M. Principles and applications of electrochemical capacitors. *Electrochim. Acta* **2000**, *45*, 2483–2498. [CrossRef]
166. Simon, P.; Gogotsi, Y. Materials for electrochemical capacitors. *Nat. Mater.* **2008**, *7*, 845–854. [CrossRef] [PubMed]
167. Inagaki, M.; Konno, H.; Tanaike, O. Carbon materials for electrochemical capacitors. *J. Power Sources* **2010**, *195*, 7880–7903. [CrossRef]
168. Saliger, R.; Fischer, U.; Herta, C.; Fricke, J. High surface area carbon aerogels for supercapacitors. *J. Non. Cryst. Solids* **1998**, *225*, 81–85. [CrossRef]

169. Zu, G.; Shen, J.; Zou, L.; Wang, F.; Wang, X.; Zhang, Y.; Yao, X. Nanocellulose-derived highly porous carbon aerogels for supercapacitors. *Carbon N. Y.* **2016**, *99*, 203–211. [CrossRef]
170. Kim, Y.-J.; Masutzawa, Y.; Ozaki, S.; Endo, M.; Dresselhaus, M.S. PVDC-Based Carbon Material by Chemical Activation and Its Application to Nonaqueous EDLC. *J. Electrochem. Soc.* **2004**, *151*, E199. [CrossRef]
171. Frackowiak, E.; Béguin, F. Carbon materials for the electrochemical storage of energy in capacitors. *Carbon N. Y.* **2001**, *39*, 937–950. [CrossRef]
172. Jayalakshmi, M.; Balasubramanian, K. Simple capacitors to supercapacitors—An overview. *Int. J. Electrochem. Sci.* **2008**, *3*, 1196–1217.
173. Vix-Guterl, C.; Frackowiak, E.; Jurewicz, K.; Friebe, M.; Parmentier, J.; Béguin, F. Electrochemical energy storage in ordered porous carbon materials. *Carbon N. Y.* **2005**, *43*, 1293–1302. [CrossRef]
174. Fuertes, A.B.; Lota, G.; Centeno, T.A.; Frackowiak, E. Templated mesoporous carbons for supercapacitor application. *Electrochim. Acta* **2005**, *50*, 2799–2805. [CrossRef]
175. Zhou, H.; Zhu, S.; Hibino, M.; Honma, I.; Ichihara, M. Lithium Storage in Ordered Mesoporous Carbon (CMK-3) with High Reversible Specific Energy Capacity and Good Cycling Performance. *Adv. Mater.* **2003**, *15*, 2107–2111. [CrossRef]
176. Lufrano, F.; Staiti, P. Mesoporous carbon materials as electrodes for electrochemical supercapacitors. *Int. J. Electrochem. Sci.* **2010**, *5*, 903–916.

Review

Functionalized Reduced Graphene Oxide as a Versatile Tool for Cancer Therapy

Banendu Sunder Dash [1], Gils Jose [1], Yu-Jen Lu [2] and Jyh-Ping Chen [1,3,4,5,*]

1. Department of Chemical and Materials Engineering, Chang Gung University, Kwei-San, Taoyuan 33302, Taiwan; banendusunder@gmail.com (B.S.D.); gilsjose84@gmail.com (G.J.)
2. Department of Neurosurgery, Chang Gung Memorial Hospital, Linkou, Kwei-San, Taoyuan 33305, Taiwan; luyj@cgmh.org.tw
3. Department of Plastic and Reconstructive Surgery and Craniofacial Research Center, Chang Gung Memorial Hospital, Linkou, Kwei-San, Taoyuan 33305, Taiwan
4. Research Center for Food and Cosmetic Safety, Research Center for Chinese Herbal Medicine, College of Human Ecology, Chang Gung University of Science and Technology, Taoyuan 33305, Taiwan
5. Department of Materials Engineering, Ming Chi University of Technology, Tai-Shan, New Taipei City 24301, Taiwan
* Correspondence: jpchen@mail.cgu.edu.tw; Tel.: +886-3-2118800

Citation: Dash, B.S.; Jose, G.; Lu, Y.-J.; Chen, J.-P. Functionalized Reduced Graphene Oxide as a Versatile Tool for Cancer Therapy. *Int. J. Mol. Sci.* **2021**, *22*, 2989. https://doi.org/10.3390/ijms22062989

Academic Editor: Ana María Díez-Pascual

Received: 1 February 2021
Accepted: 11 March 2021
Published: 15 March 2021

Publisher's Note: MDPI stays neutral with regard to jurisdictional claims in published maps and institutional affiliations.

Copyright: © 2021 by the authors. Licensee MDPI, Basel, Switzerland. This article is an open access article distributed under the terms and conditions of the Creative Commons Attribution (CC BY) license (https://creativecommons.org/licenses/by/4.0/).

Abstract: Cancer is one of the deadliest diseases in human history with extremely poor prognosis. Although many traditional therapeutic modalities—such as surgery, chemotherapy, and radiation therapy—have proved to be successful in inhibiting the growth of tumor cells, their side effects may vastly limited the actual benefits and patient acceptance. In this context, a nanomedicine approach for cancer therapy using functionalized nanomaterial has been gaining ground recently. Considering the ability to carry various anticancer drugs and to act as a photothermal agent, the use of carbon-based nanomaterials for cancer therapy has advanced rapidly. Within those nanomaterials, reduced graphene oxide (rGO), a graphene family 2D carbon nanomaterial, emerged as a good candidate for cancer photothermal therapy due to its excellent photothermal conversion in the near infrared range, large specific surface area for drug loading, as well as functional groups for functionalization with molecules such as photosensitizers, siRNA, ligands, etc. By unique design, multifunctional nanosystems could be designed based on rGO, which are endowed with promising temperature/pH-dependent drug/gene delivery abilities for multimodal cancer therapy. This could be further augmented by additional advantages offered by functionalized rGO, such as high biocompatibility, targeted delivery, and enhanced photothermal effects. Herewith, we first provide an overview of the most effective reducing agents for rGO synthesis via chemical reduction. This was followed by in-depth review of application of functionalized rGO in different cancer treatment modalities such as chemotherapy, photothermal therapy and/or photodynamic therapy, gene therapy, chemotherapy/phototherapy, and photothermal/immunotherapy.

Keywords: reduced graphene oxide; chemotherapy; photothermal therapy; photodynamic therapy; gene therapy; immunotherapy

1. Introduction

Cancer, unrestrained cell growth in the human body, has severely threatened human health worldwide due to its incurability and high death rate [1]. Although reasons for this fatal disease are uncountable, the mechanism of cancer development is associated with the failure of a body's normal control mechanism, which results in the abnormal proliferation of new cells [2]. Owing to the severity of this disease, researchers and medical professionals have made huge contributions in advancing various treatment modalities—including surgery, chemotherapy, and radiation therapy—for saving human life [3,4]. Although conventional drug delivery systems and treatment approaches have provided some treatment efficacy, its effectiveness is limited by various factors, including multi-drug resistance, rapid metabolism and elimination of drugs, non-specific cytotoxicity, etc. [5–7].

Considering the drawbacks of conventional therapies, introduction of nanomaterials in biomedical research has provided a revolutionary application of biomaterials in cancer therapy [8]. Indeed, due to their unique properties, nanomaterials have gained increasing attention for new and innovative use in biomedical research, particularly as nanocarriers for delivery of therapeutic drugs in cancer therapy [9]. To date, a wide variety of nanocarriers—including liposomes, micelles, peptides, and inorganic particles—are being explored as nanovehicles for delivery of cancer therapeutics [10–18]. Among these, carbon-based nanomaterials in the graphene family have gained particular attention due to their effectiveness and versatility for cancer treatment [19–22]. The carbon-based nanomaterials have essential structural and surface features for loading and pH-sensitive release of aromatic anticancer drugs [23]. Considering certain limitations associated with other materials, these nanomaterials attracted tremendous attention for delivery of cancer therapeutics not only due to their unique physico-chemical properties such as high surface area for drug loading [24], but also due to preferred biological properties such as endosomal escape after intracellular uptake for gene delivery and gene therapy [25,26].

The lateral dimensions and thickness of graphene family nanomaterials, such as graphene oxide (GO), reduced graphene oxide (rGO), graphene quantum dots, and graphene nanoribbons can be fine-tuned from original two-dimensional (2D) structure into zero-, one-, or three-dimensional assemblies [27], which provide improved accumulation as drug vehicles and contrast agents at specific target sites [28]. Such unique and tunable features have promised their new applications in drug delivery [29]. Nonetheless, the promise that these nanomaterials have shown in nanomedicine is not only limited to drug delivery, but also in highly sensitive biosensors and high throughput bioassays, as well as scaffolds for tissue engineering [30].

Similar to GO, rGO is a 2D nanomaterial in graphene family with a single-atom-thick layer of sp2 hybridized carbon atoms arranged in a honeycomb lattice structure, which is obtained by reducing GO through chemical, thermal, or electrical methods to eliminate the oxygen-containing functional groups on the surface. Owing to the unique surface property and presence of functional groups, functionalized rGO can accommodate high loading of genes to increase the delivery efficacy of nucleic acid therapeutics in gene therapy. The high surface area also enables the loading of abundant hydrophobic aromatic anticancer drugs for chemotherapy or photosensitizers for photodynamic therapy (PDT) via π–π interaction [31]. Besides being excellent photo-absorbers with high light absorption ability in the near infrared (NIR) range, rGO is associated with pronounced photothermal effect compared with GO, rendering potential applications in cancer photothermal therapy (PTT) [32]. Indeed, with its facile synthesis, high water dispersibility, easy surface functionalization, and good biocompatibility, rGO has emerged as an excellent multifunctional nanomaterial for PTT [33]. After combining this unique characteristic with the high loading capacity of anticancer drugs, rGO reveals itself as a promising nanomaterial for chemo-photothermal therapy [34]. Surface functionalization with multiple therapeutic moieties or conjugation with targeting ligands on rGO surface further permit its use in targeted synergistic cancer therapy such as chemo-phototherapy and photothermal/immunotherapy.

2. Preparation of Reduced Graphene Oxide (rGO) by Chemical Reduction

The typical methods for preparation of rGO involve reducing GO by thermal, chemical or electrical methods. Among them, the chemical reduction method that deoxygenates GO with a reducing agent prevails over other non-chemical routes for rGO synthesis, which can produce stable dispersions of rGO with improved quality. Herein, we review the most effective chemical reagents that can act as a direct or indirect reducing agent to convert oxygenated graphene (GO) into rGO (Figure 1). The characteristics of as-produced rGO and its applications are summarized in Table 1.

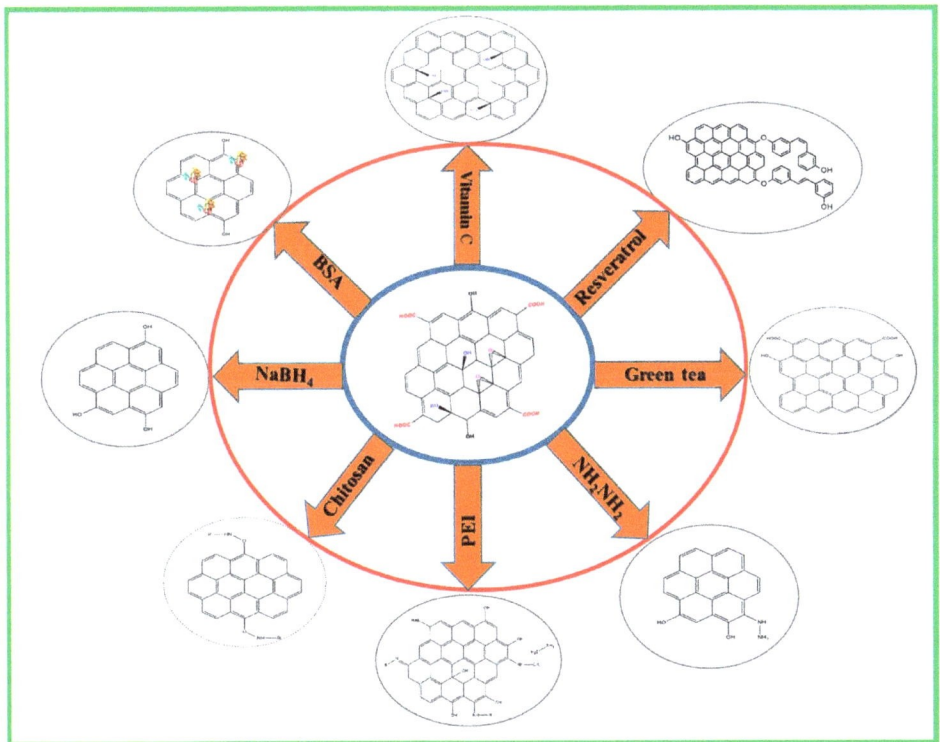

Figure 1. Schematic diagram illustrating the preparation of rGO from GO using different chemical reducing agents. BSA, bovine serum albumin; PEI, polyethyleneimine.

It should be noted that although different synthesis routes were reported for rGO synthesis, chemical reduction is widely accepted as the most promising method for large-scale production of rGO. However, considering the toxicity of many chemical reducing agents, biomedical application of rGO prefers non-toxic green chemicals for the reduction process, which could provide stability, non-toxicity, and functionality to resulting rGO products. However, as various kinds of rGO synthesized via different routes have been used successfully for cancer therapy, there is still no definite answer regarding which synthesis route could produce the most effective rGO product for application in cancer therapy.

2.1. Vitamin C

Vitamin C (L-hexuronic acid or L-ascorbic acid) is a mild reducing chemical widely used as a reducing agent for GO due to its non-toxicity. It is consider as one of the suitable choices for reducing GOs, not only because it produces highly reduced GO nanosheet suspended in water or in hydrogels at room temperature or mild temperatures, but also because it produces an environmentally friendly by-product, dehydroascorbic acid, after the reaction [35,36]. The chemical reduction of GO by vitamin C has been described in many reports and most of them describing simple mixing of GO with vitamin C using a magnetic stirrer at 60–70 °C and react for 30 min to 2 h. The reduction of GO to rGO is confirmed by color change from brown to black [37–39]. A stable suspension of highly reduced GO could be produced using vitamin C in aqueous solution as well as in organic solvent such as dimethylformamide and N-methyl-2-pyrrolidone. Moreover, as vitamin C is composed of carbon, oxygen, and hydrogen, the possibility of introducing heteroatoms to rGO could be avoided.

2.2. Hydrazine Hydrate

Hydrazine (N_2H_4) or hydrazine hydrate ($N_2H_4 \cdot H_2O$) is one of the most widely used reductants for rGO synthesis in large scale. Nonetheless, considering the explosive and toxic nature of hydrazine hydrate, the reduction of GO using this reducing agent should be performed with care [40]. Literature describes the reduction process by mixing dispersion of GO with hydrazine hydrate and ammonia solution with a weight ratio of hydrazine to GO at 7:10. After being vigorously shaken or stirred for a few minutes, the reaction was carried out in a water bath at 96 °C for 1 h. Once the reduction is complete, excess hydrazine must be removed by dialysis against 0.5% ammonia solution [41]. In another study, GO was reduced by hydrazine hydrate in the presence of poly(sodium 4-styrenesulfonate) (PSS) to produce stable PSS-coated rGO nanosheet in aqueous dispersion [42]. Treatment of GO (5 mL, 0.5 mg/mL) with hydrazine (0.50 mL, 32.1 mM) at 100 °C for 24 h is another way of reducing GO [43]. There are some other studies that used 80 °C for reaction, where aqueous solution of GO was stirred and sonicated for at least 1 h before reacting with hydrazine hydrate (weight ratio of hydrazine hydrate to GO = 1:1) with continuous stirring and sonication [44]. All studies confirm that hydrazine is a good reducing agent for producing rGO.

2.3. Resveratrol

Resveratrol is polyphenol compound used both as a reducing agent and a stabilizer. Resveratrol-guided reduction of GO could provide better biocompatibility, solubility, and selectivity compared to many other reducing agents. The reduction process involves addition of 50 µM resveratrol into GO (1 mg/mL), which was sonicated for 15 min beforehand, and reacted at 40 °C for 1 h. This was continued by cooling and sonication for 15 min, followed by continuous stirring for 1 h at 90 °C. After centrifugation and washing in distilled water, prepared rGO could be recovered from the solution [45].

2.4. Chitosan

Chitosan, a biocompatible and biodegradable polysaccharide derived from incomplete deacetylation of chitin, serves as a reducing agent in the synthesis of many nanoparticles [46,47]. Suspension of rGO in aqueous solution has been prepared by chemical reduction of GO at room temperature in the presence of chitosan. To perform the synthesis, a 1:1(w/w) mixture of GO and chitosan were heated at 37 °C for 72 h under constant stirring. Followed by this, excess chitosan in the solution was removed by centrifugation at 8000 rpm for 1 h and subsequently washed with 2% acetic acid solution. The rGO was dispersed in distilled water by sonication [48,49]. Due to the higher biocompatibility, chitosan-based reduction can enhance the potential biological and medicinal applications of rGO.

2.5. Polyethyleneimine (PEI)

Polyethyleneimine (PEI) is a widely used reducing agent and surface modifier in the fabrication of rGO [50], which is a water soluble cationic polymer containing primary, secondary, and tertiary amino groups. To reduce GO, 60 mL of GO (0.1 mg/mL) dispersion and PEI solution was mixed under vigorous stirring at 80 °C for 2 h. The transformation of yellowish-brown to black dispersion indicates the successful transformation of GO to rGO. The mixture was then centrifuged and washed with water for recovery of PEI-rGO [50–52]. The incorporation of PEI molecule into GO can act as a source of carbon and produce rGO in a one-step hydrothermal process.

2.6. Sodium Borohydride

Sodium borohydride ($NaBH_4$) is consider as one of the efficient, nontoxic, noncorrosive, inexpensive reducing agents available for reduction of GO. $NaBH_4$ has been frequently used as a reducing agent for aldehydes and ketones to produce alcohols. For synthesizing rGO from GO, a GO suspension (0.5 mg/mL) was mixed with $NaBH_4$ as well as $CaCl_2$ and stirred for 12 h at room temperature [53]. In another study, the reduction of GO by NaBH$_4$ was carried out at different temperatures, showing the highest extent of reduction when the reaction was conducted at 80 °C [54].

2.7. Bovine Serum Albumin (BSA)

Bovine serum albumin (BSA) is an affordable protein with high biocompatibility, which can act as a reductant and a stabilizer of GO due to the presence of the amino acid tyrosine (Tyr) within it [55,56]. The reduction of GO was carried out by reacting 1 mg/mL GO solution with 50 mg/mL BSA at 70 °C. After the solution pH was brought up to 12 with 1 M NaOH, the mixture was stirred at 50 °C for 24 h to observe a transition of solution color from light brown (GO) to dark black (rGO) [57]. Excess BSA was removed by centrifugal filtration with a 150 kD molecular-weight-cut-off (MWCO) membrane to obtain purified rGO/BSA hybrids suspended in water and stored at 4 °C [57,58].

Table 1. Reducing agents for producing reduced graphene oxide (rGO) from graphene oxide (GO).

Reducing Agent	Characterisitics	Applications	Reference
Vitamin C	Natural compound; non-toxic; mild reaction temperature; environment friendly byproducts; avoid introducing heteroatoms; reaction in aqueous or organic solution	Embedded in chitosan hydrogel for bone tissue engineering; functionalized with antimicrobial peptide for antibacterial activity	[37–39]
Hydrazine hydrate	Explosive; toxic; large scale production; low cost	Improve electrical conductivity; embedded in polyacrylic acid nanofiber mats for controlled release of antibiotics	[41–44]
Resveratrol	Natural phenolic compound; anti-oxidant; stabilizer; biocompatibility; solubility; green synthesis	Produce marked changes in cellular morphology and reduce cell viability of cancer cells for cancer therapy	[45]
Chitosan	Biocompatible; biodegradable; reduction at body temperature; biological and medicinal applications	Reversible change of dispersion/aggregation state with pH; pH-sensitive release of drug; loading with drug and photosensitizer for cancer chemotherapy/phototherapy	[48,49]
Polyethylenimine	Surface modifier; one-step hydrothermal reduction; high cargo loading; prevent agglomeration	Improved gas barrier property in composite films; in hemin-bovine serum albumin composite as peroxidase mimetics; gene delivery; increase strength of nylon composites	[50–52]
Sodium borohydride	Efficient; ambient conditions; reaction in aqueous solution	Decrease electrical resistance; enhance electrical conductivity	[53,54]
Bovine serum albumin	Biocompatible; stabilizer; binding by adhesion to surface; metal particle-binding platform; cell adhesive	For cancer chemo-photothermal therapy; adsorption and assembly of metal particles; create protein–metal nanocluster for detecting trypsin	[57,58]
Gree tea polyphenols	Biocompatible, biodegradable; green synthesis; good dispersion in both aqueous and organic solutions; non-toxic	Enhance thermal conductivity in chitosan polymer composites; deposite onto electrode for detection of sunset yellow in foods; reduce cytotoxicity of GO	[59–61]

2.8. Green Tea Polyphenols

Considering the harmful and hazardous natures of many commercial chemical reducing agents, researchers have focused on green methods for the production of rGO by chemical reduction. As a result, green tea extract is considered as a good option considering its easy availability, eco-friendly characteristics, and cheap price. Green tea is rich in polyphenolic compounds, with epigallocatechin gallate (EGCG) making up about 50–60% of total tea polyphenols. To reduce GO, green tea powder (2 g) was added to 100 mL of deionized water and boiled at 100 °C for 20 min, and then filtered. The GO (50 mg) was added to the green tea solution and sonicated for 30 min, followed by reflux at 90 °C under nitrogen atmosphere. After that, the solution was washed with water to remove excess green tea powder [59]. Alternatively, 10 mL green tea extract was added dropwise to 20 mL of GO aqueous suspension (0.5 mg/mL) within 45 min and the mixture was refluxed at 60 °C for 6 h before precipitation of rGO out of the solution [60,61].

3. Application of Reduced Graphene Oxide (rGO) in Cancer Therapy

rGO-based nanocomposite has emerged as a promising nanomaterial in nanomedicine. Most recent cytotoxicity studies indicate that surface functionalization of rGO could lead to enhanced biocompatibility as well as increased stability in physiological buffers. Therefore, the use of rGO-based nanomaterials for targeted pH-responsive drug delivery may overcome current challenges and provide new treatment modality in cancer therapy. Nonetheless, other than cytotoxicity study, the distribution and excretion of rGO-based nanomaterials is of paramount importance before clinical translation. The first hurdle that rGO-based cancer therapeutics faces will be the reticuloendothelial system (RES) after intravenous delivery. Overall, nanoparticles with a particle size of ~100 nm are expected to have prolonged circulating half-lives, which should be the preferred size of rGO-based nanocomposites. After escaping the RES, circulating rGO can exit tumor blood vessels and accumulate in cancerous interstitium due to the leaky tumor blood vessels by the enhanced permeation and retention (EPR) effect. For rGO-based nanomaterials to extravasate the vasculature, they should preferably possess neutral or negative charge and be within 10–100 nm in size. Indeed, nanomaterials coated with biocompatible moieties with size smaller than 100 nm are believed to be cleared from the body without noticeable toxicity after systemic administration. Even though the EPR effect may increase accumulation at the tumor site, improving the active targeting ability of rGO-based nanocomposites or using administration route other than intravenous injection should be studied. A systematic study of the biological behavior of injected rGO in vivo—such as stability, biodistribution, secretion, etc.—should be attempted for better clinical use. Furthermore, the functionalization of rGO may be difficult in large scale, which might limit the application of functionalized rGO in cancer therapy from bench to bedside.

rGO is widely accepted for application in single mode cancer therapy such as chemotherapy, photothermal therapy (PTT), photodynamic therapy (PDT), and gene therapy as well as in dual mode cancer therapy including chemotherapy/phototherapy and photothermal therapy/immunotherapy. A schematic diagram illustrating the mechanisms involved is depicted in Figure 2. In this section, we categorize most up-to-date studies using functionalized rGO-based nanocarriers in cancer therapy into several sections based on the treatment modality involved. A summary of rGO-based nanocarriers, the agents used for functionalizing rGO, cancer cell lines used in the study and the type of study is provided in Table 2.

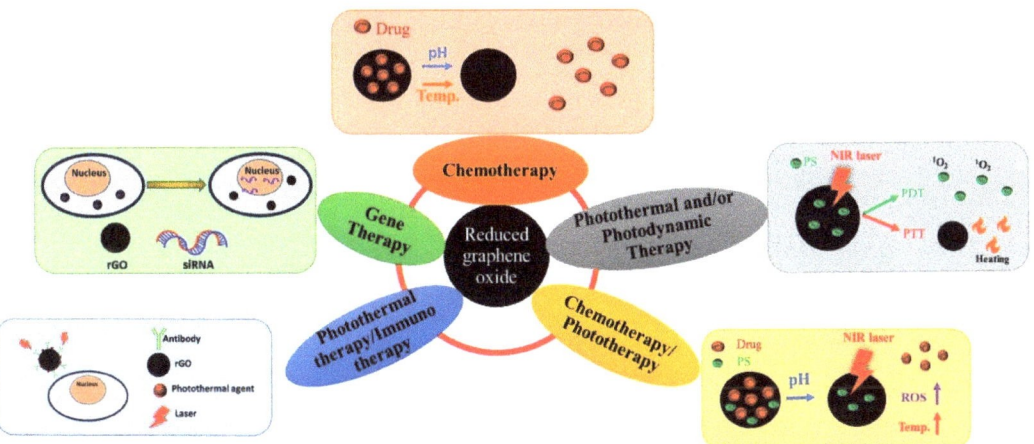

Figure 2. Applications of reduced graphene oxide (rGO) in cancer therapy.

3.1. Chemotherapy

The chemotherapy that cures malignancy by means of chemical drugs has gained worldwide acceptance due to its capability to block cell proliferation and to cause cell apoptosis [62]. The chemotherapeutic drugs like doxorubicin (DOX), cisplatin, paclitaxel (PTX), mitoxantrone (MTX), and 5-fluorouracil (5-FU) have been proved to inhibit growth rates of different cancer cells and to limit their metabolic functions. However, the associated unwanted complications with these drugs due to their cytotoxicity towards normal/healthy cells have limited the application of chemotherapy in cancer therapy. Hence, nanocarrier-based drug delivery system was introduced as an effective approach to alleviate this limitation, with the aim to deliver chemotherapeutic drugs with minimum side effects. Indeed, many nanomaterials—such as polymeric nanoparticles, liposomes, micelles, and metal nanoparticles—have been employed for this purpose to deliver different chemotherapeutic drugs due to their unique characteristics unattainable through free drug administration. Recently, researchers are turning to development of rGO-based drug delivery platforms for carrying large amount of chemotherapeutic drugs with large surface area, as well as the pH-responsive drug release behavior offered by rGO.

Ma and co-workers used a green approach to convert GO to rGO using riboflavin as a reducing agent and the resulting riboflavin-rGO nanocarrier was used for DOX loading through π–π interaction [63]. The results suggested that rGO exhibits high DOX loading, good stability, and pH-sensitive sustained drug release, which is evident from the effective cytotoxicity against MCF-7 and A549 cancer cells in vitro [64]. In another study, Wei et al. used rGO-C_6H_4-COOH for DOX loading, followed by modifying with PEI to enhance water solubility and conjugating with folic acid (FA) for targeted drug delivery. Due to specific targeting of FA to CBRH7919 cancer cells as well as pH-responsive drug release after endocytosis, the conjugation of DOX with rGO-PEI-FA could arrest cancer cells in the G2 phase and lead to cell apoptosis. There are plenty of other examples for rGO-based targeted delivery of DOX. Daysi et al. also used FA-functionalized nanocomposite consisting of chemically-reduced rGO and manganese-doped zinc sulfide quantum dots (FA-rGO/ZnS:Mn) for targeted delivery of DOX [65]. The dispersion stability, DOX loading and release efficiency, internalization, and biocompatibility of FA-rGO/ZnS:Mn resulted in excellent anti-cancer efficiency against breast cancer cells. Moreover, FA functionalization improved the selectivity of this drug delivery platform for specific targeting of folate receptor molecules overexpressed on cancer cell surface. Taken together, the FA-rGO/ZnS:Mn was suggested to be an excellent theranostic nanocomposite for breast cancer treatment. Similarly, Miao et al. explained the application of DOX-loaded cholesteryl hyaluronic acid (CHA)-coated rGO nanosheets (CHA-rGO) for the treatment of CD44-overexpressing tumors [66]. As a primary ligand of CD44, HA increases the stability and safety of rGO by enhancing the tumor-targeted distribution of DOX to CD44-overexpressing cancer cells, thereby providing better drug accumulation and substantial reduction of tumor volume from in vivo study in nude mice. He and co-workers used rGO capped by alkyl-grafted mesoporous silica (MSN-C_{18}) as a carrier of DOX for chemotherapy [67]. The in vitro cell viability assay performed using SMMC-7721 cancer cells showed enhanced DOX release upon near NIR light exposure, which led to higher cytotoxicity toward cancer cells, indicating its potential use as a nanocarrier for controlled drug release.

The application of rGO in chemotherapy is not only limited to targeted delivery of DOX, as other drugs or natural compounds with anti-cancer activity were reported to be delivered through rGO. Chen et al. used methoxypolyethylene glycol amine (mPEG-NH_2) for one-step green reduction and PEGylation of GO to synthesize rGO/PEG [68]. The rGO/PEG showed excellent water stability and two-fold increase of resveratrol loading over GO/PEG via hydrophobic interactions and π–π stacking. From in vitro experiments, NIR laser irradiation (808 nm) could enhance resveratrol release from rGO/PEG to increase the cytotoxicity against 4T1 murine breast cancer cells by lowering cell viability and inducing cell apoptosis. In animal models with subcutaneously implanted cancer cells, resveratrol-loaded rGO/PEG injected intratumorally to tumor-bearing nude mice also significantly suppress tumor growth under photothermally controlled drug delivery. In another study, rGO synthesized through reduction by Euphorbia milii plant extract was used as a carrier of the chemotherapeutic drug paclitaxel for cancer treatment [69]. The drug-loaded rGO showed high cytotoxicity toward human lung cancer cell line (A549) for potential chemotherapy of lung carcinoma.

The chemotherapy using dual drugs may be a better approach to kill cancer cells in the metastatic stage as the synergistic effect offered by dual drugs may introduce more anti-proliferative effect to cause cancer cell death. Muthoosamy et al. developed amphiphilic polymer PF-127 functionalized rGO for co-loading of anti-cancer drugs paclitaxel and curcumin on the surface through π–π interactions [70]. The drug-loaded composite showed synergistic anti-tumor efficacy towards both A549 lung cancer cells and MDA-MB-231 breast cancer cells. However, loading of an aromatic hydrophobic drug like paclitaxel on the surface of a hydrophobic carrier like rGO by π–π stacking and hydrophobic–hydrophobic interactions is challenging as hydrophobic carriers are not stable in physiological solutions. To address this problem, Hashemi et al. introduced a rGO-based nanocarrier with high paclitaxel loading capacity through functionalization and stabilization with R9 peptides, where pristine rGO sheets were found to be unstable in aqueous solutions and aggregated to decrease the surface area available for drug loading [71]. In a different study, Dhanavel et al. developed dual drug-encapsulated chitosan/rGO nanocomposite by entrapping 5-fluorouracil (5-FU) and curcumin in chitosan/sodium tripolyphosphate gel in the presence of rGO nanosheet for dual drug delivery to HT-29 colon cancer cells [72]. The synergistic cytotoxicity was observed for dual drug-loaded nanocomposite to inhibit the growth of HT-29 colon cancer cells compared with single drug therapy.

Considering the biological aspects, researchers have developed composites of rGO with nanoparticles or polymers for cancer treatment. Among these, combination of rGO with gold (Au) nanoparticle has gained considerable attention. In one study, Sanad et al. prepared rGO–gold nanocomposites (rGO-Au) by incorporating Au nanoparticles inside the rGO matrix through in situ reduction with sodium borohydride, followed by loading 5-FU by pore capping [73]. The results obtained from cytotoxicity assay determined by the reduced half maximal inhibitory concentration (IC_{50}) using MTT assay in addition to enhanced cell apoptosis from flow cytometry analysis suggested the nanocomposite can enhance targeted delivery of 5-FU as well as cytotoxicity to MCF-7 breast cancer cells. In a similar study, Jafarizad et al. prepared Au nanoparticle-loaded rGO as a covalent drug delivery system for pH-dependent release of mitoxantrone (MTX) [74]. For polymer coating, Ryu and co-workers prepared PEI-rGO nanocarrier for pH-responsive delivery of

DOX to Hela and A549 cancer cell lines [75]. The drug-loaded nanocomposite was further coated with pH-responsive charge-conversional polymer polyethyleneimine-poly-L-lysine-poly-L-glutamic acid (PKE) to endow charge-conversional property and serum stability to PEI-rGO-based drug delivery system. They found that DOX-loaded PEI-rGO after PKE coating released more DOX under low pH lysosomal condition and showed enhanced anticancer activity in HeLa and A549 cancer cells. Considering the side effects associated with chemotherapy, SreeHarsha et al. prepared hybrid nanoparticle by coating rGO with chitosan and stabilized with tripolyphosphate to produce stabilized nanocomposite for delivery of DOX to PC-3 cancer cells [76]. The sustained DOX release observed under photothermal conditions endowed this nanocarrier with improved efficacy in treating prostate cancer.

3.2. Photothermal Therapy (PTT) and/or Photodynamic Therapy (PDT)

Photothermal therapy (PTT) involves local temperature rise after exposing a photothermal agent to electromagnetic radiation such as visible or NIR light, which up converting light energy into heat, can induce death of cancer cells [77]. Many nanomaterials are effective photothermal agents in causing cancer cell apoptosis/necrosis with local hyperthermia from NIR laser exposure [78]. Graphene-based materials are good photothermal agents used for PTT considering its multifunctionality [79]. Moreover, the combination of graphene-based materials with inorganic particles, like iron oxide and gold nanoparticles, can further enhance the photothermal effect and lead to higher cancer cell death rate during PTT [80–83]. On the other hand, photodynamic therapy (PDT) is another form of phototherapy involving light and a photosensitizer (PS), which when used together with oxygen, can produce molecular oxygen or reactive oxygen species to elicit cancer cell death [84,85]. In one study, GO was fond to act both as a photothermal agent for PTT and a photosensitizer for PDT, making it an excellent candidate for synergistic phototherapy [86].

Robison et al. pioneered the use of rGO for PTT, who showed nano-sized rGO produced by chemical reduction of GO has six-fold higher NIR absorption rate than GO, endorsing its preferred use over GO as a photothermal agent [87]. The modification with targeting peptide bearing the Arg-Gly-Asp (RGD) motif further provided rGO with selective cellular uptake ability by U87MG glioma cancer cells, indicating rGO is a multi-functional photothermal agent. Their results also provided strong evidence that nano-sized rGO is highly effective as a photothermal agent when compared to other carbon-based nanomaterials as well as inorganic nanoparticles like gold or iron oxide. Similarly, Shim et al. modified rGO with clostridium perfringens enterotoxin peptide-linked chlorin e6 (Ce6) as a dual photodynamic and photothermal cancer therapeutic platform [88]. The intracellular uptake studies performed on U87 glioblastoma cells confirmed the ligand-mediated cellular uptake. The combined therapy using 660 nm light source for the PDT agent (Ce6) and 808 nm for rGO showed enhanced targeted dual phototherapy. In another study, He and co-workers synthesised palladium nanoflowers-decorated rGO (rGO/PdNFs) and explored its versatile applications in catalysts, sensor, and PTT [89]. The modification of rGO with PdNFs increased its photothermal conversion due to enhanced absorption in the NIR window. Both in vitro study using HeLa cells and in vivo animal experiments performed in tumor-bearing mice model indicates rGO/PdNFs could result in effective photothermal antitumor efficacy. The study using alanine-grafted rGO as a photothermal platform for cancer therapy confirmed that PTT using 808 nm laser irradiation produced 89% and 33% higher photothermal effect compared to GO and rGO, respectively. The conjugation of alanine to GO via π–π interactions reduced GO to rGO, not only increased the 808 nm absorbance but also acted as a targeting ligand to kill U87MG cancer cells selectively [90].

The photothermal applications rGO were further explored in combination with other nanoparticles and photosensitizers in synergistic cancer therapy. Otari et al. performed one-step reduction of GO to rGO and decoration of rGO with Au nanoparticles and thermostable antimicrobial nisin peptides to synthesize NAu-rGO [91]. After treating MCF-7 breast cancer cells with the nanocomposite followed by 800-nm diode laser (0.5 W/cm^2) treatment for 5 min, 80% cell growth inhibition was found. Similarly, Zhang et al. attached polyethylene glycol (PEG) modified Ru (II) complex (Ru-PEG) to rGO surface by hydrophobic π–π interaction, and applied the nanocomposite as a photothermal agent and a photosensitizer for PTT/PDT [92]. The A549 lung cancer cells treated with rGO-Ru-PEG and sequentially exposed to of 808 nm (for PTT) and 450 nm (for PDT) wavelength light source resulted in enhanced cytotoxicity due to combined phototherapeutic effects. The rapid reduction in relative tumor volume observed from animal experiments also confirmed the synergistic effect of dual phototherapy.

For localized combination cancer therapy, Chang and co-workers developed rGO/AE/AuNPs hydrogel containing rGO and Au nanoparticles, by using amaranth extract (AE) both as a reducing agent and a precursor, which crosslinked upon 660 nm laser exposure to form a composite hydrogel [93]. Both Au nanoparticles and rGO acted as photothermal agents while the chlorophyll derivatives in AE acted as a photosensitizer to accelerate the generation of cytotoxic singlet oxygen. Upon hydrogel formation on the surface of HeLa cancer cells in situ, the composite hydrogel was used as a combined PTT/PDT platform by repeated irradiation with 808 nm laser in multiple antitumor therapies. In another study, a PTT/PDT reagent was synthesized by conjugating tetrakis(4-carboxyphenyl) porphyrin (TCPP) to rGO–PEI, which was formed based on carboxylic acid functionalized rGO for combination PTT/PDT therapy [94]. The rGO–PEI–TCPP composite showed excellent stability in different biological solutions. The results obtained from studies with CBRH7919 cancer cells indicates induced cell apoptosis upon laser irradiation, due to the combined photothermal and photodynamic effects with the production of heat and singlet oxygen. The increase in temperature upon exposing a photothermal agent to laser light may cause side effects, and hence determining the optimum concentration of PTT agent or performing experiment at lower laser power is important in phototherapy. Jafarirad and team developed a non-invasive strategy for low-level laser induced cancer therapy. The hybrid nanocomposites (ZnO/rGO, Ag-ZnO/rGO, and Nd-ZnO/rGO) synthesized by green synthesis methods were further optimized for concentration in anti-tumor study in vitro using MCF-7 cancer cells [95]. The results confirmed a low concentration (12.5 µg/mL) of hybrid together with low irradiation doses (8–32 J/cm^2) could lead to higher cell death.

In conventional PDT, the unfavourable bioavailability, low absorption band and limitations in tissue oxygenation are considered as possible limitations. To overcome these limitations, Kapri et al. fabricated a ~5 nm thick MoS$_2$ nanoplatelet and integrated them with n-type nitrogen doped rGO for PDT [96]. The p-MoS$_2$/n-rGO-MnO$_2$-PEG composite was prepared by modifying the nanosheet with poly(ethylene glycol) (PEG) to improve biocompatibility and colloidal stability in physiological solution, which was further surface decorated with MnO$_2$ to overcome the hypoxic conditions prevalent in tumor microenvironment, by increasing intracellular O$_2$ after reaction of MnO$_2$ with endogenous H$_2$O$_2$ in cancer cells. The nanosheet reveals increased apoptosis under NIR light irradiation by alleviating hypoxia and enhances the efficacy of PDT on HeLa cells in vitro. It is difficult to selectively kill cancer cells during PTT as normal cells are also simultaneously affected by the photothermal effect, which is the most common disadvantage associated with phototherapy. To solve this problem, an interesting design based on rGO was demonstrated by synthesizing water dispersible Cu$_2$O nanocrystal-rGO nanocomposites. In contrast to the highly efficient killing of both normal and cancer cells initiated by the photothermal effect under NIR irradiation, the photocatalytic effect of this nanomaterial results in selective killing of cancer cells in contrast to unselective cell-killing under NIR light [97]. This was demonstrated from the cytotoxicity assay performed on A549, HK-2 and MDA-MB-231 cancer cell lines in vitro.

The efficacy of rGO-based PDT/PTT could be upregulated by conjugating rGO with a ligand molecule to actively targeting cancer cells. Jiang et al. used hyaluronic acid (HA) as a targeting ligand to specifically deliver a photosensitizer Ce6 to CD44 overexpressing cancer cells for PDT with NIR irradiation [98]. The nanoplatform was prepared by dopamine-reduced rGO sheet and coated with mesoporous silica to load Ce6 as well as HA. The combination of photothermal conversion and controllable Ce6 release after NIR irradiation confers the nanoplatform with enhanced singlet oxygen generation that could lead to more significant destruction of targeted cancer cells. Lima-Sousa and co-workers also functionalized rGO with HA-grafted poly-maleic anhydride-*alt*-1-octadecene (HA-g-PMAO) for targeted PTT [99]. In vitro studies confirmed internalization by CD44 overexpressing MCF-7 cells as an on-demand PTT platform to elicit cancer cell ablation for potential targeted cancer therapy. Taking advantage of the fact that polyphenol compounds in green tea (GT)-reduced rGO can act as targeting ligand, targeted delivery of the nanocomposite resulted in 20% higher photothermal destruction of the high metastatic SW48 cancer cells than that of the low metastatic HT29 cells [100]. Although the exact mechanism is still under investigation, the attachment of polyphenol-modified rGO to cancer cell surface could be confirmed from flow cytometry studies for photothermal destruction of colon cancer cell line at 0.3 mg/mL rGO and 0.25 W/cm^2 NIR laser power density. To use heparin sulphate proteoglypican-3 (GPC3) as a targeting ligand for hepatocellular carcinoma, Liu et al. conjugated biotinylated GPC3 antibody to rGO (rGO-GPC3) and bind avidinylated nanobubbles to rGO-GPC3 using the biotin-avidin bioaffinity system for PTT. Using ultrasound-targeted nanobubble destruction, the local concentration of rGO around HepG2 cell line could be increased for photothermal ablation and PTT of hepatocellular carcinoma under 808 nm NIR irradiation [101].

Indocyanine green (ICG), a NIR dye approved by the U.S. Food and Drug Administration (FDA), can combine with rGO to promote the NIR absorption ability of rGO and enhance the PTT efficacy. A novel nanoagent using ICG-loaded polydopamine (PDA)-reduced graphene oxide nanocomposites (PDA-rGO) was found to be loaded with a large amount of ICG molecules for exhibiting stronger photothermal effect and amplify the PTT efficacy for cancer theranostics [102]. After photoacoustic imaging-guided PTT treatments using 808 nm NIR laser at 0.6 W/cm^2 for 5 min, the tumors in orthotopic 4T1 breast cancer mice model were completely eradicated with no observable treatment toxicity. Also using ICG, Sharker et al. designed a pH-responsive, NIR-sensitive rGO-based nanocomposite (ICG-CPPDN/rGO), by ionic complexation of ICG with CPPDN/rGO, for local destruction of cancer cells with minimal invasiveness to surrounding normal cells [103]. The nanocomposites showed pH-dependent photothermal effect from pH 5.0 to 7.4 due to the pH response relief and quenching effects of ICG on rGO sheet, which leads to photo-thermolysis as the pH was changed from 5.0 to 7.4 in vitro. Due to acidic tumor microenvironment, the nanocomposite showed improved photothermal destruction of MDA-MB-231 cancer cells both in vitro and in vivo compared to free ICG upon local NIR laser treatment.

3.3. Gene Therapy

Inhibiting gene expression by promoting site specific cleavage of target messenger RNA, small interfering RNA (siRNA) regulates the expression of genes by RNA interference (RNAi) and represents one of the promising developments in cancer therapy [104]. Considering the limitations of delivery of naked siRNA, which include endosomal escape, rapid excretion, low stability in blood serum, non-specific accumulation in tissues, siRNAs were usually delivered by loading to nanoparticles [105]. The use nanoparticles as nano-vehicles for siRNA delivery also offers the possibility of targeted gene therapy for more effective cancer treatment outcomes [106]. Various nano-sized particulate systems, including silica and silicon-based nanoparticles, metal and metal oxide nanoparticles, carbon nanotube, graphene, dendrimer, polymers, cyclodextrin, liposome, and semiconductor nanocrystals have been developed for systematic delivery of siRNA [107]. The cationic PEI-rGO nanoparticles after reducing and modifying GO with PEI have gained particular attention in rGO-based cancer gene therapy, since PEI is widely used for non-viral transfection and offers advantages over other polycations with high endosomolytic activity and strong DNA compaction ability [108,109].

As low molecular weight PEI was proved to exhibit less cytotoxicity, Chau et al. studied the covalent functionalization of GO with triethyleneglycoldiamine or 800 Da molecular weight PEI via the epoxy ring opening reaction. The data obtained using gel electrophoresis confirmed that PEI-rGO was more efficient in complexing with siRNA to offer higher complexing capacity, making it an excellent candidate for gene silencing applications [110]. In a separate study, rGO was modified with low molecular weight branched polyethyleneimine (BPEI) via polyethylene glycol (PEG) spacer (PEG-BPEI-rGO) as a nano-vehicle for photothermally controlled gene delivery [111,112]. The nanocomposite formed stable nano-sized complex with plasmid DNA to offer high gene transfection efficiency for experiments performed with PC-3 and NIH/3T3 cell lines without significant cytotoxicity. Most importantly, PEG-BPEI-rGO demonstrated enhanced gene transfection efficiency upon NIR irradiation. After investigating with a proton sponge effect inhibitor Bafilomycin A1, the enhancement of gene transfer was found to be associated with accelerated endosomal escape of the nanocomposite, with the photothermal effect of rGO.

3.4. Chemotherapy/Phototherapy

As chemotherapy and phototherapy have produced promising outcomes in cancer therapy, researchers started combining these therapeutic modalities for synergistic cancer treatment [113]. Specifically, considering the high loading efficiency of chemotherapeutic drugs and the unique NIR laser-responsive characteristics for PTT/PDT, graphene-based nanomaterials have received attention for chemo-phototherapy [114]. By intravenous delivery of rGO-based nanocomposites, the combination of chemotherapy with phototherapy (PTT and/or PDT) also demonstrated promising anti-cancer efficacy with subcutaneously implanted cancer cells in vivo, as depicted schematically in Figure 3 [115,116].

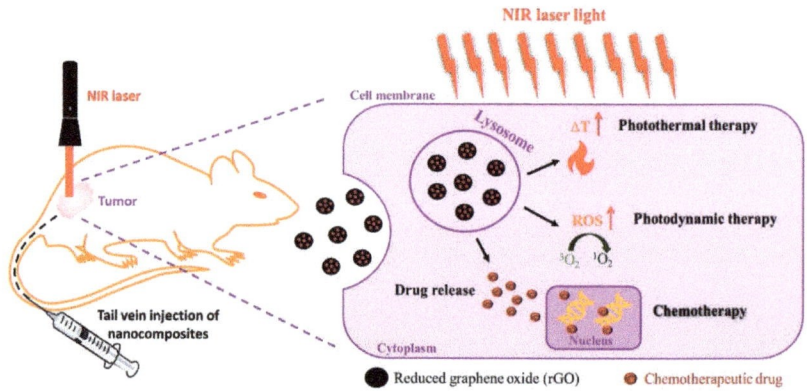

Figure 3. A schematic diagram showing the treatment of subcutaneously implanted cancer cells by combined chemotherapy/phototherapy using reduced graphene oxide (rGO).

Oz and co-workers functionalized rGO by surface anchoring maleimide-containing catechol (dopa-MAL) through noncovalent interaction. Using thiol–maleimide chemistry, they modify rGO with cyclic peptide c(RGDfC) as a targeting ligand for targeted delivery of DOX to cancer cells. The in vitro studies performed with MDA-MB-231 cell line indicated that DOX loaded rGO/dopa-MAL-c(RGDFC) was more effective than free DOX in killing the cancer cells after exposure to 980 nm laser irradiation (2 W/cm^2) for 10 min [117]. This was attributed to the enhancement of chemotherapy with PTT from the endocytosed nanocomposite upon NIR laser exposure for targeted synergistic cancer cell killing. Surface functionalization of rGO for enhancing the hydrophilicity is a strategy for effective drug delivery. Hu et al. used folic acid modified dextran-g-octadecanoic acid to decorate rGO surface with hydrophilic dextran moiety through octadecanoic acid hydrophobic anchoring and with folic acid for enhanced intracellular uptake by cancer cells. The nanocomposite was loaded with anticancer drug DOX for chemotherapy/phototherapy where in vitro analysis performed with HeLa cells at endosomal acidic environment (pH 5.3) confirms increased DOX release upon weakening of noncovalent binding between DOX and rGO. Compared with single mode chemotherapy, the combination of local chemotherapy with external NIR-induced PTT demonstrated dual therapy and offered higher therapeutic efficacy [118]. The study confirmed that by eliciting higher cytotoxicity toward cancer cells through the photothermal response of rGO under NIR irradiation, the concentration of DOX needed for chemotherapy could be significantly reduced to minimize the potential side effect of chemo drugs. Similarly, Hu and co-workers also delivered DOX using dextran-reduced rGO through direct conjugation of dextran on rGO surface by hydrogen bonds. This was followed by self-assembly to form rGO/Dex nanoparticles. After conjugating with RGD peptide for recognition by $\alpha_v\beta_3$ integrin on cancer cell surface to enhance intracellular uptake, the in vitro chemo-phototherapy performed with external NIR laser on B16F10 cell line resulted in higher anti-cancer efficacy [119].

Considered as one of the most widely used chemotherapeutic drugs, DOX has been loaded to rGO synthesized with various reducing agents. For this purpose, functionalized rGO with bovine serum albumin (BSA) as a reduced agent was used as a carrier of DOX [57]. Brain tumor cells (U87MG) treated with BSA-rGO revealed that combination of photo-chemo treatment enhanced the treatment efficacy as compared to single mode phototherapy using BSA-rGO or DOX (chemotherapy). Similarly, Zaharie-Butucel and team reduced GO using chitosan to combine PTT and PDT, followed by using chitosan-rGO as a carrier for DOX in synergetic therapy of colon cancer [49]. Targeted delivery of DOX using polydopamine-functionalized rGO (pRGO) is another example in this category [120]. pRGO modified with CD44 targeting ligand hyaluronic acid (HA) was used in combination with DOX-loaded mesoporous silica (MS) (pRGO@MS-HA) to generate both pH and NIR-triggered DOX release from the multifunctional nanosystems, showing excellent combined effect in multimodal cancer therapy. In another example, Hao et al. used tea polyphenol to produce rGO as a nanocarrier for delivery of DOX [121]. After exposure to 808 nm NIR laser irradiation at 3 W/cm^2 for 5 min, combined chemo-PTT can enhance cytotoxicity of DOX toward human tongue squamous cancer cells, CAL27.

To improve photothermal properties, Ma et al. added gold (Au) clusters to rGO surface by electrostatic interaction. Afterward, rGO was functionalized with 3-(3-phenylureido) propanoic acid–polyethylene glycol (PPEG) via π–π bond interaction for improving the biocompatibility. Using DOX for chemotherapy, rGO/Au/PPEG elicited effective phototherapy/chemotherapy effect on HeLa cell line [122]. In a separate study, Yang et al. introduced Au nanorods as well as hydroxyapatite to rGO surface (RGO/AuNR/HA) for delivery of the anticancer drug 5-fluorouracil (5-FU). The nanocomposite was designed for synergistic dual therapy, in which hydroxyapatite was used to enhance 5-FU loading, while RGO and Au nanorods (AuNR) offer enhanced photothermal effect under NIR laser irradiation [123]. Due to sequential drug release with the pH-sensitive drug release behavior of hydroxyapatite in the first stage and photothermal conversion from RGO/AuNR after NIR laser irradiation in the second stage, the designed nanocomposite exhibits greater antitumor activity from the chemo-photo effect. A hybrid with ultra-small plasmonic gold nanorods vesicles (rGO-AuNRVe) loaded on the surface of rGO was shown be to be endowed with amplified photothermal effect. This hybrid could provide a high loading capacity of DOX, provided by the cavity of the vesicle and the large surface area of rGO. Furthermore, the release of DOX was sequential with DOX release first from the vesicular cavity under NIR photothermal heating and followed by release from rGO surface induced by the intracellular acidic environment [124]. Intravenous injection of rGO-AuNRVe-DOX followed by low power 808 nm NIR laser irradiation (0.25 W/cm^2) leads to effective inhibition of tumor growth of subcutaneously implanted U87MG human glioblastoma cells in nude mice from combinatory chemo- and photothermal therapies.

Of course, the application of rGO in chemo-phototherapy is not limited to DOX as there are reports employing different chemotherapeutic drugs in combination with rGO to enhance the cytotoxicity of the chemo drug through combination with PTT. In one study, a nanocookie prepared by coating amorphous carbon on a mesoporous silica support (PSS) and self-assembled on rGO nanosheet, was used as a photo-responsive drug carrier for delivery of a hydrophobic anticancer drug camptothecin (CPT). Other than providing a large payload of CPT, this nanocomposite provided a burst-like drug release and intense photothermal effect upon NIR exposure [125]. The tumor volume change observed from MDA-MB231 tumor-bearing nude mice confirmed chemo-phototherapy due to synergistic photothermal and chemo therapeutic effects (nanocookie–CPT + NIR) was more effective than chemotherapy alone (nanocookie-CPT) or photothermal effect alone (nanocookie + NIR). In a separate study, Vinothini at el. developed magnetic iron oxide nanoparticles functionalized rGO for loading CPT. Furthermore, a photosensitizer 4-hydroxycoumarin (4-HC) was bound to rGO via allyl amine (AA) linker. The cytotoxicity study using MCF-7 human breast cancer cells upon 365 nm laser irradiation at 20 mW/cm^2 for 3 min indicated CPT-loaded MrGO-AA-g-4-HC could produce reactive oxygen species (ROS) for killing of MCF-7 cancer cells for PDT. Unfortunately, with the limited penetration depth of light source in the visible wavelength range (365 nm) for inducing PDT effect of 4-HC, the in vivo results only demonstrated significant tumor growth suppression for CPT-loaded MrGO-AA-g-4-HC without laser treatment [126].

3.5. Photothermal Therapy/Immunotherapy

Immunotherapy, which kills cancer cells by improving immunity, is a new approach for cancer therapy. The success of immunotherapy is determined by two main tools, checkpoint inhibitors (CPIs) and chimeric antigen receptor (CAR) T cells [127,128]. Although immunotherapies have achieved promising results against metastatic cancers, traditional immunotherapies are often expensive and can have toxic side effects. During the process of PTT, the heat generated by the photothermal agent not only ablates the tumor but also produces tumor-associated antigens by causing immunogenic cell death, which can lead to antitumor immunity in the body. Hence, the combination of PTT and immunotherapy (photo-immunotherapy) has shown great promise in cancer therapy recently [129,130]. Although many nanoparticles have been used for photo-immunotherapy, rGO stands out as one of the best choice among them [131]. Wang et al. prepared a nanocomposite consisting of PEGylated rGO hybridized with iron oxide nanoparticles through electrostatic interaction for photothermal-immunotherapy of metastatic cancer. This nanocomposite was an excellent photothermal agent for direct killing of cancer cells by PTT, which also stimulated immune responses by triggering the maturation of dendritic cells as well as the secretion of cytokines to cause immunogenic cell death of tumor cells [131]. In vivo antitumor studies revealed the nanocomposite to be an excellent photothermal agent for PTT when exposed to NIR laser to destroy primary tumor effectively. After NIR laser treatment of 4T1 orthotopic mouse breast tumor, the intratumorally injected nanocomposites could significantly increase the survival time of tumor-bearing mice by eliciting strong antitumor immunological response of the treated animal.

Table 2. Summary of functionalized rGO-based nanocomposites used for cancer therapy.

Nanocarrier	Functionalization Agent	Cancer Cell Line	Type of Study	Reference
Chemotherapy				
Riboflavin-rGO	DOX, riboflavin	MCF-7, A549	In vitro	[63]
rGO-PEI-FA	DOX, folic acid (FA)	CBRH7919	In vitro	[64]
FA-rGO/ZnS:Mn	DOX, folic acid (FA), Mn-doped ZnS quantum dots	MDA-MB-231	In vitro	[65]
CHA-rGO	DOX, cholesteryl hyaluronic acid (CHA)	KB	In vitro, in vivo	[66]
PEG-BPEI-rGO	DOX, branched polyethylenimine (BPEI), polyethylene glycol (PEG)	PC-3	In vitro	[112]

Table 2. Cont.

Nanocarrier	Functionalization Agent	Cancer Cell Line	Type of Study	Reference
NrGO/PEG	Resveratrol, PEG	4T1	In vitro, in vivo	[68]
MSN-C_{18}-rGO	DOX, mesoporous silica grafted with alkyl chains (MSN-C_{18})	SMMC-7721	In vitro	[67]
GP	PF-127 polymer, curcumin, paclitaxel	A549, MDA-MB-231	In vitro	[70]
CS/rGO	Chitosan (CS), 5-FU, curcumin	HT-29	In vitro	[72]
R9-rGO	R9 peptide, paclitaxel	HeLa, MCF-7	In vitro	[71]
rGO-Au	5-FU, gold (Au)	MCF-7	In vitro	[73]
MPA-AuNPs/rGO	MTX, SMTX-gold nanoparticles (AuNPs)	MCF-7	In vitro	[74]
PK_5E_7(PEI-rGO)	DOX, PK_5E_7 polymer, PEI	Hela, A549	In vitro	[75]
rGOD-hNP	DOX, chitosan	PC-3	In vitro	[76]
RGO	Leaf extract, paclitaxel	A549	In vitro	[69]
Photothermal and/or Photodynamic Therapy				
rGO-RGD	RGD peptide	U87MG	In vitro	[87]
CPC/rGO	Chlorin (Ce6), claudin 4-binding peptide	U87, HeLa	In vitro	[88]
rGO/PdNFs	Palladium nanoflowers (PdNFs)	HeLa	In vitro, in vivo	[89]
Ag(Nd)-ZnO/rGO	Ag(Nd)/ZnO	MCF-7	In vitro	[95]
ARGO	Alanine	U87MG	In vitro	[90]
ICG-CPPDN/rGO	Catechol, PPDN polymer, ICG	MDA-MB-231	In vitro, in vivo	[103]
rGO/AE/AuNPs	Amaranth extract (AE), gold nanoparticles (AuNPs)	HeLa	In vitro	[93]
rGO-Ru-PEG	PEG, Ru(II)	A549	In vitro, in vivo	[92]
NAu-rGO	Nisin peptides, gold nanoparticles (AuNPs)	MCF-7, HeLa	In vitro	[91]
p-MoS_2/n-rGO-MnO_2-PEG	p-type molybdenum sulfide (p-MoS_2), MnO_2, PEG	HeLa, HEK293	In vitro	[96]
Cu_2O-rGO	Cu_2O	HK-2, MDA-MB-231, A549	In vitro	[97]
rGO-PEI-TCPP	Polyethyleneimine (PEI), tetrakis(4-carboxyphenyl) porphyrin (TCPP)	CBRH7919	In vitro	[94]
rGO-PDA@MS/HA	Mesoporous silica (MS), hyaluronic acid (HA), polydopamine (PDA), Ce6	HT-29, HCT-116	In vitro	[98]
rGO/HA-g-PMAO	Hyaluronic acid (HA) grafted PMAO	MCF-7, NHDF	In vitro	[99]
GT-rGO	Green tea	SW48, HT29	In vitro	[100]
NBs-GPC3-rGO	GPC3 antibody, nanobubbles	HepG2	In vitro	[101]
ICG-PDA-rGO	ICG, polydopamine	4T1	In vitro, in vivo	[102]
Gene Therapy				
rGO-PEI	PEI, siRNA	None	None	[110]
PEG-BPEI-rGO	Low molecular-weight branched polyethylenimine (BPEI)	PC-3	In vitro	[111]
Chemotherapy/Phototherapy				
rGO/dopa-MAL-c(RGDfC)	Catechol, DOX, c(RGDfC) peptide	HeLa, MDA-MB-231	In vitro	[117]

Table 2. *Cont.*

Nanocarrier	Functionalization Agent	Cancer Cell Line	Type of Study	Reference
rGO/C18D	DOX, octadecanic acid conjugated on dextran (C18D)	HeLa	In vitro	[118]
rGO@PSS	Camptothecin (CPT), mesoporous silica	MDA-MB-231	In vitro, in vivo	[125]
rGO/Dex	DOX, dextran, RGD peptide	B16F10	In vitro	[119]
BSA-rGO	DOX, bovine erum albumin (BSA)	U87MG	In vitro	[57]
rGO/Au/PPEG	DOX, 3-(3-phenylureido) propanoic acid (PPA)-PEG (PPEG), Au	HeLa	In vitro	[122]
Chit-rGO-IR-820	DOX, chitosan, IR-820	C26	In vitro	[49]
pRGO@MS-HA	DOX, hyaluronic acid (HA), mesoporous silica, polydopamine	HeLa	In vitro, in vivo	[120]
TPDL1-rGO	DOX, tea polyphenol, anti-PDL1 antibody	CAL-27, PDLCs	In vitro	[121]
MrGO-AA-g-4-HC	CPT, 4-hydroxycoumarin (4-HC), magnetic nanoparticles, camptothecin	MCF-7	In vitro, in vivo	[126]
rGO/AuNR/HAP	5-FU, gold nanorod (AuNR), hydroxyapatite	HeLa	In vitro	[123]
rGO-AuNRVe	DOX, gold nanorod vesicle	U87MG	In vitro, in vivo	[124]
Photothermal Therapy/Immunotherapy				
FNPs/rGO-PEG	Fe_3O_4 nanoparticles, PEG	4T1	In vitro, in vivo	[131]
PEG-rGO-FA-IDOi	IDO inhibitor (IDOi), folic acid, PEG	CT26	In vitro, in vivo	[132]

Yan et al. also combined immunotherapy with PTT with folic acid as a targeting ligand by conjugating indoleamine-2,3-dioxygenase (IDO) inhibitor to rGO to induce IDO inhibition and programmed cell death-ligand 1 (PD-L1) blockade for synergistic antitumor immunity. After laser irradiation, the nanocomposite can directly kill tumor cells due to PTT and trigger antitumor immune response synergistically by IDO inhibition as well as PD-L1 blockade in CT26 colon cancer cells. By combining PTT, IDO inhibition, and PD-L1 blockade, the growth of irradiated tumor in distant sites without PTT treatment can be effectively inhibited by targeting multiple antitumor immune pathways to induce synergistic antitumor immunity [132].

4. Conclusions and Outlook

Due to improved photothermal response by absorbing light in the NIR range and the potential for high loading of chemotherapeutic drugs, photosensitizers and siRNA, rGO synthesized by means of various reducing agents is well suited for applications in single or multi-mode cancer therapy. Based on the reducing agent used for rGO synthesis, and the moieties conjugated with it, rGO-based nanocomposite is endowed with triggered drug release capability after intracellular uptake, by pH change or hyperthermia. This temperature-dependent and pH-responsive drug release, when combined with PTT and/or PDT, can lead to pronounced cytotoxicity from in vitro and in vivo studies performed with various cancer cells. rGO can also act as a good vehicle for gene delivery after modification/conjugation with cationic polymers, especially PEI, which can act alone for RNAi or combined with rGO-induced PTT for combination therapy involving immunotherapy. Overall, this review concludes that rGO is a promising and versatile tool after functionalization for cancer therapy, especially in combination cancer therapy such as PTT/PDT, chemotherapy/phototherapy and photothermal therapy/immunotherapy to elicit synergistic anti-tumor efficacy. Undoubtedly, despite remarkable therapeutic efficacy demonstrated from combination cancer therapy using rGO, rationally combining the therapeutic modalities into rGO-based platform for 'smart' drug delivery will be desirable. In addition, for successful cancer therapy, the designed rGO-based nanocomposites should preferably be endowed with both therapeutic and diagnostic functions for precision nanomedicine. Additionally, the function of combined cancer therapeutics offered by rGO may be required to be programmed for realizing the synergistic effects. Moreover, it would be helpful to develop better PDT/PTT cancer therapeutics using rGO, which can alleviate the limit of penetration depth of NIR laser for effective eradication of tumors located deep in the body.

Author Contributions: Conceptualization, B.S.D., Y.-J.L., and J.-P.C.; Writing—original draft preparation, B.S.D.; Writing—review and editing, G.J. and J.-P.C.; Visualization, B.S.D.; Supervision, J.-P.C.; Funding acquisition, J.-P.C. All authors have read and agreed to the published version of the manuscript.

Funding: This research was funded by the Ministry of Science and Technology, Taiwan, ROC (MOST106-2221-E-182-056-MY3) and Chang Gung Memorial Hospital, Taiwan, ROC (BMRP249, CMRPD2I0041 and CMRPD2I0042). The APC was funded by Chang Gung University.

Institutional Review Board Statement: Not applicable.

Informed Consent Statement: Not applicable.

Acknowledgments: We acknowledge the technical support by the Microscope Core Laboratory, Chang Gung Memorial Hospital, Linkou and the Microscopy Center, Chang Gung University.

Conflicts of Interest: The authors declare no conflict of interest. The funders had no role in the design of the study; in the collection, analyses, or interpretation of data; in the writing of the manuscript, or in the decision to publish the results.

References

1. Siegel, R.L.; Miller, K.D.; Jemal, A. Cancer statistics, 2020. *A Cancer J. Clin.* **2020**, *70*, 7–30. [CrossRef]
2. Soto, A.M.; Sonnenschein, C. Environmental causes of cancer: Endocrine disruptors as carcinogens. *Nat. Rev. Endocrinol.* **2010**, *6*, 363–370. [CrossRef]
3. Feng, S.-S.; Chien, S. Chemotherapeutic engineering: Application and further development of chemical engineering principles for chemotherapy of cancer and other diseases. *Chem. Eng. Sci.* **2003**, *58*, 4087–4114. [CrossRef]
4. Patel, S.C.; Lee, S.; Lalwani, G.; Suhrland, C.; Chowdhury, S.M.; Sitharaman, B. Graphene-based platforms for cancer therapeutics. *Ther. Deliv.* **2016**, *7*, 101–116. [CrossRef] [PubMed]
5. Skalickova, S.; Loffelmann, M.; Gargulak, M.; Kepinska, M.; Docekalova, M.; Uhlirova, D.; Stankova, M.; Fernandez, C.; Milnerowicz, H.; Ruttkay-Nedecky, B.; et al. Zinc-Modified Nanotransporter of Doxorubicin for Targeted Prostate Cancer Delivery. *Nanomaterials* **2017**, *7*, 435. [CrossRef] [PubMed]
6. Baskar, R.; Lee, K.A.; Yeo, R.; Yeoh, K.-W. Cancer and radiation therapy: Current advances and future directions. *Int. J. Med. Sci.* **2012**, *9*, 193–199. [CrossRef] [PubMed]
7. Arruebo, M.; Vilaboa, N.; Sáez-Gutierrez, B.; Lambea, J.; Tres, A.; Valladares, M.; González-Fernández, A. Assessment of the evolution of cancer treatment therapies. *Cancers* **2011**, *3*, 3279–3330. [CrossRef] [PubMed]
8. Mozafari, M.R.; Pardakhty, A.; Azarmi, S.; Jazayeri, J.A.; Nokhodchi, A.; Omri, A. Role of nanocarrier systems in cancer nanotherapy. *J. Liposome Res.* **2009**, *19*, 310–321. [CrossRef] [PubMed]
9. Li, Z.; Tan, S.; Li, S.; Shen, Q.; Wang, K. Cancer drug delivery in the nano era: An overview and perspectives (Review). *Oncol. Rep.* **2017**, *38*, 611–624. [CrossRef]
10. Deshpande, P.P.; Biswas, S.; Torchilin, V.P. Current trends in the use of liposomes for tumor targeting. *Nanomed. Lond. Engl.* **2013**, *8*, 1509–1528. [CrossRef]
11. Gong, Z.; Chen, M.; Ren, Q.; Yue, X.; Dai, Z. Fibronectin-targeted dual-acting micelles for combination therapy of metastatic breast cancer. *Signal. Transduct. Target. Ther.* **2020**, *5*, 12. [CrossRef]
12. Thundimadathil, J. Cancer Treatment Using Peptides: Current Therapies and Future Prospects. *J. Amino Acids* **2012**, *2012*, 967347. [CrossRef]
13. Peng, X.-H.; Qian, X.; Mao, H.; Wang, A.Y.; Chen, Z.G.; Nie, S.; Shin, D.M. Targeted magnetic iron oxide nanoparticles for tumor imaging and therapy. *Int. J. Nanomed.* **2008**, *3*, 311–321. [CrossRef]
14. Jain, S.; Hirst, D.G.; O'Sullivan, J.M. Gold nanoparticles as novel agents for cancer therapy. *Br. J. Radiol.* **2012**, *85*, 101–113. [CrossRef]
15. Yuan, Y.-G.; Zhang, S.; Hwang, J.-Y.; Kong, I.-K. Silver Nanoparticles Potentiates Cytotoxicity and Apoptotic Potential of Camptothecin in Human Cervical Cancer Cells. *Oxidative Med. Cell. Longev.* **2018**, *2018*, 6121328. [CrossRef] [PubMed]
16. Huang, D.; He, B.; Mi, P. Calcium phosphate nanocarriers for drug delivery to tumors: Imaging, therapy and theranostics. *Biomater. Sci.* **2019**, *7*, 3942–3960. [CrossRef] [PubMed]

17. Yi, X.; Chen, L.; Chen, J.; Maiti, D.; Chai, Z.; Liu, Z.; Yang, K. Biomimetic Copper Sulfide for Chemo-Radiotherapy: Enhanced Uptake and Reduced Efflux of Nanoparticles for Tumor Cells under Ionizing Radiation. *Adv. Funct. Mater.* **2018**, *28*, 1705161. [CrossRef]
18. Tanaka, M.; Kataoka, H.; Yano, S.; Ohi, H.; Kawamoto, K.; Shibahara, T.; Mizoshita, T.; Mori, Y.; Tanida, S.; Kamiya, T.; et al. Anticancer effects of newly developed chemotherapeutic agent, glycoconjugated palladium (II) complex, against cisplatin-resistant gastric cancer cells. *BMC Cancer* **2013**, *13*, 1–9. [CrossRef] [PubMed]
19. Pei, X.; Zhu, Z.; Gan, Z.; Chen, J.; Zhang, X.; Cheng, X.; Wan, Q.; Wang, J. PEGylated nano-graphene oxide as a nanocarrier for delivering mixed anticancer drugs to improve anticancer activity. *Sci. Rep.* **2020**, *10*, 2717. [CrossRef]
20. Lu, Y.-J.; Lin, P.-Y.; Huang, P.-H.; Kuo, C.-Y.; Shalumon, K.T.; Chen, M.-Y.; Chen, J.-P. Magnetic Graphene Oxide for Dual Targeted Delivery of Doxorubicin and Photothermal Therapy. *Nanomaterials* **2018**, *8*, 193. [CrossRef]
21. Wang, Y.; Qiu, M.; Won, M.; Jung, E.; Fan, T.; Xie, N.; Chi, S.-G.; Zhang, H.; Kim, J.S. Emerging 2D material-based nanocarrier for cancer therapy beyond graphene. *Coord. Chem. Rev.* **2019**, *400*, 213041. [CrossRef]
22. Chen, Y.-W.; Su, Y.-L.; Hu, S.-H.; Chen, S.-Y. Functionalized graphene nanocomposites for enhancing photothermal therapy in tumor treatment. *Adv. Drug Deliv. Rev.* **2016**, *105*, 190–204. [CrossRef]
23. Patel, K.D.; Singh, R.K.; Kim, H.-W. Carbon-based nanomaterials as an emerging platform for theranostics. *Mater. Horiz.* **2019**, *6*, 434–469. [CrossRef]
24. Mousavi, S.M.; Low, F.W.; Hashemi, S.A.; Samsudin, N.A.; Shakeri, M.; Yusoff, Y.; Rahsepar, M.; Lai, C.W.; Babapoor, A.; Soroshnia, S.; et al. Development of hydrophobic reduced graphene oxide as a new efficient approach for photochemotherapy. *RSC Adv.* **2020**, *10*, 12851–12863. [CrossRef]
25. Degors, I.M.S.; Wang, C.; Rehman, Z.U.; Zuhorn, I.S. Carriers Break Barriers in Drug Delivery: Endocytosis and Endosomal Escape of Gene Delivery Vectors. *Acc. Chem. Res.* **2019**, *52*, 1750–1760. [CrossRef] [PubMed]
26. Fortuni, B.; Inose, T.; Ricci, M.; Fujita, Y.; Van Zundert, I.; Masuhara, A.; Fron, E.; Mizuno, H.; Latterini, L.; Rocha, S.; et al. Polymeric Engineering of Nanoparticles for Highly Efficient Multifunctional Drug Delivery Systems. *Sci. Rep.* **2019**, *9*, 2666. [CrossRef] [PubMed]
27. Kostarelos, K.; Novoselov, K.S. Exploring the Interface of Graphene and Biology. *Science* **2014**, *344*, 261–263. [CrossRef] [PubMed]
28. Kostarelos, K.; Novoselov, K.S. Graphene devices for life. *Nat. Nanotechnol.* **2014**, *9*, 744–745. [CrossRef] [PubMed]
29. Chau, N.D.Q.; Ménard-Moyon, C.; Kostarelos, K.; Bianco, A. Multifunctional carbon nanomaterial hybrids for magnetic manipulation and targeting. *Biochem. Biophys. Res. Commun.* **2015**, *468*, 454–462. [CrossRef]
30. Reina, G.; González-Domínguez, J.M.; Criado, A.; Vázquez, E.; Bianco, A.; Prato, M. Promises, facts and challenges for graphene in biomedical applications. *Chem. Soc. Rev.* **2017**, *46*, 4400–4416. [CrossRef]
31. Karki, N.; Tiwari, H.; Tewari, C.; Rana, A.; Pandey, N.; Basak, S.; Sahoo, N.G. Functionalized graphene oxide as a vehicle for targeted drug delivery and bioimaging applications. *J. Mater. Chem. B* **2020**, *8*, 8116–8148. [CrossRef] [PubMed]
32. Li, D.; Zhang, W.; Yu, X.; Wang, Z.; Su, Z.; Wei, G. When biomolecules meet graphene: From molecular level interactions to material design and applications. *Nanoscale* **2016**, *8*, 19491–19509. [CrossRef] [PubMed]
33. Mun, S.G.; Choi, H.W.; Lee, J.M.; Lim, J.H.; Ha, J.H.; Kang, M.-J.; Kim, E.-J.; Kang, L.; Chung, B.G. rGO nanomaterial-mediated cancer targeting and photothermal therapy in a microfluidic co-culture platform. *Nano Converg.* **2020**, *7*, 10. [CrossRef]
34. Bao, Z.; Liu, X.; Liu, Y.; Liu, H.; Zhao, K. Near-infrared light-responsive inorganic nanomaterials for photothermal therapy. *Asian J. Pharm. Sci.* **2016**, *11*, 349–364. [CrossRef]
35. Zhang, J.; Yang, H.; Shen, G.; Cheng, P.; Zhang, J.; Guo, S. Reduction of graphene oxide vial-ascorbic acid. *Chem. Commun.* **2010**, *46*, 1112–1114. [CrossRef] [PubMed]
36. Fernández-Merino, M.J.; Guardia, L.; Paredes, J.I.; Villar-Rodil, S.; Solís-Fernández, P.; Martínez-Alonso, A.; Tascón, J.M.D. Vitamin C Is an Ideal Substitute for Hydrazine in the Reduction of Graphene Oxide Suspensions. *J. Phys. Chem. C* **2010**, *114*, 6426–6432. [CrossRef]
37. Habte, A.T.; Ayele, D.W. Synthesis and Characterization of Reduced Graphene Oxide (rGO) Started from Graphene Oxide (GO) Using the Tour Method with Different Parameters. *Adv. Mater. Sci. Eng.* **2019**, *2019*, 5058163. [CrossRef]
38. Kosowska, K.; Domalik-Pyzik, P.; Krok-Borkowicz, M.; Chłopek, J. Synthesis and Characterization of Chitosan/Reduced Graphene Oxide Hybrid Composites. *Materials* **2019**, *12*, 2077. [CrossRef] [PubMed]
39. Joshi, S.; Siddiqui, R.; Sharma, P.; Kumar, R.; Verma, G.; Saini, A. Green synthesis of peptide functionalized reduced graphene oxide (rGO) nano bioconjugate with enhanced antibacterial activity. *Sci. Rep.* **2020**, *10*, 9441. [CrossRef]
40. Iskandar, F.; Hikmah, U.; Stavila, E.; Aimon, A.H. Microwave-assisted reduction method under nitrogen atmosphere for synthesis and electrical conductivity improvement of reduced graphene oxide (rGO). *RSC Adv.* **2017**, *7*, 52391–52397. [CrossRef]
41. Li, D.; Müller, M.B.; Gilje, S.; Kaner, R.B.; Wallace, G.G. Processable aqueous dispersions of graphene nanosheets. *Nat. Nanotechnol.* **2008**, *3*, 101–105. [CrossRef] [PubMed]
42. Cong, H.-P.; He, J.-J.; Lu, Y.; Yu, S.-H. Water-Soluble Magnetic-Functionalized Reduced Graphene Oxide Sheets: In situ Synthesis and Magnetic Resonance Imaging Applications. *Small* **2010**, *6*, 169–173. [CrossRef] [PubMed]
43. Altinbasak, I.; Jijie, R.; Barras, A.; Golba, B.; Sanyal, R.; Bouckaert, J.; Drider, D.; Bilyy, R.; Dumych, T.; Paryzhak, S.; et al. Reduced Graphene-Oxide-Embedded Polymeric Nanofiber Mats: An "On-Demand" Photothermally Triggered Antibiotic Release Platform. *ACS Appl. Mater. Interfaces* **2018**, *10*, 41098–41106. [CrossRef]

44. Ren, P.-G.; Yan, D.-X.; Ji, X.; Chen, T.; Li, Z.-M. Temperature dependence of graphene oxide reduced by hydrazine hydrate. *Nanotechnology* **2010**, *22*, 055705. [CrossRef] [PubMed]
45. Gurunathan, S.; Han, J.W.; Kim, E.S.; Park, J.H.; Kim, J.-H. Reduction of graphene oxide by resveratrol: A novel and simple biological method for the synthesis of an effective anticancer nanotherapeutic molecule. *Int. J. Nanomed.* **2015**, *10*, 2951–2969. [CrossRef]
46. Bhumkar, D.R.; Joshi, H.M.; Sastry, M.; Pokharkar, V.B. Chitosan Reduced Gold Nanoparticles as Novel Carriers for Transmucosal Delivery of Insulin. *Pharm. Res.* **2007**, *24*, 1415–1426. [CrossRef]
47. Fang, M.; Long, J.; Zhao, W.; Wang, L.; Chen, G. pH-Responsive Chitosan-Mediated Graphene Dispersions. *Langmuir* **2010**, *26*, 16771–16774. [CrossRef] [PubMed]
48. Justin, R.; Chen, B. Body temperature reduction of graphene oxide through chitosan functionalisation and its application in drug delivery. *Mater. Sci. Eng. C* **2014**, *34*, 50–53. [CrossRef]
49. Zaharie-Butucel, D.; Potara, M.; Suarasan, S.; Licarete, E.; Astilean, S. Efficient combined near-infrared-triggered therapy: Phototherapy over chemotherapy in chitosan-reduced graphene oxide-IR820 dye-doxorubicin nanoplatforms. *J. Colloid Interface Sci.* **2019**, *552*, 218–229. [CrossRef]
50. Liu, H.; Kuila, T.; Kim, N.H.; Ku, B.-C.; Lee, J.H. In situ synthesis of the reduced graphene oxide–polyethyleneimine composite and its gas barrier properties. *J. Mater. Chem. A* **2013**, *1*, 3739–3746. [CrossRef]
51. Zhang, X.; Yu, Y.; Shen, J.; Qi, W.; Wang, H. Fabrication of polyethyleneimine-functionalized reduced graphene oxide-hemin-bovine serum albumin (PEI-rGO-hemin-BSA) nanocomposites as peroxidase mimetics for the detection of multiple metabolites. *Anal. Chim. Acta* **2019**, *1070*, 80–87. [CrossRef]
52. Roy, S.; Tang, X.; Das, T.; Zhang, L.; Li, Y.; Ting, S.; Hu, X.; Yue, C.Y. Enhanced Molecular Level Dispersion and Interface Bonding at Low Loading of Modified Graphene Oxide To Fabricate Super Nylon 12 Composites. *ACS Appl. Mater. Interfaces* **2015**, *7*, 3142–3151. [CrossRef]
53. Yang, Z.-Z.; Zheng, Q.-B.; Qiu, H.-X.; Li, J.; Yang, J.-H. A simple method for the reduction of graphene oxide by sodium borohydride with CaCl2 as a catalyst. *New Carbon Mater.* **2015**, *30*, 41–47. [CrossRef]
54. Guex, L.G.; Sacchi, B.; Peuvot, K.F.; Andersson, R.L.; Pourrahimi, A.M.; Ström, V.; Farris, S.; Olsson, R.T. Experimental review: Chemical reduction of graphene oxide (GO) to reduced graphene oxide (rGO) by aqueous chemistry. *Nanoscale* **2017**, *9*, 9562–9571. [CrossRef]
55. Liu, J.; Fu, S.; Yuan, B.; Li, Y.; Deng, Z. Toward a Universal "Adhesive Nanosheet" for the Assembly of Multiple Nanoparticles Based on a Protein-Induced Reduction/Decoration of Graphene Oxide. *J. Am. Chem. Soc.* **2010**, *132*, 7279–7281. [CrossRef]
56. Dasgupta, N.; Ranjan, S.; Patra, D.; Srivastava, P.; Kumar, A.; Ramalingam, C. Bovine serum albumin interacts with silver nanoparticles with a "side-on" or "end on" conformation. *Chem. Biol. Interact.* **2016**, *253*, 100–111. [CrossRef]
57. Cheon, Y.A.; Bae, J.H.; Chung, B.G. Reduced Graphene Oxide Nanosheet for Chemo-photothermal Therapy. *Langmuir* **2016**, *32*, 2731–2736. [CrossRef]
58. Griep, M.H.; Demaree, J.D.; Cole, D.P.; Henry, T.C.; Karna, S.P. Protein-Mediated Synthesis of Au Nanocluster Decorated Reduced Graphene Oxide: A Multifunctional Hybrid Nano-Bio Platform. *Plasmonics* **2020**, *15*, 897–903. [CrossRef]
59. Wang, Y.; Shi, Z.; Yin, J. Facile Synthesis of Soluble Graphene via a Green Reduction of Graphene Oxide in Tea Solution and Its Biocomposites. *ACS Appl. Mater. Interfaces* **2011**, *3*, 1127–1133. [CrossRef] [PubMed]
60. Vatandost, E.; Ghorbani-HasanSaraei, A.; Chekin, F.; Naghizadeh Raeisi, S.; Shahidi, S.-A. Green tea extract assisted green synthesis of reduced graphene oxide: Application for highly sensitive electrochemical detection of sunset yellow in food products. *Food Chem.* **2020**, *6*, 100085. [CrossRef] [PubMed]
61. Abdullah, M.F.; Zakaria, R.; Zein, S.H.S. Green tea polyphenol–reduced graphene oxide: Derivatisation, reduction efficiency, reduction mechanism and cytotoxicity. *RSC Adv.* **2014**, *4*, 34510–34518. [CrossRef]
62. Jose, G.; Lu, Y.-J.; Hung, J.-T.; Yu, A.L.; Chen, J.-P. Co-Delivery of CPT-11 and Panobinostat with Anti-GD2 Antibody Conjugated Immunoliposomes for Targeted Combination Chemotherapy. *Cancers* **2020**, *12*, 3211. [CrossRef]
63. Ma, N.; Zhang, B.; Liu, J.; Zhang, P.; Li, Z.; Luan, Y. Green fabricated reduced graphene oxide: Evaluation of its application as nano-carrier for pH-sensitive drug delivery. *Int. J. Pharm.* **2015**, *496*, 984–992. [CrossRef] [PubMed]
64. Wei, G.; Yan, M.; Dong, R.; Wang, D.; Zhou, X.; Chen, J.; Hao, J. Covalent Modification of Reduced Graphene Oxide by Means of Diazonium Chemistry and Use as a Drug-Delivery System. *Chem. A Eur. J.* **2012**, *18*, 14708–14716. [CrossRef] [PubMed]
65. Diaz-Diestra, D.; Thapa, B.; Badillo-Diaz, D.; Beltran-Huarac, J.; Morell, G.; Weiner, B.R. Graphene Oxide/ZnS:Mn Nanocomposite Functionalized with Folic Acid as a Nontoxic and Effective Theranostic Platform for Breast Cancer Treatment. *Nanomaterials* **2018**, *8*, 484. [CrossRef] [PubMed]
66. Miao, W.; Shim, G.; Kang, C.M.; Lee, S.; Choe, Y.S.; Choi, H.-G.; Oh, Y.-K. Cholesteryl hyaluronic acid-coated, reduced graphene oxide nanosheets for anti-cancer drug delivery. *Biomaterials* **2013**, *34*, 9638–9647. [CrossRef]
67. He, D.; Li, X.; He, X.; Wang, K.; Tang, J.; Yang, X.; He, X.; Yang, X.; Zou, Z. Noncovalent assembly of reduced graphene oxide and alkyl-grafted mesoporous silica: An effective drug carrier for near-infrared light-responsive controlled drug release. *J. Mater. Chem. B* **2015**, *3*, 5588–5594. [CrossRef]
68. Chen, J.; Liu, H.; Zhao, C.; Qin, G.; Xi, G.; Li, T.; Wang, X.; Chen, T. One-step reduction and PEGylation of graphene oxide for photothermally controlled drug delivery. *Biomaterials* **2014**, *35*, 4986–4995. [CrossRef]

69. Lin, S.; Ruan, J.; Wang, S. Biosynthesized of reduced graphene oxide nanosheets and its loading with paclitaxel for their anti cancer effect for treatment of lung cancer. *J. Photochem. Photobiol. B Biol.* **2019**, *191*, 13–17. [CrossRef] [PubMed]
70. Muthoosamy, K.; Abubakar, I.B.; Bai, R.G.; Loh, H.-S.; Manickam, S. Exceedingly Higher co-loading of Curcumin and Paclitaxel onto Polymer-functionalized Reduced Graphene Oxide for Highly Potent Synergistic Anticancer Treatment. *Sci. Rep.* **2016**, *6*, 32808. [CrossRef]
71. Hashemi, M.; Yadegari, A.; Yazdanpanah, G.; Jabbehdari, S.; Omidi, M.; Tayebi, L. Functionalized R9–reduced graphene oxide as an efficient nano-carrier for hydrophobic drug delivery. *RSC Adv.* **2016**, *6*, 74072–74084. [CrossRef]
72. Dhanavel, S.; Revathy, T.A.; Sivaranjani, T.; Sivakumar, K.; Palani, P.; Narayanan, V.; Stephen, A. 5-Fluorouracil and curcumin co-encapsulated chitosan/reduced graphene oxide nanocomposites against human colon cancer cell lines. *Polym. Bull.* **2020**, *77*, 213–233. [CrossRef]
73. Sanad, M.F.; Shalan, A.E.; Bazid, S.M.; Abu Serea, E.S.; Hashem, E.M.; Nabih, S.; Ahsan, M.A. A graphene gold nanocomposite-based 5-FU drug and the enhancement of the MCF-7 cell line treatment. *RSC Adv.* **2019**, *9*, 31021–31029. [CrossRef]
74. Jafarizad, A.; Aghanejad, A.; Sevim, M.; Metin, Ö.; Barar, J.; Omidi, Y.; Ekinci, D. Gold Nanoparticles and Reduced Graphene Oxide-Gold Nanoparticle Composite Materials as Covalent Drug Delivery Systems for Breast Cancer Treatment. *Chem. Sel.* **2017**, *2*, 6663–6672. [CrossRef]
75. Ryu, K.; Park, J.; Kim, T.-I. Effect of pH-Responsive Charge-Conversional Polymer Coating to Cationic Reduced Graphene Oxide Nanostructures for Tumor Microenvironment-Targeted Drug Delivery Systems. *Nanomaterials* **2019**, *9*, 1289. [CrossRef] [PubMed]
76. SreeHarsha, N.; Maheshwari, R.; Al-Dhubiab, B.E.; Tekade, M.; Sharma, M.C.; Venugopala, K.N.; Tekade, R.K.; Alzahrani, A.M. Graphene-based hybrid nanoparticle of doxorubicin for cancer chemotherapy. *Int. J. Nanomed.* **2019**, *14*, 7419–7429. [CrossRef] [PubMed]
77. Nomura, S.; Morimoto, Y.; Tsujimoto, H.; Arake, M.; Harada, M.; Saitoh, D.; Hara, I.; Ozeki, E.; Satoh, A.; Takayama, E.; et al. Highly reliable, targeted photothermal cancer therapy combined with thermal dosimetry using a near-infrared absorbent. *Sci. Rep.* **2020**, *10*, 9765. [CrossRef]
78. Doughty, A.C.V.; Hoover, A.R.; Layton, E.; Murray, C.K.; Howard, E.W.; Chen, W.R. Nanomaterial Applications in Photothermal Therapy for Cancer. *Materials* **2019**, *12*, 779. [CrossRef]
79. de Melo-Diogo, D.; Lima-Sousa, R.; Alves, C.G.; Correia, I.J. Graphene family nanomaterials for application in cancer combination photothermal therapy. *Biomater. Sci.* **2019**, *7*, 3534–3551. [CrossRef]
80. Alegret, N.; Criado, A.; Prato, M. Recent Advances of Graphene-based Hybrids with Magnetic Nanoparticles for Biomedical Applications. *Curr. Med. Chem.* **2017**, *24*, 529–536. [CrossRef]
81. Vines, J.B.; Yoon, J.-H.; Ryu, N.-E.; Lim, D.-J.; Park, H. Gold Nanoparticles for Photothermal Cancer Therapy. *Front. Chem.* **2019**, *7*, 167. [CrossRef]
82. Modugno, G.; Ménard-Moyon, C.; Prato, M.; Bianco, A. Carbon nanomaterials combined with metal nanoparticles for theranostic applications. *Br. J. Pharm.* **2015**, *172*, 975–991. [CrossRef]
83. Bai, L.-Z.; Zhao, D.-L.; Xu, Y.; Zhang, J.-M.; Gao, Y.-L.; Zhao, L.-Y.; Tang, J.-T. Inductive heating property of graphene oxide–Fe3O4 nanoparticles hybrid in an AC magnetic field for localized hyperthermia. *Mater. Lett.* **2012**, *68*, 399–401. [CrossRef]
84. Gazzi, A.; Fusco, L.; Khan, A.; Bedognetti, D.; Zavan, B.; Vitale, F.; Yilmazer, A.; Delogu, L.G. Photodynamic Therapy Based on Graphene and MXene in Cancer Theranostics. *Front. Bioeng. Biotechnol.* **2019**, *7*, 295. [CrossRef]
85. Li, W.; Yang, J.; Luo, L.; Jiang, M.; Qin, B.; Yin, H.; Zhu, C.; Yuan, X.; Zhang, J.; Luo, Z.; et al. Targeting photodynamic and photothermal therapy to the endoplasmic reticulum enhances immunogenic cancer cell death. *Nat. Commun.* **2019**, *10*, 3349. [CrossRef] [PubMed]
86. Luo, S.; Yang, Z.; Tan, X.; Wang, Y.; Zeng, Y.; Wang, Y.; Li, C.; Li, R.; Shi, C. Multifunctional Photosensitizer Grafted on Polyethylene Glycol and Polyethylenimine Dual-Functionalized Nanographene Oxide for Cancer-Targeted Near-Infrared Imaging and Synergistic Phototherapy. *ACS Appl. Mater. Interfaces* **2016**, *8*, 17176–17186. [CrossRef] [PubMed]
87. Robinson, J.T.; Tabakman, S.M.; Liang, Y.; Wang, H.; Sanchez Casalongue, H.; Vinh, D.; Dai, H. Ultrasmall Reduced Graphene Oxide with High Near-Infrared Absorbance for Photothermal Therapy. *J. Am. Chem. Soc.* **2011**, *133*, 6825–6831. [CrossRef] [PubMed]
88. Shim, G.; Kim, M.-G.; Jin, H.; Kim, J.; Oh, Y.-K. Claudin 4-targeted nanographene phototherapy using a Clostridium perfringens enterotoxin peptide-photosensitizer conjugate. *Acta Pharmacol. Sin.* **2017**, *38*, 954–962. [CrossRef]
89. He, Y.; Cao, W.; Cong, C.; Zhang, X.; Luo, L.; Li, L.; Cui, H.; Gao, D. Rationally Designed Multifunctional Carbon–Palladium Nanohybrids for Wide Applications: From Electrochemical Catalysis/Nonenzymatic Sensor to Photothermal Tumor Therapy. *ACS Sustain. Chem. Eng.* **2019**, *7*, 3584–3592. [CrossRef]
90. Chen, X.; Li, C.; Wang, X.; Zhao, X. Infrared heating of reduced graphene oxide nanosheets as photothermal radiation therapeutic agents for tumor regressions. *Mater. Res. Express* **2019**, *6*, 085080. [CrossRef]
91. Otari, S.V.; Kumar, M.; Anwar, M.Z.; Thorat, N.D.; Patel, S.K.S.; Lee, D.; Lee, J.H.; Lee, J.-K.; Kang, Y.C.; Zhang, L. Rapid synthesis and decoration of reduced graphene oxide with gold nanoparticles by thermostable peptides for memory device and photothermal applications. *Sci. Rep.* **2017**, *7*, 10980. [CrossRef] [PubMed]
92. Zhang, D.-Y.; Zheng, Y.; Tan, C.-P.; Sun, J.-H.; Zhang, W.; Ji, L.-N.; Mao, Z.-W. Graphene Oxide Decorated with Ru(II)–Polyethylene Glycol Complex for Lysosome-Targeted Imaging and Photodynamic/Photothermal Therapy. *ACS Appl. Mater. Interfaces* **2017**, *9*, 6761–6771. [CrossRef] [PubMed]

93. Chang, G.; Wang, Y.; Gong, B.; Xiao, Y.; Chen, Y.; Wang, S.; Li, S.; Huang, F.; Shen, Y.; Xie, A. Reduced Graphene Oxide/Amaranth Extract/AuNPs Composite Hydrogel on Tumor Cells as Integrated Platform for Localized and Multiple Synergistic Therapy. *ACS Appl. Mater. Interfaces* **2015**, *7*, 11246–11256. [CrossRef] [PubMed]
94. Wei, G.; Yan, M.; Ma, L.; Wang, C. Photothermal and photodynamic therapy reagents based on rGO–C6H4–COOH. *RSC Adv.* **2016**, *6*, 3748–3755. [CrossRef]
95. Jafarirad, S.; Hammami Torghabe, E.; Rasta, S.H.; Salehi, R. A novel non-invasive strategy for low-level laser-induced cancer therapy by using new Ag/ZnO and Nd/ZnO functionalized reduced graphene oxide nanocomposites. *Artif. Cells Nanomed. Biotechnol.* **2018**, *46*, 800–816. [CrossRef]
96. Kapri, S.; Bhattacharyya, S. Molybdenum sulfide–reduced graphene oxide p–n heterojunction nanosheets with anchored oxygen generating manganese dioxide nanoparticles for enhanced photodynamic therapy. *Chem. Sci.* **2018**, *9*, 8982–8989. [CrossRef]
97. Hou, C.; Quan, H.; Duan, Y.; Zhang, Q.; Wang, H.; Li, Y. Facile synthesis of water-dispersible Cu2O nanocrystal–reduced graphene oxide hybrid as a promising cancer therapeutic agent. *Nanoscale* **2013**, *5*, 1227–1232. [CrossRef]
98. Jiang, W.; Mo, F.; Jin, X.; Chen, L.; Xu, L.J.; Guo, L.; Fu, F. Tumor-Targeting Photothermal Heating-Responsive Nanoplatform Based on Reduced Graphene Oxide/Mesoporous Silica/Hyaluronic Acid Nanocomposite for Enhanced Photodynamic Therapy. *Adv. Mater. Interfaces* **2017**, *4*, 1700425. [CrossRef]
99. Lima-Sousa, R.; de Melo-Diogo, D.; Alves, C.G.; Costa, E.C.; Ferreira, P.; Louro, R.O.; Correia, I.J. Hyaluronic acid functionalized green reduced graphene oxide for targeted cancer photothermal therapy. *Carbohydr. Polym.* **2018**, *200*, 93–99. [CrossRef]
100. Abdolahad, M.; Janmaleki, M.; Mohajerzadeh, S.; Akhavan, O.; Abbasi, S. Polyphenols attached graphene nanosheets for high efficiency NIR mediated photodestruction of cancer cells. *Mater. Sci. Eng. C* **2013**, *33*, 1498–1505. [CrossRef]
101. Liu, Z.; Zhang, J.; Tian, Y.; Zhang, L.; Han, X.; Wang, Q.; Cheng, W. Targeted delivery of reduced graphene oxide nanosheets using multifunctional ultrasound nanobubbles for visualization and enhanced photothermal therapy. *Int. J. Nanomed.* **2018**, *13*, 7859–7872. [CrossRef]
102. Hu, D.; Zhang, J.; Gao, G.; Sheng, Z.; Cui, H.; Cai, L. Indocyanine Green-Loaded Polydopamine-Reduced Graphene Oxide Nanocomposites with Amplifying Photoacoustic and Photothermal Effects for Cancer Theranostics. *Theranostics* **2016**, *6*, 1043–1052. [CrossRef]
103. Sharker, S.M.; Lee, J.E.; Kim, S.H.; Jeong, J.H.; In, I.; Lee, H.; Park, S.Y. pH triggered in vivo photothermal therapy and fluorescence nanoplatform of cancer based on responsive polymer-indocyanine green integrated reduced graphene oxide. *Biomaterials* **2015**, *61*, 229–238. [CrossRef] [PubMed]
104. Mahmoodi Chalbatani, G.; Dana, H.; Gharagouzloo, E.; Grijalvo, S.; Eritja, R.; Logsdon, C.D.; Memari, F.; Miri, S.R.; Rad, M.R.; Marmari, V. Small interfering RNAs (siRNAs) in cancer therapy: A nano-based approach. *Int. J. Nanomed.* **2019**, *14*, 3111–3128. [CrossRef] [PubMed]
105. Zhi, D.; Zhao, Y.; Cui, S.; Chen, H.; Zhang, S. Conjugates of small targeting molecules to non-viral vectors for the mediation of siRNA. *Acta Biomater.* **2016**, *36*, 21–41. [CrossRef]
106. Keles, E.; Song, Y.; Du, D.; Dong, W.-J.; Lin, Y. Recent progress in nanomaterials for gene delivery applications. *Biomater. Sci.* **2016**, *4*, 1291–1309. [CrossRef] [PubMed]
107. Draz, M.S.; Fang, B.A.; Zhang, P.; Hu, Z.; Gu, S.; Weng, K.C.; Gray, J.W.; Chen, F.F. Nanoparticle-mediated systemic delivery of siRNA for treatment of cancers and viral infections. *Theranostics* **2014**, *4*, 872–892. [CrossRef]
108. Zakeri, A.; Kouhbanani, M.A.J.; Beheshtkhoo, N.; Beigi, V.; Mousavi, S.M.; Hashemi, S.A.R.; Karimi Zade, A.; Amani, A.M.; Savardashtaki, A.; Mirzaei, E.; et al. Polyethylenimine-based nanocarriers in co-delivery of drug and gene: A developing horizon. *Nano Rev. Exp.* **2018**, *9*, 1488497. [CrossRef] [PubMed]
109. Feng, L.; Yang, X.; Shi, X.; Tan, X.; Peng, R.; Wang, J.; Liu, Z. Polyethylene glycol and polyethylenimine dual-functionalized nano-graphene oxide for photothermally enhanced gene delivery. *Small* **2013**, *9*, 1989–1997. [CrossRef] [PubMed]
110. Chau, N.D.Q.; Reina, G.; Raya, J.; Vacchi, I.A.; Ménard-Moyon, C.; Nishina, Y.; Bianco, A. Elucidation of siRNA complexation efficiency by graphene oxide and reduced graphene oxide. *Carbon* **2017**, *122*, 643–652. [CrossRef]
111. Kim, H.; Kim, W.J. Photothermally controlled gene delivery by reduced graphene oxide-polyethylenimine nanocomposite. *Small* **2014**, *10*, 117–126. [CrossRef]
112. Kim, H.; Lee, D.; Kim, J.; Kim, T.-i.; Kim, W.J. Photothermally Triggered Cytosolic Drug Delivery via Endosome Disruption Using a Functionalized Reduced Graphene Oxide. *ACS Nano* **2013**, *7*, 6735–6746. [CrossRef]
113. Cao, J.; Chen, Z.; Chi, J.; Sun, Y.; Sun, Y. Recent progress in synergistic chemotherapy and phototherapy by targeted drug delivery systems for cancer treatment. *Artif. Cells Nanomed. Biotechnol.* **2018**, *46*, 817–830. [CrossRef] [PubMed]
114. Biagiotti, G.; Fedeli, S.; Tuci, G.; Luconi, L.; Giambastiani, G.; Brandi, A.; Pisaneschi, F.; Cicchi, S.; Paoli, P. Combined therapies with nanostructured carbon materials: There is room still available at the bottom. *J. Mater. Chem. B* **2018**, *6*, 2022–2035. [CrossRef] [PubMed]
115. Shen, J.-M.; Gao, F.-Y.; Guan, L.-P.; Su, W.; Yang, Y.-J.; Li, Q.-R.; Jin, Z.-C. Graphene oxide–Fe3O4 nanocomposite for combination of dual-drug chemotherapy with photothermal therapy. *RSC Adv.* **2014**, *4*, 18473–18484. [CrossRef]
116. Liu, Z.; Robinson, J.T.; Tabakman, S.M.; Yang, K.; Dai, H. Carbon materials for drug delivery & cancer therapy. *Mater. Today* **2011**, *14*, 316–323. [CrossRef]

117. Oz, Y.; Barras, A.; Sanyal, R.; Boukherroub, R.; Szunerits, S.; Sanyal, A. Functionalization of Reduced Graphene Oxide via Thiol–Maleimide "Click" Chemistry: Facile Fabrication of Targeted Drug Delivery Vehicles. *ACS Appl. Mater. Interfaces* **2017**, *9*, 34194–34203. [CrossRef] [PubMed]
118. Hu, Y.; Sun, D.; Ding, J.; Chen, L.; Chen, X. Decorated reduced graphene oxide for photo-chemotherapy. *J. Mater. Chem. B* **2016**, *4*, 929–937. [CrossRef]
119. Hu, Y.; He, L.; Ding, J.; Sun, D.; Chen, L.; Chen, X. One-pot synthesis of dextran decorated reduced graphene oxide nanoparticles for targeted photo-chemotherapy. *Carbohydr. Polym.* **2016**, *144*, 223–229. [CrossRef] [PubMed]
120. Shao, L.; Zhang, R.; Lu, J.; Zhao, C.; Deng, X.; Wu, Y. Mesoporous Silica Coated Polydopamine Functionalized Reduced Graphene Oxide for Synergistic Targeted Chemo-Photothermal Therapy. *ACS Appl. Mater. Interfaces* **2017**, *9*, 1226–1236. [CrossRef]
121. Hao, L.; Song, H.; Zhan, Z.; Lv, Y. Multifunctional Reduced Graphene Oxide-Based Nanoplatform for Synergistic Targeted Chemo-Photothermal Therapy. *ACS Appl. Biol. Mater.* **2020**, *3*, 5213–5222. [CrossRef]
122. Ma, W.; Hu, Y.; Yang, H.; Zhang, Y.; Ding, J.; Chen, L. Au-aided reduced graphene oxide-based nanohybrids for photo-chemotherapy. *Mater. Sci. Eng. C* **2019**, *95*, 256–263. [CrossRef] [PubMed]
123. Yang, Y.; Wang, Y.; Zhu, M.; Chen, Y.; Xiao, Y.; Shen, Y.; Xie, A. RGO/AuNR/HA-5FU nanocomposite with multi-stage release behavior and efficient antitumor activity for synergistic therapy. *Biomater. Sci.* **2017**, *5*, 990–1000. [CrossRef]
124. Song, J.; Yang, X.; Jacobson, O.; Lin, L.; Huang, P.; Niu, G.; Ma, Q.; Chen, X. Sequential Drug Release and Enhanced Photothermal and Photoacoustic Effect of Hybrid Reduced Graphene Oxide-Loaded Ultrasmall Gold Nanorod Vesicles for Cancer Therapy. *ACS Nano* **2015**, *9*, 9199–9209. [CrossRef]
125. Chen, Y.-W.; Chen, P.-J.; Hu, S.-H.; Chen, I.-W.; Chen, S.-Y. NIR-Triggered Synergic Photo-chemothermal Therapy Delivered by Reduced Graphene Oxide/Carbon/Mesoporous Silica Nanocookies. *Adv. Funct. Mater.* **2014**, *24*, 451–459. [CrossRef]
126. Vinothini, K.; Rajendran, N.K.; Rajan, M.; Ramu, A.; Marraiki, N.; Elgorban, A.M. A magnetic nanoparticle functionalized reduced graphene oxide-based drug carrier system for a chemo-photodynamic cancer therapy. *New J. Chem.* **2020**, *44*, 5265–5277. [CrossRef]
127. Kruger, S.; Ilmer, M.; Kobold, S.; Cadilha, B.L.; Endres, S.; Ormanns, S.; Schuebbe, G.; Renz, B.W.; D'Haese, J.G.; Schloesser, H.; et al. Advances in cancer immunotherapy 2019–latest trends. *J. Exp. Clin. Cancer Res.* **2019**, *38*, 268. [CrossRef]
128. Liu, M.; Guo, F. Recent updates on cancer immunotherapy. *Precis Clin. Med.* **2018**, *1*, 65–74. [CrossRef]
129. Wieder, T.; Eigentler, T.; Brenner, E.; Röcken, M. Immune checkpoint blockade therapy. *J. Allergy Clin. Immunol.* **2018**, *142*, 1403–1414. [CrossRef]
130. Joshi, S.; Durden, D.L. Combinatorial Approach to Improve Cancer Immunotherapy: Rational Drug Design Strategy to Simultaneously Hit Multiple Targets to Kill Tumor Cells and to Activate the Immune System. *J. Oncol.* **2019**, *2019*, 5245034. [CrossRef] [PubMed]
131. Wang, L.; Wang, M.; Zhou, B.; Zhou, F.; Murray, C.; Towner, R.A.; Smith, N.; Saunders, D.; Xie, G.; Chen, W.R. PEGylated reduced-graphene oxide hybridized with Fe3O4 nanoparticles for cancer photothermal-immunotherapy. *J. Mater. Chem. B* **2019**, *7*, 7406–7414. [CrossRef] [PubMed]
132. Yan, M.; Liu, Y.; Zhu, X.; Wang, X.; Liu, L.; Sun, H.; Wang, C.; Kong, D.; Ma, G. Nanoscale Reduced Graphene Oxide-Mediated Photothermal Therapy Together with IDO Inhibition and PD-L1 Blockade Synergistically Promote Antitumor Immunity. *ACS Appl. Mater. Interfaces* **2019**, *11*, 1876–1885. [CrossRef] [PubMed]

Article

Printed Graphene Layer as a Base for Cell Electrostimulation—Preliminary Results

Lucja Dybowska-Sarapuk [1,*], Weronika Sosnowicz [1], Jakub Krzeminski [1,2], Anna Grzeczkowicz [3], Ludomira H. Granicka [3], Andrzej Kotela [4] and Malgorzata Jakubowska [1,2]

1. Faculty of Mechatronics, Warsaw University of Technology, Andrzeja Boboli 8, 02-525 Warsaw, Poland; sosnowicz.weronika@gmail.com (W.S.); krzem@mchtr.pw.edu.pl (J.K.); m.jakubowska@mchtr.pw.edu.pl (M.J.)
2. Centre for Advanced Materials and Technologies CEZAMAT, Poleczki 19, 02-822 Warsaw, Poland
3. The Maciej Nalecz Institute of Biocybernetics and Biomedical Engineering, Polish Academy of Sciences, Ksiecia Trojdena 4, 02-109 Warsaw, Poland; agrzeczkowicz@ibib.waw.pl (A.G.); ludomira.granicka@ibib.waw.pl (L.H.G.)
4. Faculty of Medicine. Collegium Medicum, Cardinal Stefan Wyszynski University in Warsaw, Dewajtis 5, 01-815 Warsaw, Poland; andrzejkotela@gmail.com
* Correspondence: Lucja.Sarapuk@pw.edu.pl

Received: 9 October 2020; Accepted: 21 October 2020; Published: 23 October 2020

Abstract: Nerve regeneration through cell electrostimulation will become a key finding in regenerative medicine. The procedure will provide a wide range of applications, especially in body reconstruction, artificial organs or nerve prostheses. Other than in the case of the conventional polystyrene substrates, the application of the current flow in the cell substrate stimulates the cell growth and mobility, supports the synaptogenesis, and increases the average length of neuron nerve fibres. The indirect electrical cell stimulation requires a non-toxic, highly electrically conductive substrate material enabling a precise and effective cell electrostimulation. The process can be successfully performed with the use of the graphene nanoplatelets (GNPs)—the structures of high conductivity and biocompatible with mammalian NE-4C neural stem cells used in the study. One of the complications with the production of inks using GNPs is their agglomeration, which significantly hampers the quality of the produced coatings. Therefore, the selection of the proper amount of the surfactant is paramount to achieve a high-quality substrate. The article presents the results of the research into the material manufacturing used in the cell electrostimulation. The outcomes allow for the establishment of the proper amount of the surfactant to achieve both high conductivity and quality of the coating, which could be used not only in electronics, but also—due to its biocompatibility—fruitfully applied to the cell electrostimulation.

Keywords: graphene nanoplatelets; surfactants; cell electrostimulation; tissue engineering

1. Introduction

Currently, the possibility to differentiate stem cells from other cells, especially nerve cells, is attracting great attention within academia [1–3]. Nerve and glial cells form the nerve tissue; compared to other mammalian body tissues, the nerve tissue has poor regenerative properties. Neurons are made of neural stem cells (NSCs) in the process of neurogenesis [1]; the process takes place in the brain's neurogenic areas. Therefore, obtaining stem cells for research and cell therapy is limited. The connective tissue—Wharton's jelly or umbilical cord blood—appears a safe and easily accessible source of adult stem cells which can be used to differentiate into neural cells. Due to the appropriate use of mesenchymal stem cells in vitro, cells with a neural phenotype are obtainable [1]. Owning to the appropriately selected cell stimulation and its parameters in the in vitro culture, it is conceivable not only to initiate the process of cell differentiation into neural cells, but also to control their viability and proliferation [4–9]. The tissue engineering to stimulate the growth and differentiation of cells uses cell scaffolds and electrostimulation, using both with direct and alternating electric currents [10–12]. As the substrates created with the graphene materials are rough and stiff, the sufficient adhesion of cells to the scaffold is ensured [5,6,13,14].

Printed electronics is one of the fastest growing field of the electronics industry. The development of printed electronics has led to the improvement of a variety of pattern and layer application techniques, making it possible to efficiently coat both stiff and flexible substrates [15,16]. Spray coating is one of the printing techniques which allows for a non-contact and cost-effective way to apply layers and patterns to a wide variety of substrates, especially substrates with a complex geometric structure. The indubitable asset of this technique is also the purity of the process, which eliminates the risk of contamination of the printed material and the layer formed, making it possible to apply created substrates in medical applications. The use of graphene materials in printed electronics methods can enable the production of highly conductive layers and coatings that are non-toxic to mammalian cells, which can be used, for example, in regenerative medicine for cellular electrostimulation [5]. Numerous scientific publications prove the expediency of the graphene-coated platforms for the proliferation and differentiation of induced pluripotent stem cells (iPSC), human mesenchymal stem cells (hMSC), human neural stem cells (hNSC) and preosteoblasts into osteoblasts [6,14,17–19]. One of the publications also indicates better proliferation and stronger polarisation of hMSC in the presence of fluorinated graphene; hence, it is possible to control the differentiation of cells to generate neurons [20].

The cellular electrostimulation development, especially on the graphene substrates, is of paramount importance for the regenerative medicine development. However, there is currently no literature describing research utilising highly conductive graphene materials such as graphene nanoflakes. Hence, the application of graphene to biomedical issues is still under discussion among researchers [4]. To date, reports related to the usage of graphene in tissue engineering have mainly focused on carbon materials, e.g., non-conductive graphene oxide (GO) or the reduced graphene oxide (rGO), the reduction of which often complicates the technological process [13,16]. Consequently, there have appeared great expectations associated with the graphene nanoplatelets (GNP)—materials with a sheet structure, the dimensions greater than 100 nm and a thickness of several nanometres—and their use in biomedicine [21]. GNPs are biocompatible with mammalian cells; simultaneously, they maintain graphene's unique electrical properties [5,22]. GNPs also have beneficial properties, e.g., good thermal conductivity and mechanical strength. They are used to create heterophasic materials in the form of inks and pastes. Our to-date studies confirmed that the developed inks and graphene layers are non-toxic to eukaryotic cells [5]. In the literature, graphene materials have been described as excellent substrates for cell cultures, enhancing the proliferation or differentiation of cells [6,14].

One of the problems emerging in the production of nanocomposite materials based on graphene nanoplatelets is the agglomeration of particles, which has a direct impact on the quality and usability of the created composites. The phenomenon is particularly noticeable in the case of unstable, two-dimensional nanoparticles such as graphene flakes [21,23]. The agglomerates develop as a consequence of Van der Waals forces and physical interactions between the structure layers; accordingly, solution sedimentation takes place [21,24]. In the production of polymer composites, it is imperative to obtain a suspension with the greatest possible homogeneity and stability [13,16]. Due to the homogeneous dispersion of particles in the solution, homogenous surfaces are obtainable [5]. This can be achieved by using an appropriately selected (for a given application) surfactant [25]. Both the type of solvent (whether polar or nonpolar) and the type of dispersed phase (hydrophilic or lipophilic) affect the process of dispersion. The amphiphilic structure of the surfactant may generate all the desired effects described, jet the excess amount, and above the critical concentration of micellization, it will generate the formation of micelles, which are structures that disturb the formed suspension in an undesirable way. To eliminate the unintended effects of the use of the amphiphilic agent, there is a need to properly dose surfactants and check the viscosity, printability and electrical parameters of the resulting suspension [16]. The presence of a surfactant affects some of the rheological properties of the inks, e.g., the surface tension. However, to some extent, it influences the solution viscosity. The selection of an appropriate surfactant and its amount significantly improve the conductivity of the graphene coatings, hence improving the electrostimulation process. Moreover, the addition of the surfactant has an significant effect on the sedimentation of the solution, which affects the stability of the coating depositing in the spray coating process [5].

In the production of inks for biomedical applications, the selection of the right surfactant plays a very crucial role. The surfactant ensures good dispersion of particles in the solution and positively affects ink rheology, but, at the same time, it has to be non-toxic to cells [5]. The surfactant is responsible for the solution homogeneity, layer electrical conductivity and cell behaviour on the coating. Such an approach to the cell stimulation and combining it with printing electronics and nanotechnology (particularly the use of graphene nanoplatelets) is an innovative solution, novelty in the field and was not published before.

In this study, the authors present the methodology of production of highly conductive, biocompatible graphene coatings for applications in cell electrostimulation. The study investigated the influence of surfactant content on the rheological properties of ink, as well as on the conductivity and microgeometry of the layers. The described ink manufacturing methodology and selection of the appropriate component quantity allows for the obtention of inks with the desired viscosity for the spray coating technique (viscosity in the range of 1.3–2.0 mPas). At the same time, the ink forms highly electrical conductive layers (13.5 Ω/□). The printed graphene layers are also characterised with homogeneity, which significantly affects the quality of cellular electrostimulation. The obtained results enable the selection of the ink with the best properties that allow for its effective use, not only in electronics, but, due to its biocompatibility, also for successful use in electrostimulation of cells.

2. Results and Discussion

2.1. Rheology of Graphene Inks with Different Surfactant Contents

The graphene inks have to meet the rheological requirements determined by the applied printing technique. It is necessary to assess the inks' viscosity, which affects the quality of the printed surface. On the basis of the obtained results, the relationship between the shear rate and the viscosity of the tested inks was determined. The inks' viscosity results are shown in Figure 1. The obtained viscosity values of the graphene inks, measured at the shear rate of 487.5 $\frac{1}{s}$, are presented in Table 1.

The serviceability of a material for use in spray coating technology is defined by the viscosity value at the highest shear rate, which should remain within the 0.7–2.0 mPas range [5,9]. The obtained viscosity values of the graphene inks, measured for the shear rate of 487.5 $\frac{1}{s}$, were in the range of 1.4–2 mPas, while for the base ink, the result was 1.3 mPas. The obtained values of viscosity are generally comparable. The viscosity value for the base ink is the lowest, which results from the lack of dispersion of the graphene flakes in the ink solution. Not only does that result in a poor quality of the coating, but it also contributes to its low conductivity. The further viscosity values—for 2%, 5% and 10% surfactant content—are akin to each other. Then, for the inks with 15% and 20% of surfactant, the viscosity value decreases. This indicates a deterioration of the flake dispersion triggered by the too-large amount of the surfactant. The viscosity values of the inks with the surfactants are within the recommended range, which enables their effective application in spray coating technology.

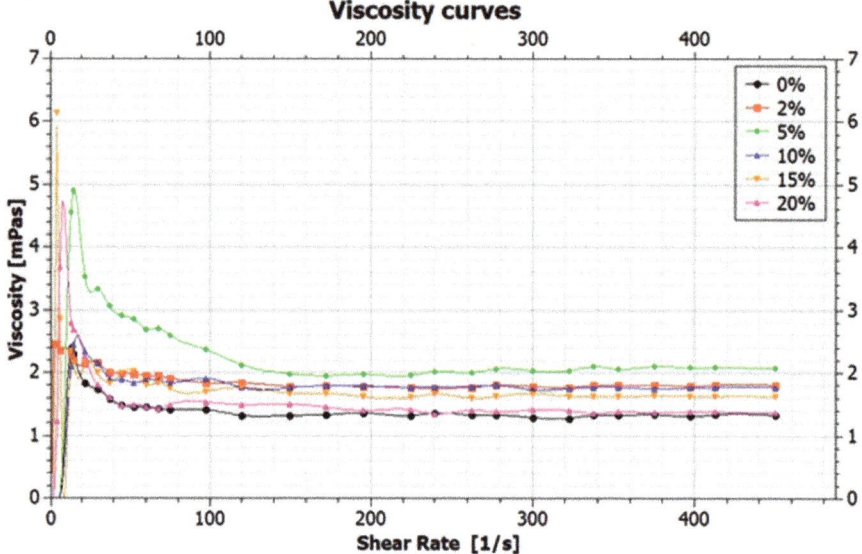

Figure 1. Graphene viscosity curves for graphene inks with 0, 2, 5, 10, 15, 20% surfactant content (in regard to GNP content).

Table 1. The obtained values of the viscosity of graphene inks with 0, 2, 5, 10, 15, 20% surfactant content (in regard to GNP content), measured at the highest value of shear rate of 487.5 $\frac{1}{s}$.

Ink (Graphene Nanoplatelets/ Surfactant Content)	Viscosity Value at Shear Rate 487.5 $\frac{1}{s}$
GNP D2 0%	1.3
GNP D2 2%	1.8
GNP D2 5%	2.0
GNP D2 10%	1.8
GNP D2 15%	1.6
GNP D2 20%	1.4

2.2. Production of Conductive Graphene Layers by Spray Coating Techniques

Spray coating allows one to place the graphene layers on variously shaped surfaces. It is also possible to achieve the desired thickness of the layers in a relatively easy manner. Additionally, the contactless application and high cleanliness of the coating are important, particularly for the biomedical applications.

During the application of the coatings, significant differences between several inks were established. The base ink, containing no surfactant, contained the agglomerates of the graphene flakes which clogged the airbrush nozzle; this significantly hindered the entire process and resulted in the production of nonhomogeneous coatings. Moreover, the coatings with such an ink required more layers to achieve satisfactory homogeneity. The printing process of the remaining—surfactant enriched—inks went smoothly regardless of the surfactant content.

Figures 2 and 3 show examples of the Kapton film substrates and polystyrene culture plates: (a) without a layer; (b) with a layer made with the base ink; and (c) ink with a surfactant. Figure 3 shows the flexible Kapton film substrates before and after the graphene coatings made with inks containing 0% and 5% surfactant. It can be noticed that the substrate created with the base ink is of poor quality, and produced agglomerates are clearly visible.

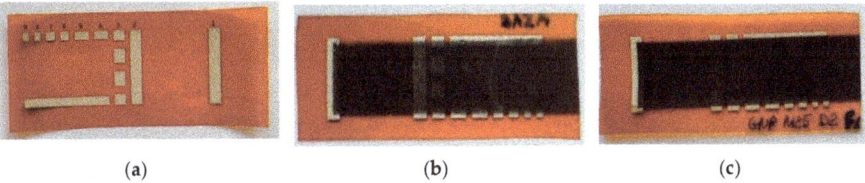

Figure 2. Kapton film substrates (**a**) without a layer; (**b**) with the layer made with the base ink with a visible agglomerate; and (**c**) with the homogeneous layer made with an ink with 5% surfactant content.

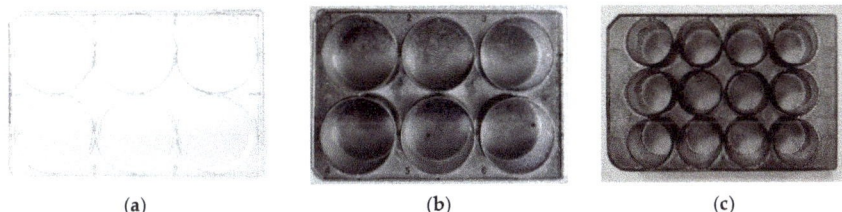

Figure 3. Culture plate (**a**) without coating; (**b**) covered with a base ink coating (without surfactant); and (**c**) covered with an ink coating containing 5% surfactant.

The quality of the obtained layers proves the surfactant's effectiveness. Analysing the above photos, it is evident how great the impact of the surfactant content in the ink is on the coating process. Not only does the surfactant component enable the correctness of the printing, but also, it unquestionably improves the quality of the produced layers, visible already during the layer application.

2.3. Layer Micro- and Macro-Geometry

From the biological perspective, due to their impact on cells and their stimulation, both micro- and macro-geometry are central. Figures 4 and 5 display the enlargements of the created graphene layers to assess the surfaces' micro- and macro-geometry, depending on the surfactant content.

The analysis of the micro-geometric images of the coatings proves that their porosity decreases with the increase in the surfactant content. The base ink coating has numerous defects, perceptible without high magnification. The addition of a little amount of surfactant significantly improves the coating homogeneity and reduces porosity. However, increasing the surfactant content from 5% to 20% provided no significant improvement in the coating quality. All the produced layers are equally homogeneous and lack pores. The increasing surfactant content insignificantly improves the quality of the microgeometry. The exception is the coating produced with the 15% surfactant ink, which has poor quality areas. Yet, in that case, it is more likely an effect of the distortion of the substrate rather than the poor quality of the produced ink itself.

Taking into account the cytocompatibility of the AKM-0531 surfactant (D2), confirmed in our previous work [25], the inks made with it can successfully be used in the cell cultures. The examination of the coatings' micro- and macro-geometry showed no significant variances for coatings made with inks containing more than 5% of surfactant. The most significant differences in the coatings' quality can be observed when comparing the SEM photographs of the coatings made with the base ink and the ink containing with the 5% surfactant. The addition of even a small amount of surfactant has a profound impact on the coatings' uniformity and quality and, thus, on their electrical conductivity.

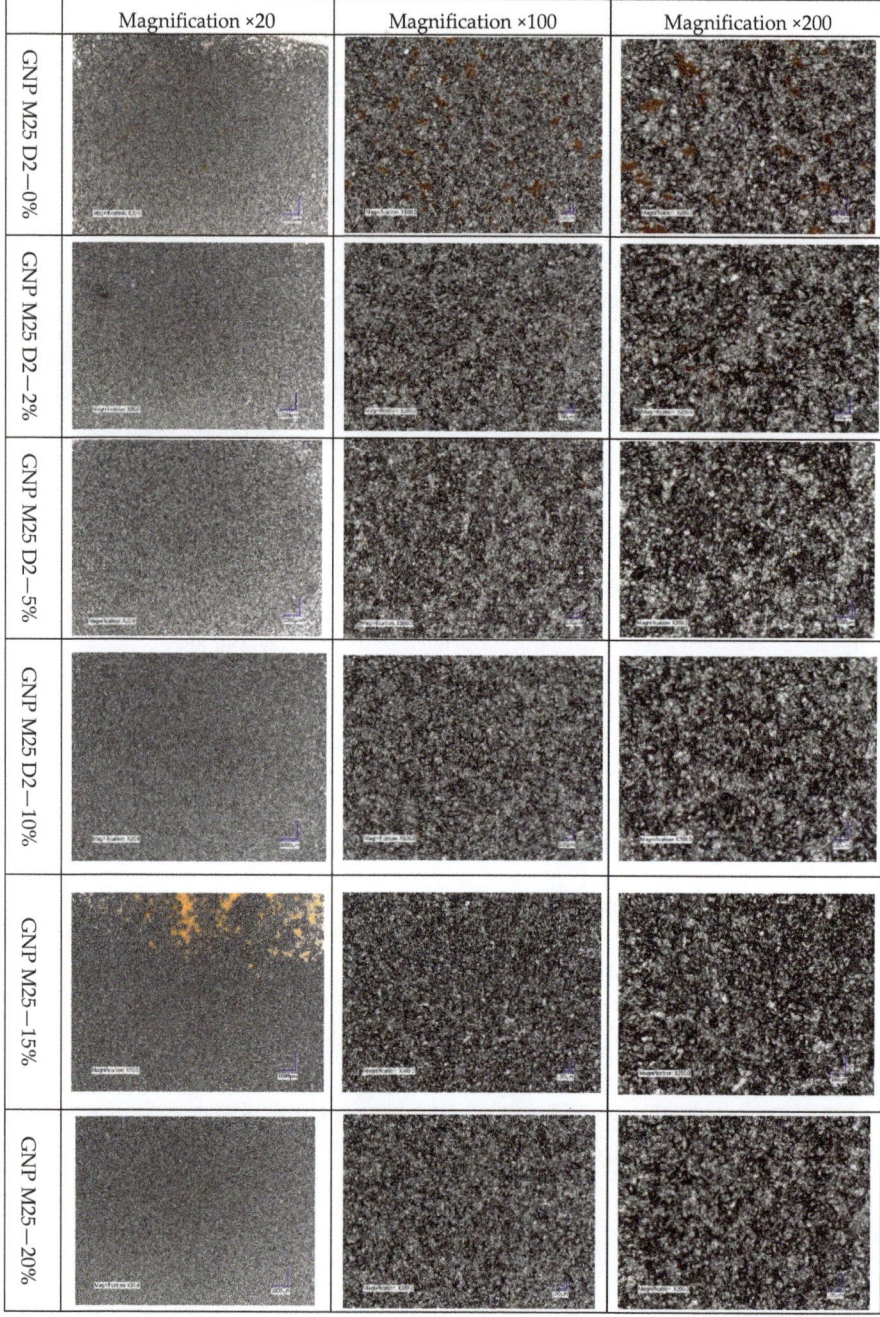

Figure 4. Digital microscope photographs showing the quality of coatings made of the inks with various surfactant content.

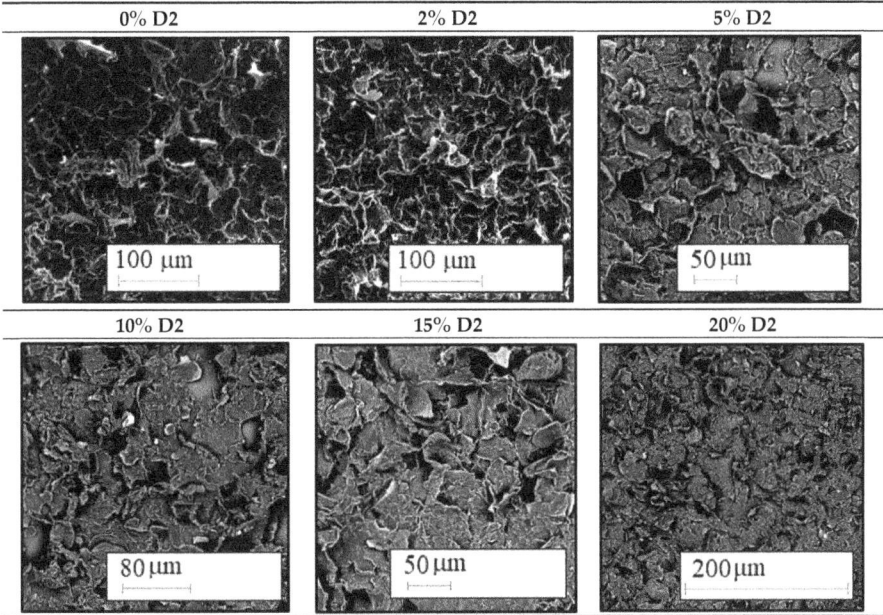

Figure 5. SEM photographs showing the quality of coatings made of the inks with various surfactant content.

2.4. Conductivity of Graphene Layers

To enable the application of the produced graphene coatings in the cellular electrostimulation, it is crucial to obtain the best current conductivity. The conductivity of the layers was measured depending on the amount of the surfactant used. The results of sheet resistivity measurements and their representation in figures are presented in Table 2 and Figure 6.

The graph's analysis indicates significant differences among the sheet resistivity values for various surfactant contents. When the surfactant content increases from 0% to 10%, the resistivity decreases successively; then, for the 15% and 20% content, their increase is visible. The lowest resistivity values were established for the 10% surfactant ink. Interestingly, the values are even five times lower than those obtained for the base ink (without any surfactant). Obtaining such low resistance proves the paramount influence that creating a homogeneous ink with a well-dispersed functional phase has on the layers' conductivity. The obtained viscosity results straightforwardly influence the coatings' conductivity (described in Section 2.1). The addition of more than 10% caused no further decrease in the resistance values; contrary, the values actually increased. The reason may be the bigger distance between the graphene nanoflakes, as a result of which, the conductivity of the layers decreased.

Choosing the spray coating method for applying the coatings allowed for the desired surface quality and conductivity to be achieved, while maintaining high purity of the process. The spray coating method allows for the obtention of graphene rough substrates, which can positively affect the behaviour of cells and their adhesion.

Table 2. The results of resistivity of the graphene layer measurements.

Average Sheet Resistivity [Ω/□]					
0% D2	2% D2	5% D2	10% D2	15% D2	20% D2
50.0	24.0	28.0	13.5	19.0	17.5

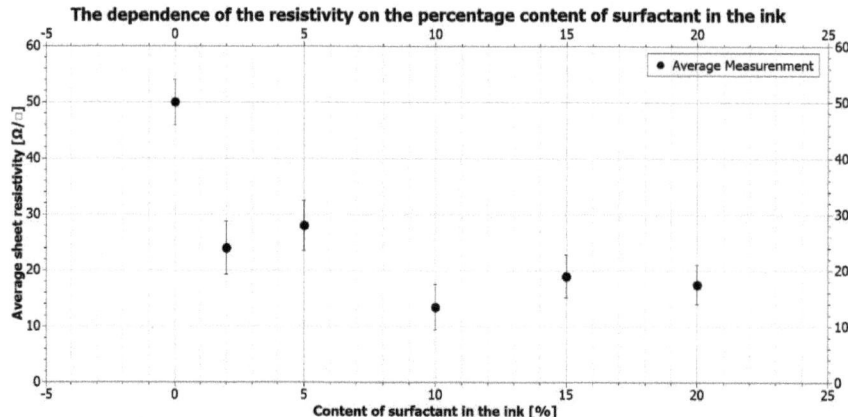

Figure 6. The dependence of the resistivity on the percentage content of the surfactant in the ink.

2.5. Cell Electrostimulation—Preliminary Results

For the applications of the graphene coatings in tissue engineering, cytocompatibility of the inks towards the cells is essential. There is a potential risk of the ink's toxicity resulting from the use of certain surfactants in the heterophasic inks [5]. The low content surfactant inks show no toxicity to the cells. Still, increasing the amount up to 20% could, potentially, have a negative effect on the cell behaviour. Accordingly, it is recommended to use less surfactant, which reduces the danger of cytotoxicity; simultaneously, it ensures good electrical conductivity. Therefore, the initial electrostimulation was performed on the layers made of the 5% surfactant content ink.

The preliminary electrostimulation process, carried out with the use of the produced substrates, resulted in the formation of new connections and cellular extensions. The contact of the cells with the graphene substrate did not cause their death; however, further research is necessary to determine more precise substrate–cell interactions. The SEM observations have made it possible to assess the ability of the proper adherence and elongation of the cells cultured on graphene layers for a seven-day culture. Figure 7 presents the results for both the current-stimulated and unstimulated control cells.

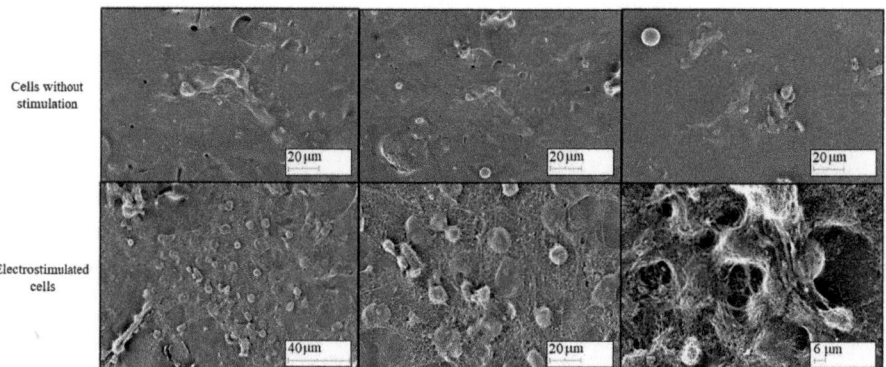

Figure 7. The SEM microscope images presenting the preliminary electrostimulation results.

The SEM pictures show that the electrostimulation greatly increased the number of correctly flattened cells. The cells from the control culture properly adhere to the substrate, flatten out and sprout. Still, when the current is applied, the well-developed network, created by the cell appendages, can be observed. Moreover, the electrostimulated cells released appendages, which, in the case of neurons, create ovules of dendrites and axons.

Based on microscopic observation, it can be concluded that there is an observable effect of electrostimulation on the behaviour of the neuronal stem cells embedded in the graphene layer. Electrostimulated cells not only multiply decidedly faster, but also form numerous cylindrical cytoplasmic appendages, which are extensions of the nerve cell body.

3. Materials and Methods

The composition of the inks was selected on the basis of the results obtained in our previous research [25]. The graphene nanoplatelets manufactured by XG Science (XG Science, Inc.; Lansing, MI, USA) (GNP M25) with average particle dimensions of 20–25 µm and a thickness of 19 sheets of graphene were used as the material of the functional phase [26]. The material was also used and characterised in our previous studies [27,28]. The percentage content of graphene flakes was 0.5 wt.%. The matrix is an essential component of graphene ink. In our study, a solution of polymeric polymethyl methacrylate (PMMA) (Sigma-Aldrich Sp.zo.o.; Poznan, Poland) in butyl carbinol acetate (OKB) (Sigma-Aldrich Sp.zo.o.; Poznan, Poland) at a concentration of 8% was used. Due to the long dissolution time of PMMA in OKB (min. 24 h), the matrix solution was conducted as the first step. The process of homogenisation of the carrier was carried out on a magnetic stirrer with a heated plate model RTC produced by IKA (IKA Poland Sp. zo.o.; Warsaw, Poland) at a temperature of 50 °C. Each of the produced inks contained the same matrix material in content of 27.5 wt.% and the same solvent—acetone, in content of approximately 72 wt.%. The percentage contents of each component in the made solution are presented in Table 3 below.

Table 3. Percentage content of ink components.

Percentage Content of Surfactant in Regard to GNP Content [%]	Ink Components			
	Surfactant (%)	Graphene Nanoplatelets (%)	Matrix (%)	Solvent (%)
0	0	0.5	27.5	72.000
2	0.01	0.5	27.5	71.990
5	0.025	0.5	27.5	71.975
10	0.05	0.5	27.5	71.950
15	0.075	0.5	27.5	71.925
20	0.1	0.5	27.5	71.900

The AKM-0531 surfactant MALIALIM series from the NOF company (NOF America Corporation; White Plains, NY, USA) was used in the studies. The MALIALIM surfactant is made of functional polymers with a comb structure. According to the datasheet, the molecular weight of the surfactant is in the range of 10^4–10^5 u [29]. MALIALIM surfactants have ionicity groups on the main chains responsible for the absorption of the polymer to the powder surface and its wetting. The polyoxyalkylene groups on side chains cause the repulsive effect of the functional phase particles, improving their dispersion in suspension [30].

In regard to the percentage content of the graphene flakes, inks containing 2, 5, 10, 15 and 20 wt.% surfactant were made. Concentrations of the surfactant with respect to the whole ink solution are presented in Table 4 below. Previous studies have shown that the layers produced with the developed inks have not caused toxic effects against cells [25]; thus, it seems certain that the layers used in the current work are cytocompatible.

Table 4. Weight content of the surfactant in the ink composition for a 50 g sample.

	Ink				
	GNP D2 2%	GNP D2 5%	GNP D2 10%	GNP D2 15%	GNP D2 20%
Content of the surfactant (g)	0.005	0.0125	0.025	0.0375	0.05

The layers were printed on two different substrates: a 25-µm-thick flexible Kapton film (DuPont, Inc.; Wilmington, DE, USA) and the polystyrene cell culture plates from Nest Scientific Biotechnology (Wuxi Nest Scientific Co., Ltd.; Wuxi, Jiangsu, China). The graphene inks were made according to the technology presented in Figure 8. The ink base was a combination of the surfactant, the graphene flakes and acetone. The obtained ink base was sonicated in an InterSonic IS-10 (Intersonic Sp.zo.o.; Olsztyn, Poland), 700 W ultrasonic cleaner for 90 min carried out at 35 kHz. After sonication, a polymer was added to the ink base. Finally, the whole solution was sonicated again for 15 min.

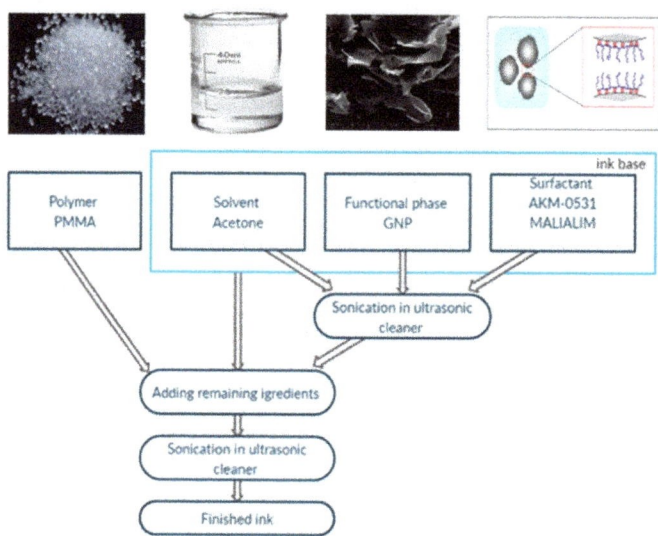

Figure 8. The technology of the graphene inks with graphene nanoflake production [31–34].

The layers were produced by spray coating. The Harder & Steenbeck Infinity CR Plus Solo Airbrush (Harder and Steenbeck, GmbH & Co. KG; Norderstedt, Germany) was used for application. The airbrush was connected to a compressed gas cylinder with 0.3 MPa pressure. The airbrush nozzle's diameter was 400 µm. The distance between the nozzle and the substrate was approximately 20 cm. After the application, the coatings were dried in a Memmert UF55 laboratory dryer with natural air circulation (Memmert GmbH & Co. KG; Schwabach, Germany). The drying of the Kapton foil substrates was carried out at 120 °C and lasted 60 min. For polystyrene culture plates, the drying temperature was 70 °C, and the drying time was 240 min.

For the viscosity measurements of the manufactured inks, a Wells-Brookfield model DV2T cone-plate viscometer (AMETEK Brookfield, Inc.; Middleboro, MA, USA) was used with a type CP-40 cone. The coatings' resistance measurements were performed using the UNI-T digital laboratory multimeter, model UT804 (Uni-Trend Technology, Co., Ltd.; Shanghai, China). The micro- and macro-geometry of the obtained layers, their quality, and homogeneity were assessed with the Keyence VHX-900F digital microscope (Keyence International (Belgium), NA/SA; Mechelen, Belgium), and the scanning electron microscope used was the Phenom Pharos Desktop Scanning Electron Microscope

(Thermo Fisher Scientific; Waltham, MA, USA). The pictures were taken at 20×, 100× and 200× magnification and an acceleration voltage of 10 kV.

The NE-4C neuroectodermal cell line was maintained in Eagle's minimum essential medium (EMEM) (ATCC) supplemented with 10% heat-inactivated foetal bovine serum (FBS) (ATCC), 2 mM L-Glutamine (Sigma-Aldrich Sp.zo.o.; Poznan, Poland) and 10^{-6} M all-trans retinoic acid (RA) (Sigma Aldrich) in the culture flasks pre-coated with 15 µg/mL poly-L-lysine (Sigma Aldrich Sp.zo.o.; Poznan, Poland). NE-4C cells from the 8th to 11th passage were used in the study. The cells were seeded on the 6-well plates at a density of 2×10^3 /cm^2 and maintained in standard cell culture conditions (37 °C, 5% CO_2 and 95% humidity). The cell stimulation was performed using an array of comb electrodes. The selected parameters were as follows: voltage amplitude $U = 10$ V; forcing frequency $f = 1$ kHz; sinusoidal; alternating current. The stem neural cells were cultured for 7 days; the stimulation started 24 h after seeding the culture [5]. For the scanning electron microscopy observation, after the seven-day culturing, the cells were fixed in 4% glutaraldehyde. Having been dehydrated, the cell surface of the samples was covered with a thin layer of gold with a sputtering system. The specimens were observed with an Merlin Zeiss scanning electron microscope (SEM) (Carl Zeiss Microscopy GmBh, Jena, Germany) with an acceleration voltage of 5 kV.

4. Conclusions

The results confirmed the effectiveness of the applied surfactant. The results for the inks with various contents of the AKM-0531 surfactant from the MALIALIM series are in the range of 11–50 Ω/□. The low resistance values of the produced layers enable their effective use not only in cell electrostimulation, but also for the production of sensors requiring high electrical conductivity.

The results prove that the created graphene inks meet both the technical and biological requirements. The use of a surfactant increases the dispersion of the GNPs particles in the inks and improves their homogeneity, which is crucial for obtaining high-quality substrates. The high quality of the substrates directly influences the layers' electrical properties, which are essential in the electrostimulation process. Obtaining the conductive, homogenous graphene coatings contributes to the tissue engineering development.

However, to entirely exclude the possibility of long-term damage of the cellular structures, further studies concerning the influence of the size and shape of graphene structures on their potential cytotoxicity are necessary [9]. Owning to the development of the research using the mesenchymal stem cells and the neural cells produced with them, in the future, it appears probable to effectively and safely treat injuries and neurodegenerative diseases [14]. In the long term, the use of cellular electrostimulation may provide a solution to the currently incurable neurological conditions, e.g., reconstruction of the spinal cord and nerve connections possibilities.

Author Contributions: Conceptualisation, L.D.-S.; methodology, A.G. and W.S.; validation, L.H.G., A.K. and M.J.; formal analysis, J.K.; investigation, W.S., L.D.-S. and A.G.; resources, L.D.-S. and W.S.; data curation, W.S., A.G. and J.K.; writing—original draft preparation, W.S.; writing—review and editing, L.D.-S.; visualisation, J.K.; supervision, M.J.; project administration, L.D.-S.; funding acquisition, L.D.-S. and M.J. All authors have read and agreed to the published version of the manuscript.

Funding: This research was funded by Warsaw University of Technology, internal grant of the Warsaw University of Technology—research activity in the discipline of biomedical engineering, grant number 504/04540/1142/43.050002.

Conflicts of Interest: The authors declare no conflict of interest.

References

1. Wroblewski, G.; Swatowska, B.; Dybowska-Sarapuk, L.; Jakubowska, M.; Stapinski, T. Optical properties of transparent electrodes based on carbon nanotubes and graphene platelets. *J. Mater. Sci. Mater. Electron.* **2016**, *27*, 12764–12771. [CrossRef]

2. Wroblewski, G.; Kielbasinski, K.; Stapinski, T.; Jaglarz, J.; Marszalek, K.; Swatowska, B.; Dybowska-Sarapuk, L.; Jakubowska, M. Graphene Platelets as Morphology Tailoring Additive in Carbon Nanotube Transparent and Flexible Electrodes for Heating Applications. *J. Nanomater.* **2015**, *2015*, 316315. [CrossRef]
3. Sanchez, V.C.; Jachak, A.; Hurt, R.H.; Kane, A.B. Biological Interactions of Graphene-Family Nanomaterials—An Interdisciplinary Review. *Chem. Res. Toxicol.* **2012**, *25*, 15–34. [CrossRef]
4. Akhavan, O. Graphene scaffolds in progressive nanotechnology/stem cell-based tissue engineering of the nervous system. *J. Mater. Chem. B* **2016**, *4*, 3169–3190. [CrossRef] [PubMed]
5. Dybowska-Sarapuk, Ł. *Development of the Production Process of Biocompatible, Conductive Graphene Ink*; Warsaw University of Technology: Warsaw, Poland, 2019.
6. Ullah, I.; Subbarao, R.B.; Rho, G.J. Human mesenchymal stem cells-Current trends and future prospective. *Biosci. Rep.* **2015**, *35*, e00191. [CrossRef] [PubMed]
7. Heo, C.; Yoo, J.; Lee, S.; Jo, A.; Jung, S.; Yoo, H.; Lee, Y.H.; Suh, M. The control of neural cell-to-cell interactions through non-contact electrical field stimulation using graphene electrodes. *Biomaterials* **2011**, *32*, 19–27. [CrossRef]
8. Gheith, M.K.; Pappas, T.C.; Liopo, A.V.; Sinani, V.A.; Shim, B.S.; Motamedi, M.; Wicksted, J.P.; Kotov, N.A. Stimulation of neural cells by lateral currents in conductive layer-by-layer films of single-walled carbon nanotubes. *Adv. Mater.* **2006**, *18*, 2975–2979. [CrossRef]
9. Ryu, S.; Kim, B.S. Culture of neural cells and stem cells on graphene. *Tissue Eng. Regen. Med.* **2013**, *10*, 39–46. [CrossRef]
10. Guo, W.; Zhang, X.; Yu, X.; Wang, S.; Qiu, J.; Tang, W.; Li, L.; Liu, H.; Wang, Z.L. Self-Powered Electrical Stimulation for Enhancing Neural Differentiation of Mesenchymal Stem Cells on Graphene-Poly(3,4-ethylenedioxythiophene) Hybrid Microfibers. *ACS Nano* **2016**, *10*, 5086–5095. [CrossRef]
11. da Silva, L.P.; Kundu, S.C.; Reis, R.L.; Correlo, V.M. Electric Phenomenon: A Disregarded Tool in Tissue Engineering and Regenerative Medicine. *Trends Biotechnol.* **2020**, *38*, 24–49. [CrossRef]
12. Thrivikraman, G.; Boda, S.K.; Basu, B. Unraveling the mechanistic effects of electric field stimulation towards directing stem cell fate and function: A tissue engineering perspective. *Biomaterials* **2018**, *150*, 60–86. [CrossRef]
13. Gurunathan, S.; Kim, J.H. Synthesis, toxicity, biocompatibility, and biomedical applications of graphene and graphene-related materials. *Int. J. Nanomed.* **2016**, *11*, 1927–1945. [CrossRef] [PubMed]
14. Szabłowska-Gadomska, I.; Bużańska, L.; Małecki, M. Stem cell properties, current legal status and medical application. *Postepy Hig. Med. Dosw.* **2017**, *71*. [CrossRef]
15. Jing, W.; Huang, Y.; Wei, P.; Cai, Q.; Yang, X.; Zhong, W. Roles of electrical stimulation in promoting osteogenic differentiation of BMSCs on conductive fibers. *J. Biomed. Mater. Res. Part A* **2019**, *107*, 1443–1454. [CrossRef] [PubMed]
16. Hu, G.; Kang, J.; Ng, L.W.T.; Zhu, X.; Howe, R.C.T.; Jones, C.G.; Hersam, M.C.; Hasan, T. Functional inks and printing of two-dimensional materials. *Chem. Soc. Rev.* **2018**, *47*, 3265–3300. [CrossRef] [PubMed]
17. Park, S.Y.; Park, J.; Sim, S.H.; Sung, M.G.; Kim, K.S.; Hong, B.H.; Hong, S. Enhanced differentiation of human neural stem cells into neurons on graphene. *Adv. Mater.* **2011**, *23*, 263–267. [CrossRef] [PubMed]
18. Zhang, Y.; Nayak, T.R.; Hong, H.; Cai, W. Graphene: A versatile nanoplatform for biomedical applications. *Nanoscale* **2012**, *4*, 3833–3842. [CrossRef] [PubMed]
19. Nayak, T.R.; Andersen, H.; Makam, V.S.; Khaw, C.; Bae, S.; Xu, X.; Ee, P.-L.R.; Ahn, J.-H.; Hong, B.H.; Pastorin, G.; et al. Graphene for Controlled and Accelerated Osteogenic Differentiation of Human Mesenchymal Stem Cells. *ACS Nano* **2011**, *5*, 4670–4678. [CrossRef]
20. Wang, Y.; Lee, W.C.; Manga, K.K.; Ang, P.K.; Lu, J.; Liu, Y.P.; Lim, C.T.; Loh, K.P. Fluorinated Graphene for Promoting Neuro-Induction of Stem Cells. *Adv. Mater.* **2012**, *24*, 4285–4290. [CrossRef]
21. Krittayavathananon, A.; Li, X.; Batchelor-McAuley, C.; Sawangphruk, M.; Compton, R.G. Comparing the effect of different surfactants on the aggregation and electrical contact properties of graphene nanoplatelets. *Appl. Mater. Today* **2018**, *12*, 163–167. [CrossRef]
22. Jiménez-Suárez, A.; Prolongo, S.G. Graphene Nanoplatelets. *Appl. Sci.* **2020**, *10*, 1753. [CrossRef]
23. Neto, H.C.; Guinea, F.; Peres, N.M.R.; Novoselov, K.S.; Geim, A.K. The electronic properties of graphene. *Rev. Mod. Phys.* **2009**, *81*, 109–162. [CrossRef]
24. Krittayavathananon, A.; Li, X.; Sokolov, S.V.; Batchelor-McAuley, C.; Sawangphruk, M.; Compton, R.G. The solution phase aggregation of graphene nanoplates. *Appl. Mater. Today* **2018**, *10*, 122–126. [CrossRef]

25. Dybowska-Sarapuk, Ł.; Rumiński, S.; Wróblewski, G.; Słoma, M.; Młożniak, A.; Kalaszczyńska, I.; Lewandowska-Szumieł, M.; Jakubowska, M. Aqueous biological graphene based formulations for ink-jet printing. *Pol. J. Chem. Technol.* **2016**, *18*, 46–52. [CrossRef]
26. Romanenko, A.I.; Anikeeva, O.B.; Kuznetsov, V.L.; Obrastsov, A.N.; Volkov, A.P.; Garshev, A.V. Quasi-two-dimensional conductivity and magnetoconductivity of graphite-like nanosize crystallites. *Solid State Commun.* **2006**, *137*, 625–629. [CrossRef]
27. Dybowska-Sarapuk, Ł.; Kotela, A.; Krzemiński, J.; Wróblewska, M.; Marchel, H.; Romaniec, M.; Łęgosz, P.; Jakubowska, M. Graphene nanolayers as a new method for bacterial biofilm prevention: Preliminary results. *J. AOAC Int.* **2017**, *100*, 900–904. [CrossRef]
28. Pepłowski, A.; Walter, P.A.; Janczak, D.; Górecka, Ż.; Święszkowski, W.; Jakubowska, M. Solventless conducting paste based on graphene nanoplatelets for printing of flexible, standalone routes in room temperature. *Nanomaterials* **2018**, *8*, 829. [CrossRef]
29. Molecular Weight of NOF Company Dispersants. Available online: https://www.nof.co.jp/english/dispersant/pdf/molecular_weight_of_dispersants.pdf (accessed on 17 October 2020).
30. MALIALIM Surfactants. Available online: https://www.nofamerica.com/store/index.php?dispatch=categories.view&category_id=131 (accessed on 17 October 2020).
31. Solvent. Available online: http://freyamakeschemistrystressfree.blogspot.com/2017/10/14-know-what-is-meant-by-terms-solvent.html (accessed on 17 October 2020).
32. Surfactant. Available online: https://www.nofamerica.com/store/images/img_malialim_02.jpg (accessed on 17 October 2020).
33. Graphene Nanoplatelets. Available online: https://nanografi.com/product_images/uploaded_images/us1059-sem-graphene-nanoplatelets.jpg (accessed on 17 October 2020).
34. Polymethyl Methacrylate. Available online: https://www.indiamart.com/proddetail/lg-pmma-injection-moulding-ig840-20230863530.html (accessed on 17 October 2020).

Publisher's Note: MDPI stays neutral with regard to jurisdictional claims in published maps and institutional affiliations.

© 2020 by the authors. Licensee MDPI, Basel, Switzerland. This article is an open access article distributed under the terms and conditions of the Creative Commons Attribution (CC BY) license (http://creativecommons.org/licenses/by/4.0/).

Review

Graphene-Based Sensors for the Detection of Bioactive Compounds: A Review

Carlos Sainz-Urruela [1], Soledad Vera-López [1,2], María Paz San Andrés [1,2] and Ana M. Díez-Pascual [1,2,*]

[1] Universidad de Alcalá, Facultad de Ciencias, Departamento de Química Analítica, Química Física e Ingeniería Química, Ctra. Madrid-Barcelona Km. 33.6, 28805 Alcalá de Henares, Madrid, España (Spain); carlos.sainz@uah.es (C.S.-U.); soledad.vera@uah.es (S.V.-L.); mpaz.sanandres@uah.es (M.P.S.)

[2] Universidad de Alcalá, Instituto de Investigación Química Andrés M. del Río (IQAR), Ctra. Madrid-Barcelona Km. 33.6, 28805 Alcalá de Henares, Madrid, España (Spain)

* Correspondence: am.diez@uah.es

Citation: Sainz-Urruela, C.; Vera-López, S.; San Andrés, M.P.; Díez-Pascual, A.M. Graphene-Based Sensors for the Detection of Bioactive Compounds: A Review. *Int. J. Mol. Sci.* **2021**, *22*, 3316. https://doi.org/10.3390/ijms22073316

Academic Editor: Vojtěch Adam

Received: 5 March 2021
Accepted: 22 March 2021
Published: 24 March 2021

Publisher's Note: MDPI stays neutral with regard to jurisdictional claims in published maps and institutional affiliations.

Copyright: © 2021 by the authors. Licensee MDPI, Basel, Switzerland. This article is an open access article distributed under the terms and conditions of the Creative Commons Attribution (CC BY) license (https:// creativecommons.org/licenses/by/ 4.0/).

Abstract: Over the last years, different nanomaterials have been investigated to design highly selective and sensitive sensors, reaching nano/picomolar concentrations of biomolecules, which is crucial for medical sciences and the healthcare industry in order to assess physiological and metabolic parameters. The discovery of graphene (G) has unexpectedly impulsed research on developing cost-effective electrode materials owed to its unique physical and chemical properties, including high specific surface area, elevated carrier mobility, exceptional electrical and thermal conductivity, strong stiffness and strength combined with flexibility and optical transparency. G and its derivatives, including graphene oxide (GO) and reduced graphene oxide (rGO), are becoming an important class of nanomaterials in the area of optical and electrochemical sensors. The presence of oxygenated functional groups makes GO nanosheets amphiphilic, facilitating chemical functionalization. G-based nanomaterials can be easily combined with different types of inorganic nanoparticles, including metals and metal oxides, quantum dots, organic polymers, and biomolecules, to yield a wide range of nanocomposites with enhanced sensitivity for sensor applications. This review provides an overview of recent research on G-based nanocomposites for the detection of bioactive compounds, providing insights on the unique advantages offered by G and its derivatives. Their synthesis process, functionalization routes, and main properties are summarized, and the main challenges are also discussed. The antioxidants selected for this review are melatonin, gallic acid, tannic acid, resveratrol, oleuropein, hydroxytyrosol, tocopherol, ascorbic acid, and curcumin. They were chosen owed to their beneficial properties for human health, including antibiotic, antiviral, cardiovascular protector, anticancer, anti-inflammatory, cytoprotective, neuroprotective, antiageing, antidegenerative, and antiallergic capacity. The sensitivity and selectivity of G-based electrochemical and fluorescent sensors are also examined. Finally, the future outlook for the development of G-based sensors for this type of biocompounds is outlined.

Keywords: bioactive compound; graphene; graphene oxide; melatonin; gallic acid; tannic acid; resveratrol; oleuropein; hydroxytyrosol; tocopherol; ascorbic acid; curcumin

1. Introduction

Carbon is an essential element on earth, which can be found in several structures in nature. The most usual and stable ones are diamond and graphite. Diamond has a rigid 3-D structure with sp^3 carbon atoms arranged in a lattice, which is a variation of the face-centered cubic crystal structure. It has superlative physical qualities, most of which originate from the strong covalent bonding between its atoms. Graphite consists of a layered structure made of hexagonal rings of carbon atoms with sp2-hybridization. These layers are linked by Van der Waals forces that generate an exfoliating structure. Due to the weak forces between the graphite layers, it is possible to isolate one of these layers to obtain graphene (G), an atomically thick 2D sheet comprising sp2 carbon atoms arranged in a honeycomb structure. It has superior electronic, thermal, optical, and mechanical properties with values that surpass those obtained in any other materials [1,2]. For instance, it has exceptional thermal conductivity, in the range of 3000–5000 W·m^{-1}·K^{-1} [3], superior to that of copper, which is around 400 W·m^{-1}·K^{-1}, very high electron mobility (25,000 cm^2 V^{-1} s^{-1}) [4], the highest electrical conductivity known at room temperature (6000 S cm^{-1}) [5], a very large specific surface area (2640 m^2 g^{-1}) [6], and it is impermeable to gases. Further, it is a zero-gap semiconductor material, is electroactive and transparent, absorbing only 2.3% of the incident light. Moreover, G presents a Young's modulus close to 1 TPa and an ultimate strength of 130 GPa, thus being the strongest material ever measured, stiffer than steel [7,8]. These unique properties make G an ideal candidate for a wide range of applications such as sensors, supercapacitors, fuel cells, photovoltaic devices, batteries, nanocomposites, flexible electronic devices, and so forth [9–13].

G is the starting point of other structures like fullerenes, nanotubes, or graphene quantum dots. It should be noticed that the term "graphene" used in the literature includes a broad range of graphene-like structures which differ in the preparation method and consequently in the chemical structure (usually the oxidation level), shape, size, and the number of layers. G synthesis is based on two general approaches, i.e., bottom-up and top-down approaches, as shown in Scheme 1. In the first approach, the starting material is graphite, and the aim is to exfoliate it via mechanical, liquid phase, or electrochemical exfoliations. Another approach in this group is to exfoliate graphite oxide to graphene oxide (GO), followed by chemical or thermal reduction. The bottom-up method is based on making graphene from molecular precursors building blocks by chemical vapor deposition (CVD) or epitaxial growth.

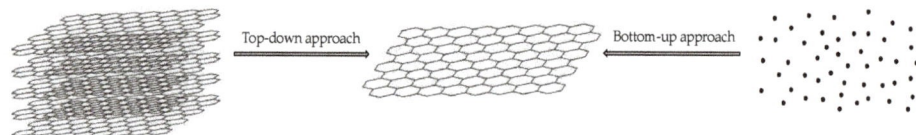

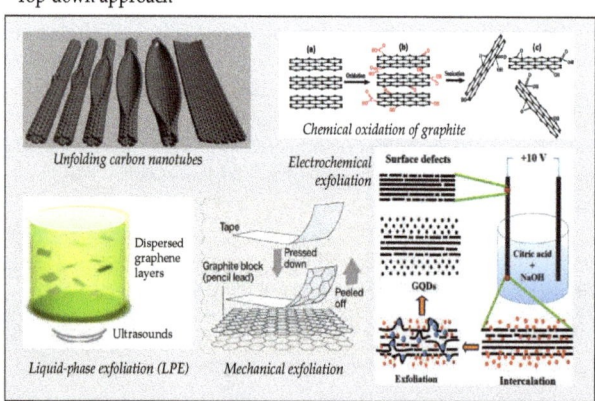
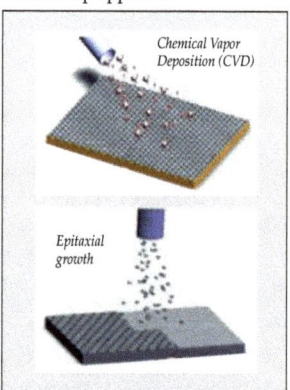

Scheme 1. Representation of bottom-up and top-down approaches for G synthesis. GDQ: graphene quantum dots.

Mechanical exfoliation is the simplest technique, in which G is isolated by peeling it off from graphite flakes using tape. The main concern of this method is the low graphene quality and the limitation for large-scale production [2]. CVD is a scalable and cost-effective technique to produce high-quality graphene films, hence it is the most used to fabricate G at a large-scale, although it is difficult to attain a proper control of the G thickness. CVD uses the saturation of carbon by a high-tempered hydrocarbon gas on a transition metal substrate [14]. When the metal becomes cold, carbon solubility decreases, and carbon atoms precipitate, forming a G layer. Epitaxial growth enables control over the layer thickness by modifying the time and temperature throughout the process. However, it is one of the most expensive synthesis methods since a silicon carbide has to be heated between 1080 °C and 1320 °C to lead the growth of G; thus, it is not affordable at a large-scale. Moreover, the face of SiC used, *Si*-terminal or *C*-terminal, highly influences the thickness, mobility, and density of graphene [15]. Chemical oxidation of graphite is frequently used to obtain a graphene derivative, GO. Different oxidizing reagents can be used, like concentrated acids, such as $KMnO_4$ or K_2CrO_4. The yield of graphene oxidation depends greatly upon the graphite source (flake size, crystallinity, purity, etc.) and can be tuned with other parameters such as oxidation time, acid ratios, mixing, washing, removal of non-oxidized material, etc. Subsequently, a reduction stage has to be applied to reduce the GO, although the resulting product differs significantly from raw G, and it is commonly named reduced graphene oxide (rGO) [16,17]. Another method to get a G layer is unfolding carbon nanotubes (CNTs) via chemical or plasma etching [18]. Other approaches include liquid-phase exfoliation (LPE), which uses specific organic molecules, or electrochemical exfoliation, which relies on the penetration of graphite by ions from the electrochemical solution using a potential [19]. Depending on the synthesis method, graphene features are different, as well as the yield and the costs [20,21]. In addition, it should be noted that due to graphene forces, it tends to fold or join with other G layers, leading to an agglomerated nanomaterial. For this reason, surfactants are usually required to attain stable G monolayers in solution [22,23].

Some researchers have discovered that GO and rGO have better properties for some applications than graphene. GO is a modified form of G that comprises carboxylic groups on the edges and epoxy and hydroxyl groups on the layer plane (Scheme 2). It can be obtained via chemical oxidation graphite [24], as we mentioned before, via oxidation of G using the Hummers' method [17,25] or by electrochemical exfoliation of graphite oxide GO [26,27], leading to a larger surface, which allows keeping compounds in the interlayer space. This property is especially useful for biomedical applications.

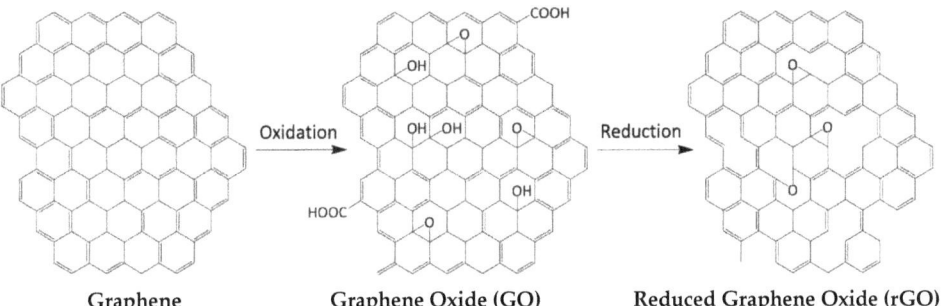

Scheme 2. Molecular structures of Graphene, Graphene Oxide, and Reduced Graphene Oxide.

On the other hand, rGO is obtained via the thermal treatment of GO to remove functional groups [24] or by chemical reduction of GO (Scheme 2). Reduction of the epoxide groups of the GO can be conducted by using hydrazine or sodium borohydride as reduction agents, but the dehydroxylation and decarboxylation need heat treatment (as an endothermic reaction). However, these components are corrosive, combustible, and highly toxic, which may be dangerous for personnel health and the environment. Hence, eco-friendly, natural reducing agents are sought, such as aminoacids (i.e., ascorbic acid) or plant extracts (i.e., Ginkgo biloba leaves).

GO is removed more easily, both kinetically and thermodynamically; rGO has improved the electric properties compared to GO, due to the reduced amount of functional groups, although its retention capability is lower [17]. The functional groups of both G and GO enable them to be functionalized and to interact with other materials [28].

On the other hand, graphene quantum dots (GQDs) have recently emerged amongst the family of carbon nanomaterials. They are 0D graphene sheets, with circular planar geometry and very small size, about 3–20 nm (Figure 1). Due to the confinement of the excitons and the quantum effect, GQDs exhibit many excellent properties such as chemical inertness, photobleaching resistance, and stable luminescence. Thus, photoluminescence can be induced in graphene material when the size, structure, and surface property are controlled properly as for traditional QDs [29]. Fluorescent GQDs have great potential for application in bioimaging, diagnosis, and drug delivery and can be used in Föster resonance energy transfer (FRET) optical sensors as donors [30]. Further, GQDs combined with conventional plasmonic material such as noble metals can significantly improve the absorption (and interaction) with visible and IR light, which is only 2.3% for bare G. Besides, GQDs exhibit low toxicity, high conductivity, and good biocompatibility, which combined with their cost-effectiveness make them suitable and efficient in both optical and electrochemical sensing applications [31,32].

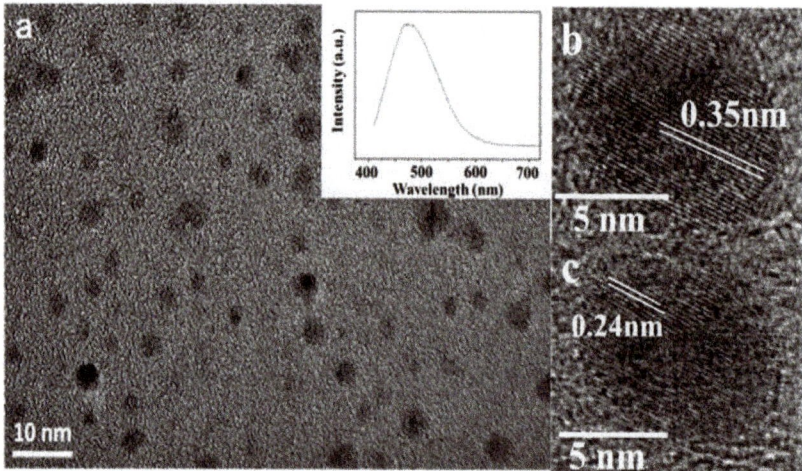

Figure 1. (a) TEM images of graphene quantum dots (GQDs) of different sizes and the photoluminescence spectrum excited at 400 nm. (b,c) High-resolution transmission electron microscopy (HRTEM) images of GQDs. Taken from Wu et al. [32].

A chemical sensor is innately defined as a device that responds to changes in the local chemical environment via an electrical, electrochemical, and/or optical signal. It transforms the chemical information of a sample (concentration of one or more of its components) into an analytically useful signal. The main parts of a chemical sensor shown in Scheme 3.

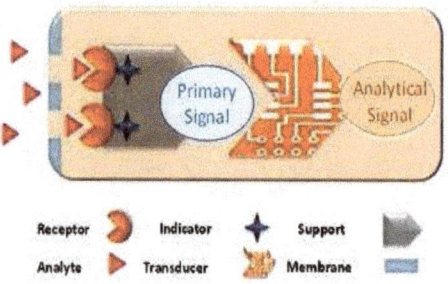

Scheme 3. Basic representation of a chemical sensor.

(1) A receptor zone has the role of transforming the chemical information into a form of primary energy or signal. It is the chemical part of the sensor and generally contains a reagent immobilized on a solid support. If the receptor is not capable of generating the primary signal by itself, it also contains an indicator. If the reagent is biological in nature and is called a biosensor, if it is synthetic, the term chemoreceptor is often used. It should be pointed out that, traditionally, the term biosensor also applies to chemoreceptors that act in a similar way to affinity bioreactants; that is, they have both the ability to recognize the analyte (molecular recognition) as well as to chemically interact with it, as is the case with Molecularly Imprinted Polymers (MIPs), or with non-natural reagents but of biological origin (aptamers or plastic antibodies). The selectivity, sensitivity, and precision of the sensor depend essentially on this component.

(2) A transducer zone, which transforms the primary signal into the final analytical signal, generally an electric current or a potential difference directly or inversely proportional to the analyte concentration. It is the most technical part of the sensor since, together with the transducer itself, it contains the necessary components for the transformation and treatment of the signals. Although there are different types of primary signals (mass, conductivity, magnetic, etc.), the most common transducers are electrochemical and optical. In the latter, the transducer is called a detector.

(3) A membrane to separate the whole sensor from the outside.

For a chemical sensor to be useful, its response must be reproducible, stable, sensitive, and selective. Over the last years, a change in sensor technology towards more sensitive elements, complex architecture, and size reduction has arisen due to the emergence of nanotechnology, that is, the science that deals with the manipulation of matter on the scale of atoms and molecules. Nanomaterials such as graphene [33], carbon nanotubes [34], metal oxide nanoparticles [35], and their polymeric composites [36] have been used to develop sensors to detect a variety of analytes.

Bioactive ingredients of natural products can protect the human body from harm, as well as prevent and treat disease. Screening bioactive compounds from natural products is attracting particular attention in a broad range of applications, including pharmacology, cosmetics, the food industry, biomedicine, and so forth. It is a very promising research area in full development, which has resulted in many studies devoted to diversify the resources of bioactive compounds and improve their synthesis or recovery pathways. However, despite all this significant research in various fields, the definition of bioactive compounds remains ambiguous and unclear. The term "bioactive" arises from bio- from the Greek (βίο-) "bios", which means life, and active from the Latin "activus", which means: dynamic or full of energy. In a strictly scientific sense, the term "bioactive" is an alternative term for "biologically active" [37]. Thus, a bioactive compound is just a substance that has biological activity [38]. These compounds contain chemicals that are found in small amounts in plants or certain foods such as fruits, vegetables, nuts, oils, and whole grains; further, they typically have an associated beneficial effect on health. It should be noted that in addition to natural bioactive substances [39], there are also synthetic bioactive molecules [40].

The benefits of G-based sensors can be summarized as follows: the high specific surface area and atomic thickness of G layers enable direct contact between the carbon atoms and the analytes; as a result, graphene-based sensors have better sensitivity compared to silicon [41]. In addition, the flexibility combined with the high optical transparency and electrical conductivity of G enables the acquisition of high-quality signals without motion artifacts or visual disturbances. Furthermore, a high signal-to-noise ratio can be achieved in electrophysiological signals due to the efficient signal transmission attained by the high electrical conductivity [42]. Besides, the large specific surface area, the feasibility of functionalization, and a high electron transfer rate enable receptors such as enzymes, antibodies, and DNA to be effectively immobilized on a G surface [43]. Consequently, many sensors based on G nanocomposites and their derivatives have been reported, including wearable sensors and implantable devices, that enable performing quick, real-time analysis [44] and are also applied in biomedicine for diagnosis, prognosis, prediction, and treatment methods. It is a novel technology, increasingly employed, in which it is completely essential to work at the nanoscale.

Although a number of reviews dealing with G-based sensors for the detection of biomolecules have been published, most of them are too general and deal only with electrochemical sensors. Further, they are devoted mainly to target biological molecules such as DNA, urea, glucose, and so forth. In this review, an attempt is made to show the state-of-art in the synthesis and performance of G-based fluorescent and electrochemical sensors for the specific detection of bioactive compounds, providing insights into the unique advantages offered by graphene and its derivatives. Their synthesis process (including functionalization routes), novel structures, and main properties are summarized, and the main challenges are also discussed. The reported works try to improve the conventional procedures via the use of cheaper, faster, more effective, or more eco-friendly methods.

2. Bioactive Compounds: Properties and Applications

Bioactive compounds are substances that typically occur in small quantities in food and can be beneficial for health. They are intensively studied to evaluate their effects, including antioxidant, antiallergic, antimicrobial, antithrombotic, antiatherogenic, hypoglycaemic, anti-inflammatory, antitumor, cytostatic, immunosuppressive properties, and hepatoprotective activities [45]. Unlike essential macro- and micronutrients, they are not essential for life, and the body can function properly without them.

Among the most common bioactive compounds found in our diets are polyphenols, which are well known for their antioxidant properties [46]. They inhibit or interfere with the process of free radical formation, preventing the oxidation of cells. Dietary antioxidants can protect the body from oxidative damage that may result over time in many pathologies such as cancer or cardiovascular disease. It is believed that polyphenols may exert a cardio protective effect via several paths. They may enhance the functioning of the inner lining of blood vessels, hinder platelet aggregation (preventing blood accumulations in the arteries), and positively influence blood lipids and insulin sensitivity [47]. Typical dietary antioxidants are vitamin C, vitamin E, and carotenoids, and most of them come from vegetables like olive, rice, broccoli, eggplant, ginger, onion, citrus, coffee, tomatoes, etc.

The antioxidant potential of phenolic compounds depends on the number and arrangement of the hydroxyl groups [48]. Phenolic antioxidants can give hydrogen atoms to lipid radicals and produce lipid derivatives and antioxidant radicals (Equation (1)), which are more stable and less readily available to promote autoxidation. The antioxidant free radical could then interfere with the chain-propagation reactions (Equations (2) and (3)).

$$R*/RO*/ROO* + AH \rightarrow A* + RH/ROH/ROOH \qquad (1)$$

$$RO*/ROO* + A* \rightarrow ROA/ROOA \qquad (2)$$

$$ROO* + RH \rightarrow ROOH + R* \qquad (3)$$

As the hydrogen bond energy in a free radical scavenger (FRS) decreases, the hydrogen transfer to the free radical is energetically more favorable, hence faster. Any compound that has a reduction potential lower than that of a free radical (or oxidized species) is able to donate its hydrogen atom to that of the free radical unless the reaction is kinetically unfeasible. For instance, FRS including α-tocopherol ($E^{\circ\prime}$ = 500 mV) that have a reduction potential lower than that of peroxyl radicals ($E^{\circ\prime}$ = 1000 mV) is able to donate their hydrogen to the peroxyl radical to form a hydroperoxide [49]. The phenoxyl radical is then stabilized by the delocalization of its unpaired electron around the aromatic ring.

The influence of the antioxidant concentration on the autoxidation rate is conditioned by many factors such as the antioxidant structure, oxidation conditions, and nature of the sample being oxidized [50]. Frequently phenolic antioxidants lose their activity at high concentrations and are more effective in extending the induction period when added to any oil that has not deteriorated to any great extent. Thus, to attain the best protection against oxidation, antioxidants should be added to foodstuffs as early as possible during processing and storage. Nonetheless, the effect of polyphenols on humans is still not clear, and many studies claim that the beneficial effect arises from the polyphenol-rich foods rather than the isolated polyphenols. Up to date, the European Food Safety Authority (EFSA), which evaluates health statements made on food products, has rejected all claims on polyphenols except for olive oil, which contains hydroxytyrosol, and contributes to the protection of blood lipids from oxidative stress.

Among natural phenolic antioxidants are phenolic acids, flavonoids, coumarins, stilbenes, hydrolyzable and condensed tannins, lignans, and lignins, and contain one or more hydroxyl groups attached directly to an aromatic ring. The antioxidants selected for this review were red wine polyphenolic compounds, including gallic acid, tannic acid, resveratrol, oleuropein, hydroxytyrosol, tocopherol, and ascorbic acid (the last two found in lower amounts). Further, melatonin, an indole amine also present in wine that shows a wide range of anticancer activities, and curcumin, a polyphenol with multiple health benefits, has also been addressed. All these antioxidants are either autofluorescent or can be derivatized to form a fluorescent compound. Their chemical structure is displayed in Scheme 4.

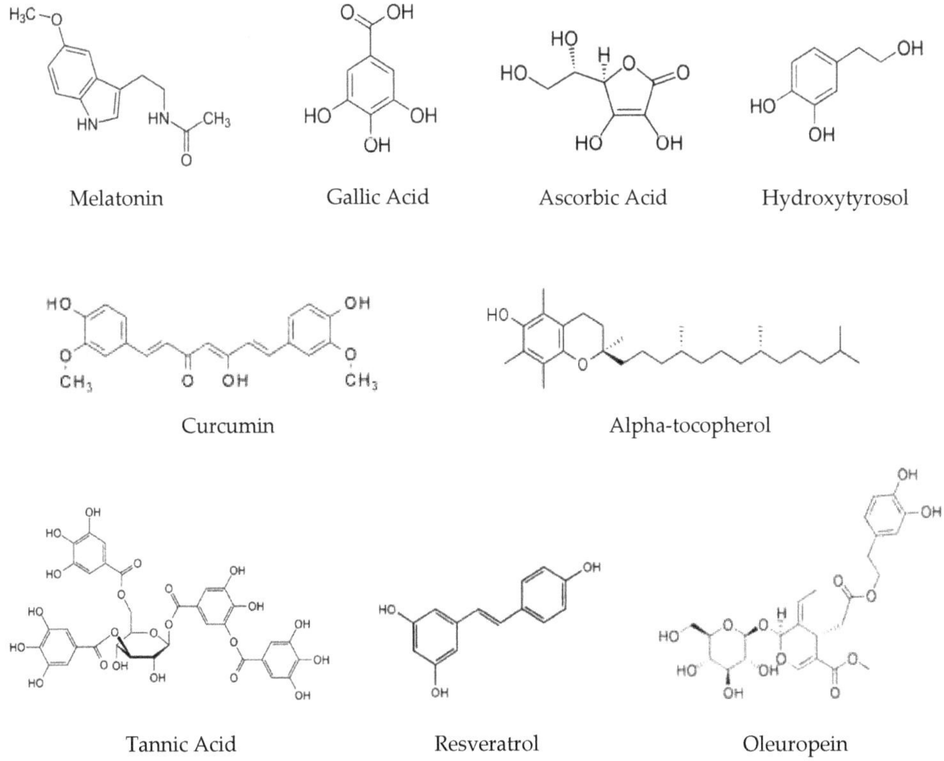

Scheme 4. Chemical structure of the selected antioxidants.

2.1. Melatonin

Melatonin, *N*-acetyl-5-methoxytryptamine (MLT), is a pleiotropic neurotransmitter with cellular and physiological actions, widely distributed in nature [51]. MLT is a hormone released primarily by the pineal gland that regulates the sleep-wake cycle. As a dietary supplement, it often is used in the short-term treatment of insomnia. This molecule is involved in circadian rhythms, hematopoiesis, angiogenesis, metastasis, sexual behavior, hypertension, oxidative stress, metabolic syndrome, and immune function. Furthermore, one of its most important activities is that it is a potent free radical detoxifier and regulator of redox-active enzymes [52]. The hydrophilic and lipophilic character of MLT has allowed it to cross the blood-brain barrier and to enter into cells, which have given this hormone the capacity of developing all the mentioned implications in animals [53]. Moreover, MLT improves the antioxidant reaction stimulating other antioxidative molecules such as glutathione peroxidase, glutathione reductase, SOD, and catalase [54]. Recent studies have shown that MLT deficiency causes a gradual acceleration of aging [55]. Furthermore, it presents immunomodulatory, thermoregulatory, and antitumor properties and appears to be an excellent candidate for the prevention and treatment of various cancers such as skin, breast, prostate, and colon [56].

2.2. Gallic Acid

Gallic acid (GA, 3,4,5-trihydroxybenzoic acid) belongs to the group of hydrolyzable tannins, and to the subclass, gallotannins. It is considered one of the strongest natural antioxidants. Gallic acid name is derived from oak galls, which are historically used to get tannins. It is distributed in different families of the vegetable kingdom, such Anacardiaceae, Myrtaceae, and Fabaceae, as well as the fungi genus such *Aspergillus*, *Penicillium*, and *Termitomyces* [57,58].

Its synthesis can be performed via three possible routes. One starts with the conversion of phenylalanine in caffeic acid, later in 3,4,5-trihydroxycinnamic acid, and then in gallic acid. Another starts from the artificial production of 3,4,5-trihydroxycinnamic acid, and its side chain comes from the formation of protocatechuic acid, which is derived from caffeic acid [57]. The third route uses 3-dehydroshikimate, in which the action of the enzyme shikimate dehydrogenase produces 3,5-didehydroshikimate. This compound tautomerizes to form a redox gallic acid equivalent, which is converted into GA due to its spontaneous aromatization [59,60]. Several beneficial effects are reported for gallic acid, including antioxidant, anti-inflammatory, and antineoplastic properties. Further, this compound has been reported to have therapeutic activities in gastrointestinal, neuropsychological, metabolic, and cardiovascular disorders.

2.3. Tannic Acid

Tannic acid (TA, penta-m-digallolyl glucose) is one of the main tannins in plants. It has multiple applications in medicine due to its antioxidant, antimutagenic, antiallergenic, and anticarcinogenic capacity [61]. Further, it is widely used in the wine and food industries. However, in Europe is not considered a food additive due to its toxicity in large amounts in animals [62]. Conversely, TA with magnesium was used as a treatment for many toxic substances in the early twentieth century. During that time, tannic acid was also applied to treat burn injuries. Nowadays, it is used in pharmacology to develop numerous applications.

It has bactericide action since it reacts with proteins irreversibly, thus complexing within bacterial membranes, neutralizing their activity [63], and has effective anticaries properties. Its antiviral effectiveness is also well documented. Besides, it is used for the treatment of intestinal problems, owed to its complexation ability with other molecules and antioxidant behavior, and has been proven to be effective against ulcers by functioning as a protective coating of the gastrointestinal tract [64]. TA is extracted from several plant species like *Caesalpinia spinosa*, *Rhus semialata*, *Quercus infectoria*, and *Rhus coriaria*.

2.4. Resveratrol

Resveratrol (3,5,4′-trihydroxystilbene, RSV), a natural member of the stilbene family, has been reported to have numerous health benefits. It is present in several plants and fruits, such as peanuts, mulberries, blueberries, and, above all, in grapes, specifically in their skin, seeds, and woody parts [65–67]. Drinking red wine, unlike white wine where these rich parts in resveratrol are removed, could be responsible for health-promoting properties [68,69]. It is a phytoalexin, that is, an antimicrobial and antioxidative substance synthesized by a plant in response to environmental stresses, and it is used for the treatment of various diseases, including dermatitis, gonorrhea, fever, hyperlipidemia, arteriosclerosis, and inflammation. Furthermore, the antiproliferative and proapoptotic effects of resveratrol in tumor cell lines have recently been documented in vitro [70]. For all these reasons, RSV is one of the most studied antioxidants.

Its structure was not characterized until 1940 when Taraoka isolated it from *Veratrum grandiflorum* roots [71]. However, its properties have been used in medicinal preparations for 2000 years [72]. RSV has two isomers, one *trans* and another *cis*, being the first more stable and, thus, more abundant in nature. The RSV synthesis starts with phenylalanine, which after being modified by three enzymes, it is turned to cumaril-CoA. This product interacts with three molecules of malonil-CoA thanks to a stilbene synthetase, the resveratrol synthetase, generating the *trans*-resveratrol molecule [73]. A remarkable fact is that the synthesis of resveratrol decreases with the grape maturation, because of the lower genetic expression induction.

Some of the species with more quantity of resveratrol, besides the grapes, are *Polygonum cuspidatum* [74], *Bauhinia racemose* [75], *Veratrum grandiflorum* [76], *Veratrum formosanum* [77], *Pterolobium hexapetallum* [78], eucalyptus [79,80], and fir species [81]. RSV protects them from infections, UV radiation, chemical substances or stress [73,82,83].

2.5. Hydroxytyrosol and Oleuropein

Hydroxytyrosol (2-(3,4-dihydroxyphenyl)ethanol, HT) is a diphenolic compound naturally occurring in olives and olive oil, frequently occurring in the Mediterranean diet, with proven benefits for health [84–86]. It was first recognized by Stoll et al. as a component of echinacoside, a phenolic glycoside showing antibiotic activity against *Staphylococcus aureus* extracted from the roots of *Echinacea angustifolia* [85,87]. HT is considered the most powerful antioxidant after gallic acid. It has been most studied and used in food, cosmetic, and pharmaceutical industries [88] after being demonstrated that HT produces a wide range of biological properties besides antioxidant such as hepatoprotective [89], cytoprotective [90], neuroprotective [91], cardioprotective [92], anti-inflammatory [93], antiviral [94], anticancer [85], and anti-obesity effect [95]. Some studies affirm that the composition of olive leaves extracts rich in several phenols and flavonoids and that it varies with the harvesting season, the leaves maturity, storage conditions, and extraction method [96].

The antioxidant activity of HT in vivo is directly linked to its high bioavailability, and several works have reported a high degree of absorption, fundamental to exert its metabolic and pharmacokinetics properties [97]. HT behaves as an antioxidant acting as a free radical scavenger and radical chain breaking as well as a metal chelator. With its catecholic structure, it is able to scavenge the peroxyl radicals and break peroxidative chain reactions, leading to very stable resonance structures. On the other hand, the protection against the genotoxic action of reactive oxygen species (ROS) is one of the mechanisms explaining the anticancer effects of HT [98]. In addition, it may also act via the modulation of pro- and anti-oncogenic signaling pathways, resulting in cell apoptosis and growth restriction of tumor cells, which may be mediated by their capability to induce the accumulation of hydrogen peroxide in the culture medium [99]. A decrease in ROS production, derived by iron or copper-induced oxidation of low-density lipoproteins, has also been found, suggesting a chelating action on such metals.

HT can be obtained in several ways differing in the starting molecule and the synthetic route. The first synthesis path was reported by Schöpf et al. in 1949. It started with 2-(3,4-dimethoxyphenyl)ethanol that was demethylated by 48% HBr, leading to the corresponding bromide, which was converted into the triacetoxy derivative and subsequently hydrolyzed by methanolic NH_3 to give hydroxytyrosol [85,100].

One of the derivates of hydroxytyrosol is oleuropein, which presents very similar properties: antioxidant, antimicrobial, antitumoral, and cardioprotective. Oleuropein (2-(3,4-

dihydroxyphenyl)ethyl(2S,3E,4S)-3-ethylidene-2-(β-D-glucopyranosyloxy)-5-methoxycarbonyl)-3,4-dihydro-2H-pyran-4-yl)acetate) arises from the union of hydroxytyrosol and an oleosidic skeleton. It is a glycosylated seco-iridoid that can be found in green olive skin, flesh, seeds, and leaves [101]. During its maturation, the β-glucosidases present inside transforms the oleuropein to extra virgin olive oil, a form in which it is protected from natural oxidation. Similar to HT, it shows a beneficial effect on the reduction in cholesterol and the decrease in coronary heart diseases [102]. Besides, its consumption, mostly in the Mediterranean diet, reduces the risk of breast, prostate, and colon cancer, owed to its potential antitumoral activity [103–108].

2.6. Tocopherol

Tocopherol (TCP or TOH) is the name of a class of organic chemical compounds having some vitamin E activity. It is one of the 13 essential vitamins in humans that is not produced by the organism. Four forms of tocopherol with vitamin E activity and different degrees of methylation have been described: alpha-, beta-, gamma- and delta-tocopherol. α-tocopherol is a component integrated into the European diet because it is present in olive and sunflower oils [109]. Conversely, γ-tocopherol is more common in the American diet, where soybean and corn oil are intaken [109,110]. Fortunately for Europeans, α-tocopherol is the most absorbed form of vitamin E in humans [111].

One of the main functions of TCP is the capacity of donating an H atom to free radicals that could generate damage [112], resulting in a less reactive tocopheryl radical. Besides, TCP has anti-inflammatory, antiageing, anticancer, and cardioprotective properties, and its aids in preventing macular degeneration, platelet aggregation, and Alzheimer's disease. It is widely used in moisturizers, creams, and as a food additive.

Unlike vitamins A and D, α-tocopherol (which is also a fat-soluble vitamin) does not accumulate to "toxic" levels in the liver or extrahepatic tissues; however, an excess of vitamin E arises few problems if the doses overcome 400 mg/day. It can be a topic allergic, normally seen in cosmetic products, although the incidence is quite low. It could generate a drug interaction such as tamoxifen, an anti-breast cancer drug, or cyclosporine A, an immune-suppressant drug, or aspirin and warfarin, potentiating an anti-blood-clotting action [113]. For some of these interactions, its co-administration together with a chemotherapy drug is completely forbidden.

The natural biosynthesis of TCP in plants comes from the reaction of homogentisate (HGA) with homogentisate phytyltransferase (HPT), giving a molecule that reacts with tocopherol cyclase (TC) to get δ-tocopherol or with methyltransferase (MT) and then tocopherol cyclase (TC) to get γ-tocopherol. Subsequently, if these two tocopherols react with γ-tocopherol methyltransferase (γ-TMT), β-tocopherol and α-tocopherol are obtained, respectively [114].

α-tocopherol can be extracted and purified from seed oils. In addition, γ-tocopherol can be extracted, purified, and methylated to produce α-tocopherol. It has been found that rats can convert γ-tocopherol to α-tocopherol, methylating it in their tissues. On the other hand, α-tocopherol can be obtained synthetically, although with only 50% of the effectiveness of natural α-tocopherol.

2.7. Ascorbic Acid

Ascorbic acid ((R)-3,4-dihydroxy-5-((S)-1,2-dihydroxyethyl)furano-2(5H)-one, AA) is an organic acid with antioxidant properties. The S enantiomer (L-ascorbic acid) has vitamin C activity. AA is very common in the diet, and it is usually sold as a dietary supplement. It prevents illnesses such as scurvy [115], which gave it the name of ascorbic acid or ascorbate. Vitamin C is an essential nutrient in humans that acts as a cofactor of several enzymes, promotes carnitine and collagen synthesis, and plays a crucial role in cell division and growth regulation, in the maintenance of the immune system, in the reparation of skin and tissues and in the production of neurotransmitters. Further, it presents antiageing, anticancer, neuroprotective, and wound-healing properties. Vegetables and fruits represent a great source of vitamin C, and its recommended ingestion is established as 90 mg/day for males and 75 mg/day for females [116].

The history of AA started with James Lind demonstrating that citrus fruit consumption had positive effects on scurvy prevention and treatment, and it was named as an antiscorbutic factor. Its structure was elucidated by Albert Szent-Györgyi in 1932 [117], who was awarded the Nobel Prize for Medicine in 1937.

The ascorbic acid biosynthesis has many routes. In mammals, it can be generated from D-glucose, which is converted to D-glucuronic acid, and later to L-gulonic acid by glucuronate reductase. This molecule is then turned to gulono-1,4-lactone by aldonolactonase. Finally, ascorbic acid is converted through the action of gulono-1,4-lactone oxidase in gulono-1,4-lactone, producing 2-keto-gulono-γ-lactone, which spontaneously converts to l-ascorbic acid or vitamin C [118]. On the other hand, plants synthesize L-ascorbic acid from L-galactose, which comes from D-mannose. L-galactose is converted to L-Galactono-1,4-lactone via oxidation by NAD-dependent L-galactose dehydrogenase. L-ascorbic acid is finally formed via oxidation of L-Galactono-1,4-lactone by L-galactono-1,4-lactone dehydrogenase [119,120].

2.8. Curcumin

Curcumin ((1E,6E)-1,7-Bis(4-hydroxy-3-methoxyphenyl, CU) hepta-1,6-diene-3,5-dione, also known as diferuloylmethane) is a non-polar diarylheptanoid, natural, bright and yellow polyphenol from the *Curcuma* genus. It is the principal curcuminoid of turmeric (*Curcuma longa*), a member of the ginger family, Zingiberaceae, that is obtained mainly from its rhizome [121,122]. Curcuma has been used for years due to its beneficial properties for human health. Nowadays, it is an authorized food additive, labeled as E-100i (curcumin) and E-100ii (curcuma). It is used as an herbal supplement, cosmetics ingredient, food flavoring, and food coloring [123]. CU is present in at least two forms, the keto and enol tautomers, which are solid and liquid, respectively.

CU has anti-cancer [124,125], anti-arthritis, anti-inflammatory [126], and neuroprotective capacity, besides its antioxidant properties [127]. It aids in the management of oxidative and inflammatory conditions, metabolic syndrome, arthritis, anxiety, and hyperlipidemia. It may also help in the treatment of exercise-induced inflammation and muscle soreness, thus promoting recovery and subsequent performance in active people [128]. However, there are some factors that limit the bioactivity of curcumin, such as chemical instability, insolubility in water, absence of potent and selective activity, low bioavailability, limited tissue distribution, and extensive metabolism [129,130]. The bioavailability of curcumin depends on the delivery format, age, health condition, and human gender [131]. Some studies have shown its increased biodisponibility and its reduced degradation upon encapsulation [132,133].

The biosynthesis of curcumin remains still unclear. Peter J. Roughley and Donald A. Whiting proposed two possible mechanisms for its biosynthesis in 1973. One involves a chain extension reaction by cinnamic acid and 5-malonyl-CoA, that arylize into a curcuminoid. The second possible mechanism involves two cinnamate units coupled by malonyl-CoA. On the other hand, it is believed that plants start with p-coumaric acid instead of cinnamic acid [134].

A summary of the most important properties of all the selected antioxidants is provided in Table 1.

Given that G is a zero-gap semiconductor and an electroactive and transparent material, there are many possibilities for its application in biosensing applications. Its outstanding electrical conductivity and large specific surface have demonstrated the accurate, rapid, sensitive, and selective sensing ability of bioactive compounds. It presents enhanced sensitivity for a wide range of biomolecules when compared with other carbon materials such as CNTs, fullerenes, or amorphous carbon.

Table 1. Biological and physical properties of the antioxidants, their nutrient sources, and medical uses.

Antioxidant	Biological Properties	Physical Properties	Nutrient Sources	Functions and Medical Uses	Ref.
Melatonin	Immunomodulatory Thermoregulatory Anti-aging Anticancer Photo-and radioprotective Cardioprotective Antiarrhythmic agent	Off-white powder M_w = 232.28 g/mol $d_{25\,°C}$ = 1.175 g/cm^3 T_m = 117 °C b.p. = 512.8 °C $S_{20\,°C}$ = 2.0 g/L	Coffee, Tea Red wine, Beer Banana Tomatoes Rice, Wheat Corn, Oat	Control of hypertension, obesity, and metabolic syndrome Modulation of inflammatory markers Modulation of oxidative stress Sleep disorders/Insomnia treatment Parkinson and Alzheimer diseases	[52]
Gallic Acid	Antimicrobial Anticancer Antifungal Antiviral Astringent Antiallergic Antiinflammatory Antimelanogenic Antiulcerogenic	Crystalline white powder M_w = 170.12 g/mol $d_{25\,°C}$ = 1.694 g/cm^3 T_m = 260 °C b.p. = 501 °C T_d = 237.5 °C pK_a = 4.40 $S_{20\,°C}$ = 11.9 g/L	Blueberries Apples Flax seeds Tea, Coffee Walnuts Watercress Grapes, Wine Grenade	Control of periodontal disease Cell death in e human cancer cells Regulation of the genes involved in the cell cycle Prevention of degenerative diseases Prevention of cardiovascular diseases Inhibitor of diabetes dysfunction Inflammation suppressor	[57]
Tannic Acid	Astringent Chemotherapy drug enhancer Antiallergic Anticarcinogenic, Antimutagenic Antiinflammatory	Light yellow amorphous powder M_w = 1701.19 g/mol $d_{25\,°C}$ = 2.12 g/cm^3 T_m = 218 °C T_d = 199 °C pK_a = 10 b.p. = 218 °C $S_{20\,°C}$ = 250 g/L	Red wine Coffee, Tea Guava Spinach Black raisins Oaks Nuts Persimmon	Inhibitor of NO$_2$ production Clarifying agent in wine and beer Flavoring agent in foods Treatment of diarrhea Topical to dress skin burns Treatment of rectal disorders.	[135]
Resveratrol	Anticancer Antiallergic Antiinflammatory Cardioprotective Inmunostimulatory Antimicrobial Antiplatelet agent Antifungal	White to yellow powder M_w = 228.25 g/mol $d_{25\,°C}$ = 1.40 g/cm^3 T_m = 263 °C T_d = 222 °C b.p. = 449 °C $S_{20\,°C}$ = 0.03 g/L	Peanuts Pistachios Grapes, Wine Blueberries, Cranberries Cocoa Chocolate	Natural reducing agent Prevention of cardiovascular disease Parkinson and Alzheimer diseases Regulation of triglycerides Inhibitor of platelet aggregation Inhibitor of DNA duplication in cancer cells	[21,25,136]
Hydroxytyrosol	Immunostimulant Antimicrobial Antifungal Cardioprotective Anticancer Antiinflammatory Hepatoprotective Neuroprotective	White powder M_w = 154.16 g/mol $d_{25\,°C}$ = 1.30 g/cm^3 T_m = 55 °C T_d = 361 °C b.p. = 355 °C $S_{20\,°C}$ = 50 g/L	Olive leaves Olive oil Wine	Prevention of sexual dysfunctions Prevention of atherosclerosis Inhibitor of platelet aggregation Inhibitor of human LDL oxidation Stabilizer and antioxidant in foods	[88,137]
Tocopherol (Vitamin E)	Antiageing Anticancer Cardioprotective Antiinflammatory	Yellow-brown liquid M_w = 430.71 g/mol $d_{25\,°C}$ = 0.95 g/cm^3 T_m = 2 °C b.p. = 220 °C $S_{20\,°C}$ = 0 g/L	Nuts Avocado Salmon Mango Tomato Spinach Seed oils	Prevention of macular degeneration Prevention of Alzheimer's disease Prevention of cardiovascular diseases Inhibitor of platelet aggregation Moisturizers/creams	[138]
Ascorbic acid (Vitamin C)	Antiageing Wound healing Anticancer Immunostimulant Neuroprotective	white powder M_w = 176.12 g/mol $d_{25\,°C}$ = 1.65 g/cm^3 T_m = 190 °C b.p. = 553 °C $S_{20\,°C}$ = 330 g/L)	Guava Pepper Citrus Broccoli Grape Cauliflower Strawberry Mango	Prevention of Hepatitis Promotion of collagen synthesis Prevention of Alzheimer's disease Reparation and maintenance of skin, blood vessels, scars, tendons, ligaments, etc. Cofactor in many enzymes Natural reducing agent Cell division and growth regulation Colitis and stomach ulcer protection Inhibitor of diabetes dysfunction	[118]
Curcumin	Anticancer Antiarthritis Antiinflammatory Neuroprotective	yellow crystalline solid (keto-) or liquid (enol-) M_w = 368.38 g/mol $d_{25\,°C}$ = 1.3 g/cm^3 T_m = 183 °C b.p. = 591 °C $S_{20\,°C}$ = 0 g/L	Curcuma Curry Tea	Inflammation suppressor Treatment for viruses and pulmonary fibrosis Prevention of cancers Cosmetics ingredient Food flavoring and coloring	[128]

$S_{20\,°C}$ = solubility in water at 20 °C; M_w = Molecular weight; $d_{25\,°C}$ = density at 25 °C; T_m = melting temperature; b.p. = boiling point; T_d = decomposition temperature.

3. Graphene Functionalization Approaches

Pristine G sheets are hydrophobic in nature, so they cannot be dissolved in polar solvents. To make it soluble in common solvents, avoiding stacking between adyacent sheets, and hence to expand its range of applications, it can be functionalized via interaction with other molecules or polymers. Noncovalent functionalization by π-interactions is an attractive synthetic method because it offers the possibility of attaching functional groups to G without disturbing the electronic system. These include H-π, π–π, anion-π and cation-π interactions [139]. Complexes exhibiting the H−π interaction (190, 201–207) are of great interest since this type of interaction is also one of hydrogen bonds [140]. On the other hand, the π–π interaction is one of the most important driving forces for supramolecular self-assembly. When the countermolecule is a metal cation, a combination of electrostatic and induction energies dominates the cation−π interaction. Recently, anion−π interactions have been reported as a novel approach towards a new type of anion recognition, host architecture, and supramolecular self-assembly [141]. Overall, by controlling the relationships of several noncovalent interactions, novel organic nanostructures can be designed.

Typically, non-covalent strategies include solution mixing or in situ polymerization. The first requires both G and the functional molecule or polymer to be stably dispersed in a common solvent; it involves the dispersion of G in the appropriate solvent, the adsorption of the molecule (or polymer) to delaminated G sheets in solution, and the elimination of the solvent, resulting in sandwich-like nanocomposite [12]. On the other hand, in in-situ polymerization, G is first swollen within the liquid monomer, the initiator is subsequently added, and the polymerization begins either by heat or radiation. Nanocomposites with conductive polymers can also be produced via in situ electrochemical polymerization [142], which yields mechanically stable composite films that can be directly used as electrodes or energy devices.

The key aim of the covalent functionalization of pristine G with organic molecules or polymers is to improve its dispersibility in common organic solvents. Furthermore, groups such as chromophores provide novel properties that could be combined with G properties such as conductivity. In most cases, when organic molecules are covalently linked to the G surface, their aromaticity is perturbed, enabling the control of its electronic properties. The functionalization reactions include two general approaches: (a) formation of covalent linkages between free radicals or dienophiles and C=C bonds of G and (b) formation of covalent bonds between organic functional groups and the oxygenated groups of GO. On the one hand, free radicals and dienophiles can react with sp^2 carbons of G via 1,3 dipolar cycloaddition, aryne, or nitrene addition [143]. On the other hand, organic functional groups can be anchored to the epoxy, carboxylic acid, ketone, or hydroxyl groups onto a GO surface. In particular, chromophores, including azobenzenes, porphyrins, and phthalocyanines, with outstanding optoelectronic properties, have been covalently attached to G nanoplatelets. [144]. Thus, GO can be functionalized with porphyrins through the formation of amide bonds between amine-functionalized porphyrins and carboxylic groups of GO. Besides, GO can be grafted to polymeric chains that have reactive species like hydroxyls and amines, in particular poly(ethylene glycol), polylysine, polyallylamine, and poly(vinyl alcohol). The polymer provides improved dispersibility in certain solvents and morphological characteristics, while G offers electrical and thermal conductivity and reinforcement of the stiffness and strength.

Grafting of polymeric chains onto G nanosheets can be carried out via "grafting-to", "grafting-from", and "grafting-through" approaches (Figure 2) [145]. The first consists of synthesizing G and polymers individually and connecting them. Here, physical and chemical interactions play a role in modifying the G layers. In the "grafting-from" method, the polymer chains grow in situ from an initiator that has been previously anchored to the G surface. In the "grafting-through" approach, the polymerizable groups are anchored onto the G surface. Then, the polymerization begins in the solution that contains an initiator, monomers, and G. Polymerization of monomers takes place, and G is incorporated inside the polymer chains.

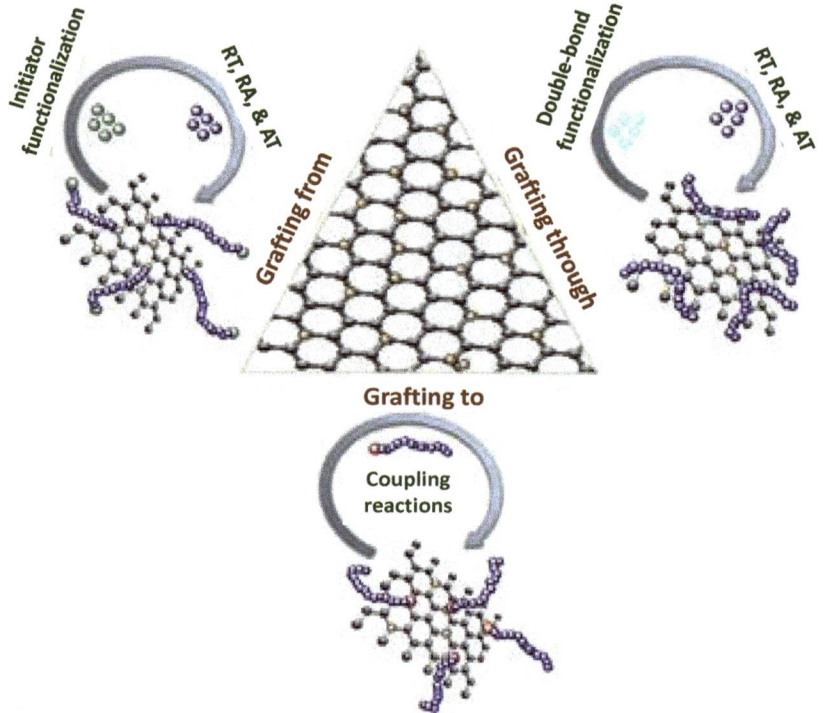

Figure 2. Representation of the anchoring of polymer chains onto graphene (G) via "grafting-to", "grafting-from", and "grafting-through" approaches. Adapted from Eskandari et al. [145].

4. Graphene-Based Sensors for Bioactive Compounds

Due to its outstanding electrical, chemical, optical, and electrochemical properties, G has excellent potential for use as a transducer in optical sensors based on fluorescence, chemiluminescence, and colorimetric detection systems [35]. G derivatives, in particular GO and rGO, have valuable characteristics to be employed in these optical sensors. On the one hand, GO has tailorable luminescent properties. On the other hand, GO and rGO have been reported to be strong fluorescence quenchers via Föster resonance energy transfer (FRET) [146]. Based on this property, two types of G-based fluorescent sensors have been described: (1) signal-on ones, in which the fluorescence intensity rises with the addition of the analyte, hence the signal can be directly correlated with the analyte concentration reaching detection limits as low as ng/L. (2) signal off sensors, based on fluorescently labeled probes that adsorb onto GO in the presence of the analyte and quench its fluorescence, though have lower sensitivity than signal on sensors [30].

Furthermore, the application of G-based materials as sensors involves two approaches: one is based on G-biomolecule interactions via van der Waals, π-π stacking, cation−π, anion−π interactions, and electrostatic forces, leading to electrical variations in the pristine G. The other is based on the chemical functionalization to immobilize the molecular receptors onto the surface of GO, rGO, or GQDs [147].

From an electrochemical viewpoint, the potential of G-based electrodes is huge given that they preserve the properties of other carbon-based materials, including chemical inertness and good electrocatalytic activities for many redox reactions, and simultaneously they offer new properties like high surface area and ultrarapid charge mobility, which guarantee high sensitivity and quick response. Further, it presents an electrochemical potential window of ~2.5 V in 0.1 M PBS (pH 7.0) [148], which is better than that of graphite, glassy carbon, and even boron-doped diamond electrodes. Besides, the charge-transfer resistance on G is considerably lower than that of graphite or glassy carbon electrodes, indicating that the electronic structure of G is beneficial for electron transfer, making it suitable for the detection of biomolecules that have high oxidation or reduction potential. The key aspects of G electrochemistry are beyond the scope of this review and have been recently reviewed [149]. Furthermore, some former reviews have focused on the interactions of G, GO, and rGO-based biosensors with their analyte targets [150–153].

In the following section, we review the most recent advances in the development of optical and electrochemical sensors based on graphene and its derivatives for the detection of bioactive compounds. We discuss the processing and detection method, the linear range and limit of detection (LOD), as well as the advantages and improvement of properties due to the presence of graphene. Although the number of papers related to this type of sensor is still scarce, very promising results have already been obtained.

4.1. Melatonin

A lot of techniques to determine low concentrations of MLT in biological samples have been reported, including HPLC with electrochemical and fluorometric detection, gas chromatography-mass spectrometry, micellar electrokinetic chromatography, spectrofluorimetry, chemiluminescence, radioimmunoassay, and colorimetry [154–158]. However, some of these techniques have many drawbacks, such as the use of expensive instruments, regular maintenance, tissue destruction, tedious and complicated processes, and the use of organic solvents which are not biocompatible and generate pollution. Researchers in their investigations try to improve at least one of these disadvantages. Niu et al. [159] focused their work on the development of an optical sensor to detect MLT via a simple, cost-effective, and sensitive method. For such a purpose, they first synthesized GO via a modified Hummer´s method. Subsequently, it was dispersed in an aqueous medium and reacted with phenyl triethoxysilane (PTEOS) and tetramethoxysilane (TMOS) via a sol-gel method to yield a GO@SiO$_2$ nanocomposite which was used as a sorbent in dispersive solid-phase extraction (dSPE). The detection of MLT was performed via HPLC combined with DAD, and a detection below 0.1 µg mL^{-1} was attained.

However, most of the graphene-based sensors for MLT detection are based on electrochemical techniques, which are typically more sensitive, accurate, faster, miniaturizable, eco-friendly, and cheaper compared with optical ones. In this regard, Apetrei et al. [160] developed a novel sensor based on graphene-coated carbon screen-printed electrode (G-CSPE) prepared via sonication of G followed by drop-casting onto the electrode. These screen-printed electrodes (SPEs) consist of a single device with three different electrodes (Figure 3): (1) Working electrode, which response is sensitive to the analyte concentration. (2) The reference electrode, which potential is constant, and the working electrode potential is measured against it. Auxiliary or counter electrode, which completes the circuit of the cell, as it allows the passage of current. The voltammetric behavior of the SPEs (unmodified and modified with G) was studied in order to evaluate the electroactive surface area of the working electrode (using K$_4$[Fe(CN)$_6$] as benchmark redox system) and to quantify the rate constant obtained from cyclic voltammetric (CV) curves. The method was applied to the analysis of commercial pharmaceutical formulations.

The G-CSPE showed better performance, with a higher degree of reversibility, lower separation between the anodic and cathodic peaks, and the ratio between the current for the catodic and anodic peak (I_c/I_a) was close to 1. Good sensitivity in small samples was obtained with a detection limit of 0.87 µM and a response time of about 4 s.

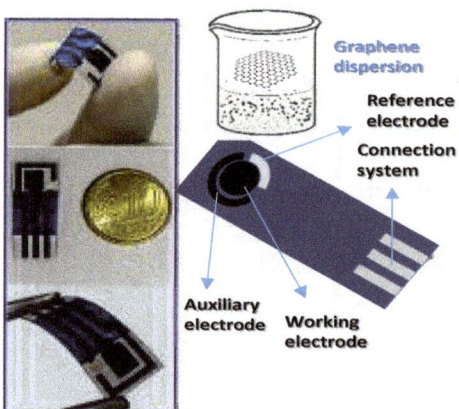

Figure 3. Graphene-coated carbon screen-printed electrode (G-CSPE) with the three-electrode system: a reference electrode, a working electrode, and an auxiliary or counter electrode.

A very similar approach was reported by Miccoli et al. [161] to detect MLT in food supplements using differential pulse voltammetry (DPV) with a G-CSPE. Graphene was suspended, and the electrodes were modified by drop-casting. Again, the electrode modified with CVD graphene led to better results than the unmodified one. Since the solution pH is crucial for the stability of melatonin molecules, studies were carried out at different pHs. They demonstrated how pH affects the relation between the signal peak and the amount of MLT. The detection limits for pHs at 6.4, 7.0, and 7.4 were 15, 30, and 60 µg/L, respectively, lower than those attained with the unmodified electrode.

Analogously, Gomez et al. [162] tested several carbon nanostructures to modify a CSPE and detect MLT and serotonin simultaneously in tablets and herb extract capsules. In particular, graphene oxide nanoribbons (GON) and graphene reduced nanoribbons (GRN) were synthesized from multi-walled carbon nanotubes (MWCNTs) via the longitudinal unzipping method, followed by chemical reduction with hydrazine, ultrasonication in aqueous media, and drop-casting onto the CSPE. The electrochemical behavior of the different electrodes was examined by DPV, reaching a LOD of 1.1 µM, with a low sample consumption (50 µL), good reproducibility, a response time of 120 s, and a recovery of 94–103%. The excellent performance obtained makes this approach promising not only in the pharmaceutical field but also in the determination of neurotransmitters in urine and other related samples.

Gupta et al. [163] prepared a sensor for the selective and sensitive determination of melatonin in human biological fluids based on the combination of rGO and a molecularly imprinted polymer (MIP). The rGO was synthesized from graphite powder via a modified Hummers' method followed by hydrazine reduction. Then, the rGO was ultrasonicated in a mixture of distilled water and DMF (1:9) and dropped on the surface of a glassy carbon electrode (GCE). Subsequently, the MIP film was prepared by the electropolymerization on the surface of the modified electrode. The synergistic effect of graphene and the MIP enlarged the number of recognition sites, resulting in an improved MLT detection, with a linear range from 0.05 to 100 µM, and a LOD of 6 nM.

Another approach to improve the sensitivity of this type of sensor and enable the simultaneous determination of related molecules coexisting in biological systems is the combination of rGO with inorganic nanoparticles. In this regard, Bagheri et al. [164] developed an electrochemical sensor to detect MLT and dopamine, based on rGO decorated with Fe_3O_4 magnetic nanoparticles on a carbon paste electrode (CPE). The nanocomposite was prepared using a modified Hummers' method followed by hydrazine reduction and then hydrothermal growth of the nanoparticles. Electrochemical studies revealed that the surface modification of the electrode considerably increased the oxidation peak currents, although it reduced the peak potentials of MLT and dopamine. The synergistic effect between the nanocomposite components enhanced the signal response, leading to a linear range of 0.02–5.80 µM and a LOD of 8.4 nM. Further, no significant effect on the recovery was found in the presence of interferences such as glucose, AA, pyridoxine, serotonin, or uric acid, among others.

Other researchers such as Zeinali et al. [165] followed a similar method to detect tryptophan and melatonin at the time. They developed an electrochemical sensor with an ionic liquid carbon paste electrode modified with rGO and SnO_2-Co_3O_4 nanoparticles. This SnO_2-Co_3O_4@rGO/IL/CPE sensor worked linearly in the range of 0.02 to 6.00 µM, with a LOD of 4.1 nM, good selectivity, stability, and repeatability, besides its cost-effectiveness and simple fabrication. Furthermore, Tadayon et al. [166], based on the previous works, synthesized a sensor to detect dopamine, melatonin, and tryptophan. It was a nanocomposite based on nitrogen-doped reduced graphene oxide (N-rGO)/$CuCo_2O_4$ nanoparticles, deposited onto a CPE via a solvothermal method. The N-rGO improved the reactivity and electrocatalytic performance of the electrode, providing multiple binding sites, as well as enhanced biocompatibility and sensitivity. Both CV and DPV studies revealed that the potential separations between the three compounds were large enough to allow their simultaneous detection (Figure 4). Hence, it was employed for their analysis in human urine, serum, and pharmaceutical samples. Recovery values ranging from 97–104% were obtained, with a linear range of 0.01–3.0 µM and a detection limit of 4.9 µM for MLT. The cost-effectiveness and easy preparative method are valuable advantages of this sensor.

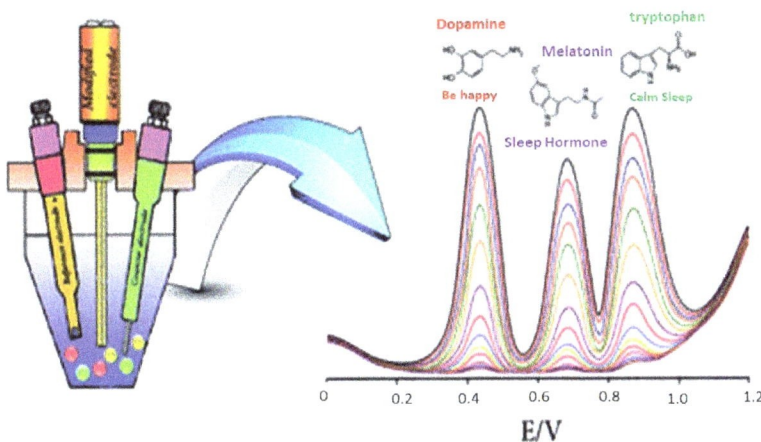

Figure 4. Differential pulse voltammetry (DPVs) for the determination of melatonin (MLT) at the *N*-reduced graphene oxide (rGO)/$CuCo_2O_4$/carbon paste electrode (CPE) in the presence of 2 mM dopamine and tryptophan. Taken from Tadayon et al. [166].

In the case of Liu et al. [167], a CuO−poly(L-lysine) (PLL)/graphene-sensing electrode for the detection in situ of MLT and pyridoxine (vitamin B_6) was prepared via electrochemical deposition. CuO and PLL acted as linkers for the bioactive molecules, which were helped by a 3D graphene, grown via CVD, that amplifies the sensitivity. MLT was detected in a concentration range of 0.016−110 µM, with a LOD of 12 nM.

There are also a few articles that used MLT to reduce GO via an eco-friendly method [168,169]. Conventionally, GO is reduced by strong chemical agents, like hydrazine; however, they are not able to be produced at a large-scale due to their toxicity and harm to the environment. MLT has many advantages; the surface of MLT-reduced GO suspension presents more amount of nitrogen, attributed to the π–π adsorption, which triggers more stability; besides, when MLT is oxidized, it cannot be reduced back to its initial state, as lots of antioxidants do, protecting the rGO from oxidation. Ultimately, the efficiency obtained by MLT is comparable to the hydrazine one, although the deoxygenation process requires 600% more time.

4.2. Gallic Acid

Most of the reported studies focused on the determination of GA via electrochemical techniques (Table 2), which are simpler and provide higher sensitivity and selectivity. For instance, Al-Ansi et al. [170] synthesized a 3D nitrogen-doped porous graphene aerogel (NPGA) via one-step hydrothermal reduction by mixing graphene oxide (GO) with p-phenylenediamine (PPD) and ammonia solution and then followed by freeze-drying. The NPGA electrode provided a new way to determinate GA and showed improved analytical behavior compared to most of the electrodes reported in previous studies. It showed a large specific surface area, excellent electrical conductivity as well as high nitrogen content, and provided a linear range of detection from 2.5 to 1000 µM and a LOD of 67 nM.

Chikere et al. [171] used amorphous ZrO_2 nanoparticles decorated onto G to modify a carbon paste electrode. ZrO_2 has a high surface area, good biocompatibility, good conductivity, and affinity for oxygen-containing groups, which resulted in an improved GA detection compared to the unmodified electrode. Thus, the mixing of zirconia and G produces an interaction that enhances the peak current of the oxidized GA. The electrode worked linearly in the range of 1 µM to 1 mM, with a LOD of 124 nM. Moreover, it was successfully applied for the determination of GA in red and white wines. A similar approach was used by Puangjan et al. [172], who synthesized an $rGO/ZrO_2/Co_3O_4$ nanocomposite by a simple reflux method, as depicted in Figure 5. Thus, GO and ZrO_2 nanoparticles were dispersed in ethylene glycol followed by the addition of $CoCl_2$ and hydrazine hydrate and then refluxed at 80 °C. The hybrid nanocomposite exhibited a synergistic catalytic effect towards oxidation of GA, caffeic acid (CA), and protocatechuic acid (PA), with LODs of 1.56, 0.62, and 1.35 nM, respectively. The modified electrode was successfully applied for the simultaneous determination of the three species in fruit juice, rice and tea samples, showing rapid response and satisfactory recoveries.

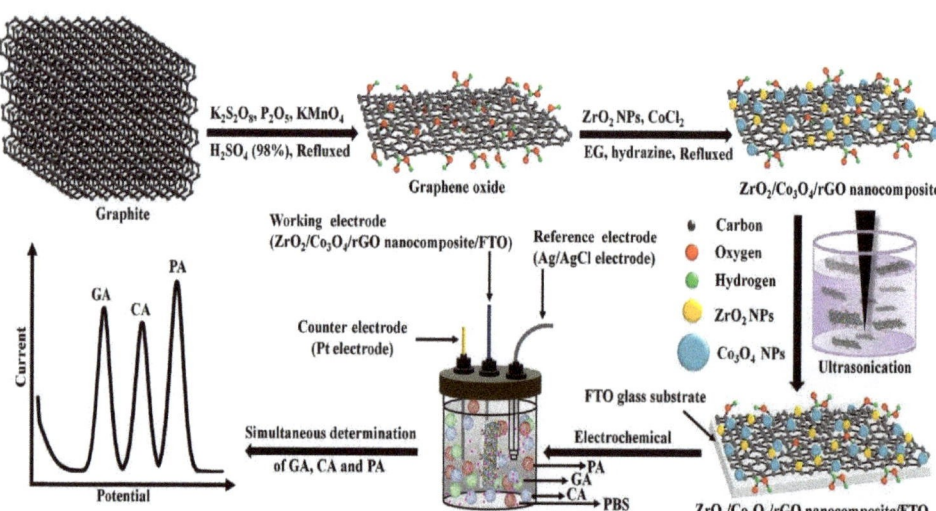

Figure 5. Schematic illustrations of the preparation of $ZrO_2/Co_3O_4/rGO$ nanocomposite and oxidation mechanisms of gallic acid (GA), caffeic acid (CA), and protochatechuic acid (PA). Taken from Puangjan et al. [172].

On the other hand, Ganesh et al. [173] developed an MWCNT-rGO nanocomposite electrode for the sensitive detection of Au nanoparticles (NPs) capped with GA. The synthesis of this type of NPs capped with GA under low-temperature sonication conditions is depicted in Figure 6. The electrode was built by drop-casting an MWCNT-GO solution onto a GCE surface, followed by UV irradiation for GO reduction. DPV measurements were performed at different AuNPs-Ga concentrations, leading to a LOD of 2.57 pM, the lowest reported for this type of sensor. Using CV and electrochemical impedance spectroscopy (EIS), it was found that this modification of the electrode surface resulted in a 10-fold increase in the current response compared to unmodified electrodes. Thus, the capping of the nanoparticles allowed very sensitive and easy detection and prevented nanoparticle agglomeration. This green approach is interesting for the progress in nanotechnology, electronic, biomedical, and material science, in which metallic nanoparticles are increasingly used.

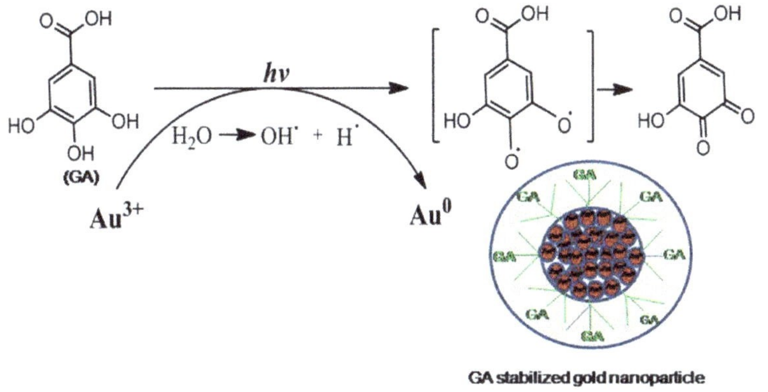

Figure 6. Diagram showing the formation of Au-nanoparticles (NPs) using GA under low-temperature sonication conditions. Adapted from Ganesh et al. [173].

Electrochemical sensors based on hybrid materials comprising polymers and graphene derivatives have also been developed. Thus, Gao et al. [174] synthesized a nanocomposite incorporating chitosan (CS), fishbone-shaped Fe_2O_3 nanoparticles, and electrochemically reduced graphene oxide (ERGO) as the sensing matrix. The NPs were prepared via a solvothermal method; then, they were mixed with GO via ultrasonication, and the dispersion was drop cast onto a GCE, followed by electrochemical reduction. The electrochemical characterization experiments showed that the modified electrode had a large surface area, excellent electronic conductivity, and high stability. A good linear relationship between the oxidation peak currents in DPV and GA concentration was found in the 1–100 µM range, with a LOD of 0.15 µM.

Ma et al. [175] developed a photoelectrochemical sensor based on polyaniline (PANI)-rGO-TiO_2 nanocomposite (Figure 7). PANI is an inexpensive and nontoxic conductive polymer with excellent stability, corrosion protection, and high mobility of charge carriers, hence highly suitable for photoelectronic materials. The nanocomposite was prepared via solvothermal synthesis of TiO_2, followed by aniline polymerization, mixing, and ultrasonication. GA was detected in the linear range of 4.17 to 250 µM, with a LOD of 1.72 µM. This sensor showed a rapid response, high sensitivity, and excellent selectivity towards GA in the presence of other species such as AA, glutathione (GSH), and L-cysteine (CYs).

An optical sensor based on GQDs obtained by pyrolysis of citric acid was reported by Benítez-Martínez et al. [176]. GA was able to quench the GQDs fluorescence via π-π stacking and non-covalent interactions. The emission band (at 474 nm) underwent a green shift when GA from real samples is added. In addition, higher quenching was observed when the polarity of the solvent tested increased. The applicability of the method was evaluated on four different types of real olive oil samples, leading to a linear response over the concentration range 2–30 mg L^{-1} and a LOD of 0.3 mg/L^{-1}. The proposed method was fast, very simple, sensitive, and reproducible.

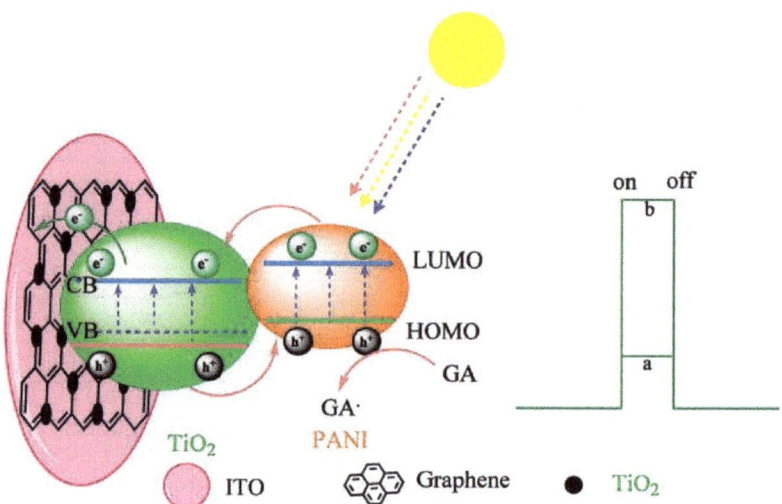

Figure 7. Schematic illustration of the photoelectrochemical process for GA oxidation at a polyaniline (PANI)–rGO–TiO$_2$ modified electrode. Taken from Ma et al. [175].

On the other hand, a few papers have been reported on the use of GA for the development of sensors for the detection of other ions. For instance, Liu et al. [177] prepared a 3D-porous graphene-based hydrogel with good mechanical strength and large surface area, fabricated by self-assembly of GO sheets reduced and modified by GA through π-π interactions, to capture toxic Cr(III) ions generated by tannery wastewater. This GA-modified hydrogel is able to capture the Cr(III) by coordination complexation with its deprotonated carboxylic groups at pH around 4.0, with an average of nearly 97% of Cr(III) in 20 min. In addition, this functionalized structure is reusable due to its desorption with HCl at pH 2.0, releasing an average of 89% in 30 min. Both adsorption and desorption processes have improved compared to the unmodified hydrogel. Otherwise, some researchers employed the anticancer ability of this antioxidant. In this regard, Croitoru et al. [178] designed a multifunctional platform based on GO, synthesized by the traditional Hummers' method, that acted as a nanocarrier where biologically active substances, such as GA could be loaded. Experimental results showed about 70% release in less than one day, 75% in four days, and 80% within 10 days. This novel nanocarrier could be useful to treat cancer or severe infections.

4.3. Tannic Acid

Although TA is a widely investigated molecule, very few papers regarding graphene-based sensors for TA determination have been published. Sinduja et al. [179] developed a colorimetric and a spectrofluorimetric sensor based on graphene quantum dots (GQDs) prepared via pyrolysis of citric acid. The mixture of GQDs and TA led to a new absorption band in the UV-Vis spectra due to the hydrogen bonding with the surface oxygen functional groups and π-π stacking interaction between aromatic groups of both compounds. On the other hand, the fluorescence intensity of GQDs linearly decreases while increasing TA concentration (Figure 8), from 0.1–1.0 μM with a LOD of 0.26 nM. Two quenching mechanisms have been proposed: (i) fluorescence resonance energy transfer (FRET), in which excited state electrons of GQDs return to the ground state via absorption of the energy emitted by the ground state electrons of TA transit to excited state resulting in a non-radiative process and (ii) simple charge transfer, in which the excited-state electron of GQDs meet TA, they transfer an electron to the Lowest Unoccupied Molecular Orbital (LUMO) of TA and then returns to ground state with a radiationless transition, which results in fluorescence quenching.

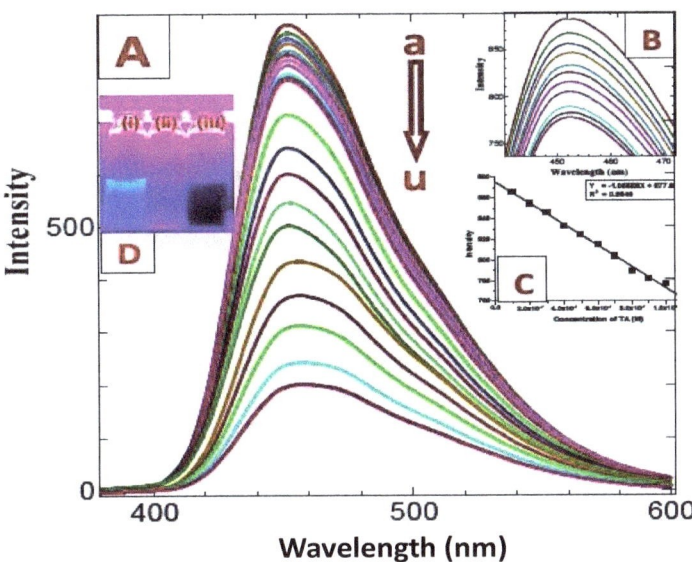

Figure 8. Fluorescence emission spectra of GQDs in the presence of different tannic acid (TA) concentrations, from 0.1 to 50 µM. The inset shows the linear plot of the intensity versus TA concentration. Taken from Sinduja et al. [179].

The electrochemical determination of TA using Zn-modified G electrodes has also been recently reported by Palisoc et al. [180]. The characterization via DPV led to a sensitive electrode, with a linear TA concentration range from 2 to 60 ppb and a LOD of 3.13 ppb.

On the other hand, several studies have been published on the use of TA as a stabilizer and reducing agent for G synthesis. For instance, Zhao et al. [181] used TA as a stabilizer for the preparation of high-quality graphene on a large scale through direct exfoliation of graphite via a green, high-efficiency, and low-cost method. Since TA acted as both dispersant and interfacial agent, G was uniformly dispersed and tightly integrated into polymer matrices for the development of high-performance and multifunctional nanocomposites. This environmentally friendly technique avoids the use of synthetic surfactants or organic solvents commonly employed for conventional G exfoliation in liquid media and prevents long reaction times. Analogously, Luo et al. [182] used TA as a reducing agent and stabilizer for GO synthesis and induced the self-assembly of rGO into a G hydrogel. The TA retained in the skeleton of 3D G also endowed the modified hydrogel with good antibacterial capability. Moreover, it showed excellent adsorption toward dyes, oils, and organic solvents; hence it is a promising candidate for efficient adsorbents in water purification. In another study, the same TA-modified hydrogel was used for the immobilization of Au-NPs. The obtained nanocomposite exhibited much higher catalytic activities than the bare NPs towards the reduction of methylene blue (MB). Overall, TA has been demonstrated to be an effective stabilizer for one-step exfoliation and noncovalent functionalization of graphene in aqueous media.

Another common application of TA is to aid in the development of several sensors. Lim et al. [183] prepared a humidity sensor based on a polyvinyl alcohol (PVA) nanocomposite filled with rGO coated with TA, which acted as a reducing and stabilizing agent and also increased the compatibility between rGO and the PVA matrix. The conductive property of rGO provides long-term stability, and the incorporation of rGO-TA into the PVA matrix enhanced mechanical strength. The PVA nanocomposite showed excellent humidity sensing properties over a wide relative humidity range. Similarly, Yoo et al. [184]

synthesized an NH$_3$ sensor based on TA-functionalized rGO, which demonstrated a high potential in gas sensing due to its high sensitivity, reversibility, and short response time.

4.4. Resveratrol

The development of optical and electrochemical sensors for resveratrol (RES) determination has been the aim of a few studies. Thus, Li et al. [185] used the quenching property of GO to prepare a fluorescent FRET-based sensor via competitive supramolecular recognition between p-sulfonated calix(6)arene (CX6)-modified reduced graphene oxide (CX6@RGO) and a probe-resveratrol complex (Figure 9). The probe molecule, Rhodamine B (RhB) or rhodamine 123 (R123), had a strong fluorescence signal, and its fluorescence was quenched by CX6@RGO. However, if RES was added, the fluorescence reappeared proportionally to the amount of RES added. This was due to the new CX6@RGO-resveratrol generated, which inhibited the quenching. Fluorescence measurements were performed in the linear range of 2–40 µM, with a RES LOD of 0.47 µM.

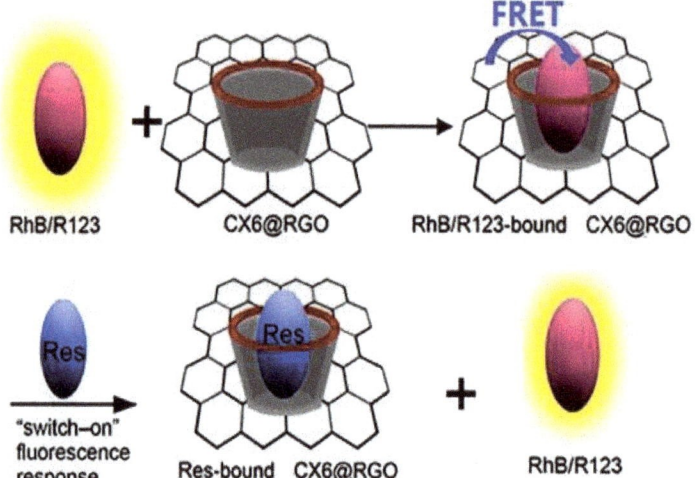

Figure 9. Displacement assay for resveratrol using calix(6)arene (CX6)-modified reduced graphene oxide (CX6@RGO) against a fluorescent dye. Taken form Li et al. [185].

Electrochemical methods have also been reported. Zhang et al. [186] used a direct laser-induced graphene (LIG) technique, which transformed the commercial Kapton/polyimide tape into 3D porous G. The prepared electrochemical sensor showed excellent repeatability, stability, reproducibility, and reliability, with an excellent linear response within the RES concentration range from 0.2 to 50 µM and a low LOD of 0.16 µM. Furthermore, the developed sensor was applied for the evaluation of RES levels in red wines and grape skins with outstanding results. On the other hand, Liu et al. [187] synthesized a sensor by one-step electrodeposition of rGO onto a GCE, which was compared with the bare GC electrode. The increased surface of rGO strongly enhanced the sensitivity of the sensor due to the π-π interaction between the rGO and RES. The response was linear in the range from 0.8 µM to 32 µM. Further, the electrode was stored at pH 2.0 and 4 °C for one month, and it retained a 95.6% of the original interaction. This approach of measuring RES using a G derivative is a cost-effective, eco-friendly, and effective technique.

On the other hand, RES has been used as a reducing agent for G derivatives. As mentioned earlier, GO is generally reduced by chemical methods; however, it results in limited solubility and an irreversible agglomeration of rGO due to the strong π-π stacking tendency between graphene layers. For that reason, surfactants are used. Another option to overcome this trouble is to employ a green reduction. In this regard, Gurunathan et al. [188]

synthesized rGO from GO using RES [189]. Then, a RES-rGO complex was prepared to study its effect against ovarian cancer, and it exhibited much more cytotoxicity than just rGO, which is also known to decrease cell viability [190,191]. The incorporation of RES causes a significant toxicity increment, inducing cell death by promoting ROS generation.

RES and G nanocomposites have also been used for the treatment of other diseases. He et al. [192] used RES for its neuroprotective effect and GO, due to its properties and cost-effectiveness, to develop a structure able to recognize amyloid β (Aβ), closely implicated in Alzheimer's disease. The Res@GO composite sensitively captured both Aβ monomers and fibers because RES is able to specifically bind with Aβ. Further, the fluorescence of RES was decreased with the GO addition via FRET. When Aβ was added, the fluorescence was restored due to RES removal. This approach based on the interaction between Aβ and the Res@GO complex was applied to detect an Alzheimer indicator via a quick and cost-effective method.

4.5. Oleuropein and Hydroxytyrosol

Oleuropein (OL) is the ester of elenolic acid and HT and is one of the most significant components of the olive leaf extract. A few studies on the development of electrochemical sensors for OL and HT detection have been reported. Gomez et al. [193] developed a novel method for OL detection in complex plant matrices based on a Graphene Oxide Pencil Graphite Electrode (GOPGE). The electrochemical behavior of OL was examined using DPV, showing a signal enhancement of 5.3 times higher than the bare electrode. A calibration curve was performed between 0.10 to 37 μM, with a LOD of 30 nM. In another study, Kurtulbas et al. [194] developed an easy, accurate and sensitive detection method to determine OL using a TiOx-modified rGO glassy carbon electrode (TiOx-RGO@GCE). The rGO was prepared using AA as a reducing agent, and the nanocomposite via sol-gel method followed by drop-casting. CV and square wave voltammetry (SWV) experiments showed that the quasi-reversible reaction is the dominant mechanism on the electrode/electrolyte interface. A linear concentration range of 1–12 μM was obtained with a LOD of 18.7 nM. The same authors developed another TiO-rGO based electrode for OL detection, optimizing the synthesis conditions (Figure 10), and a linear concentration range of 5–30 nM was obtained with a LOD of 0.57 nM [195].

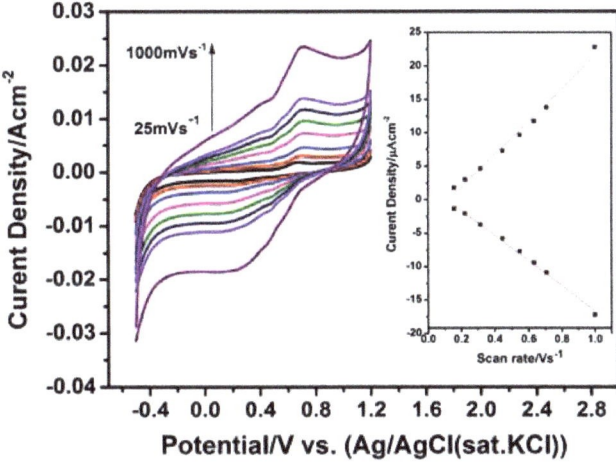

Figure 10. Cyclic voltammetric (CV) curves of 0.1 mM oleuropein (OL) on a TiO-rGO electrode at different scan rates from 25 to 1000 mV s^{-1}. Taken from Yazar et al. [195].

On the other hand, HT can be typically found in olive mill wastewater. For this reason, several studies have been reported on developing new methods for HT recovery.

Sahin et al. [196] used GO synthesized by Hummer's method as an adsorbent of HT. Adsorption of this bioactive compound from aqueous media onto GO was found to be >85% under optimum conditions. The pH of the adsorption medium was found to be a very important parameter affecting HT recovery. Increasing the pH from 3 to 9 increased the amount of adsorbed substance per GO from 0.55 to 89.46 mg. The same authors [197] also investigated a method to recover oleuropein and hydroxytyrosol from olive leaves and olive oil. For such purpose, a Zr-based metal-organic framework (UiO-66) and another based on graphene nanoplatelets (GNP/UiO-66). The use of GNP has been found to be more efficient than single-layer graphene, with excellent adsorption for organic pollutants because of its large, delocalized π-electron system. The particle size of UiO-66 was about 0.28 μm, which increased to 0.71 μm upon the addition of the GNP. Results showed that most of the hydroxytyrosol was removed from the solution, with an adsorption capacity of 142.07 mg per g UiO-66 nanoparticles at pH 10 in 180 min.

HT can also be used to reduce and stabilize GO. Baioun et al. [198] developed a green, low-cost, effective, and scalable method using an olive leaf aqueous extract rich in HT. It provided a high-efficiency removal of functional oxygen groups in the GO, generating and stabilizing rGO, which exhibited good solubility in aqueous solutions and some organic solvents.

4.6. Tocopherol

The determination of vitamin E has also been the aim of a few studies. Filik et al. [199] designed a Nafion (NF)/ERGO-modified GCE electrode. This nanocomposite provided excellent selectivity, sensitivity, stability, and reproducibility, and allowed the detection of TOH in the concentration range of 0.5 to 90 μM with a LOD of 0.06 μM. The electrode reaction of TOH is an irreversible process that takes place readily in the presence of water, free from interferences of other compounds such as AA.

MIPs have been combined with ionic liquids (ILs) to fabricate a GO/QDs nanocomposite sensor for the selective detection of traces of vitamin E in real samples. ILs are introduced on the GO surface by a one-pot room temperature synthesis strategy with reverse microemulsion polymerization (Figure 11) since they provide surface binding groups between GO and QDs, and also improve the fluorescence stability of GO due to their high thermal and chemical stability [200]. The fluorescence intensity of MIP was found to decrease with the increasing concentration of vitamin E in the range of 23–92 nM with a LOD of 3.5 nM and high precision. FRET is a possible mechanism for fluorescence quenching owing to no spectral overlap between the absorption spectrum of vitamin E and the emission spectrum of MIP.

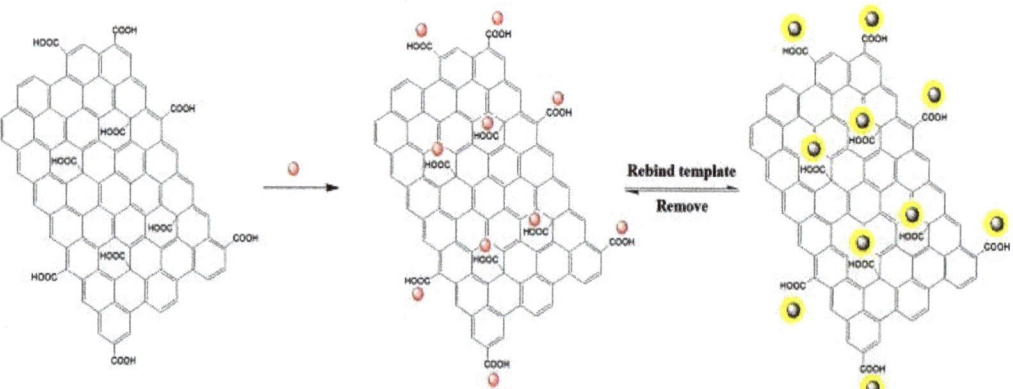

Figure 11. Representation of the synthesis of GO/QDs@molecular imprinted polymers (MIP) by a one-pot room temperature synthesis strategy with reverse microemulsion polymerization. Taken from Liu et al. [200].

4.7. Ascorbic Acid

Regarding AA determination, numerous methods have been reported in the literature; however, most of them present drawbacks such as high cost, operational complexity, laborious sample treatments, and high waste generation. Therefore, novel approaches are pursued. In particular, a few fluorescence sensors for AA determination have been reported. For instance, Liu et al. [201] developed a photoluminescent glycine (GLY)- functionalized GQDs by a simple and environmentally friendly pyrolysis method using ethylene glycol (EG) as a carbon source. The as-synthesized GLY-GQDs showed outstanding water solubility with a fluorescence quantum yield of 21.7%. The fluorescence of GLY-GQDs was quenched by Ce^{4+} via forming GLY-GQDs-Ce^{4+} non-luminescent complexes. Upon addition of AA, the fluorescence was restored due to the reduction of $Ce4+$ to $Ce3 +$. Based on this, a simple, fast, and inexpensive AA sensor was fabricated with a linear relationship in the range of 0.03–17.0 µM and a LOD of 25 nM without interference from other molecules such as uric acid dopamine, glutathione, and so on.

With regard to electrochemical methods, De Faria et al. [202] proposed a simple, sensitive, and precise approach using Flow injection analysis (FIA) with amperometric detection based on an rGO electrode prepared via simple dilution and drop-casting. The FIA system allowed a high analytical frequency, approximately 96 injections per hour, together with a linear concentration range of 65–253 µM and a LOD of 4.7 µM. Additionally, Swamy et al. [203] fabricated a sensor by decorating the surface of graphite electrode with NiO/G nanoparticles, which successfully separated the oxidation current signals of AA, dopamine, and tyrosine compared to a single, overlapped oxidative peak on a bare graphite electrode (Figure 12). The electrode has high selectivity and sensitivity (LOD of 50 µm) in addition to other factors like cost-effectiveness, convenience, and hassle-free electrochemical performance.

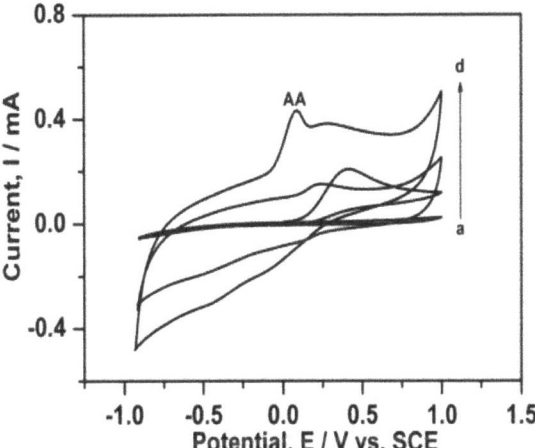

Figure 12. Cyclic voltammograms of (a) Bare graphite (b) Bare graphite with 5 mM ascorbic acid (AA) (c) NiO/G electrode in 0.1 M phosphate-buffered saline (PBS), pH 7.0 (d) NiO/G with 5 mM AA in 0.1 M buffer solution. Taken from Swamy et al. [203].

A very similar approach was applied by Kunpatee et al. [204], who used GQDs/IL-modified screen-printed carbon electrodes (SPCE) to determine AA, dopamine, and uric acid, which coexist in living systems. The GQDs/IL-SPCE exhibited excellent electrocatalytic activity for the oxidation of the three components in the mixture solution. Moreover, the anodic peak responses of the three analytes were well resolved into defined peaks. Under the optimal conditions, linear response for AA concentration was obtained in the range of 25–400 µM, with a LOD of 6.64 µM. The sensor exhibited high sensitivity, cost-effectiveness and was successfully applied for the simultaneous detection of these analytes in pharmaceutical products and biological samples. Analogously, Ji et al. [205] developed a smartphone-based integrated voltammetry system based on SPCE modified with rGO and electrochemically deposited. Experimental results corroborated that the system could be used to detect the electrochemical activity of these biomolecules with high sensitivity, linear and specific responses. Thus, AA was determined in the range of 20–375 µM, with a LOD of 1.04 µM.

Fu et al. [206] also determined these three biomolecules using a G ink-coated glass prepared via simple water immersing followed by electrochemical reaction. CV studies revealed linear calibration curves in the range of 50–1000 µM, with a LOD of 17.8 µM. This study corroborated that the elimination of additives of the G ink upon film coating is a simple and cost-effective approach for sensor applications. Similarly, Shi et al. [207] detected them using rGO/polydopamine (PDA)/AuNPs nanocomposites prepared via reduction of GO nanosheets by PDA followed by mixing with the nanoparticles. The modified nanomaterials showed a big surface area, as revealed by TEM images (Figure 13), a high level of crystallinity according to X-Ray diffraction (XRD) analysis, exceptional biocompatibility and outstanding conductivity that promoted the electrocatalytic oxidation of the biocompounds, though the sensibility was not high. Thus, AA was only detected in the linear range of 4.93–9.60 mM, with an LOD of 1.64 mM.

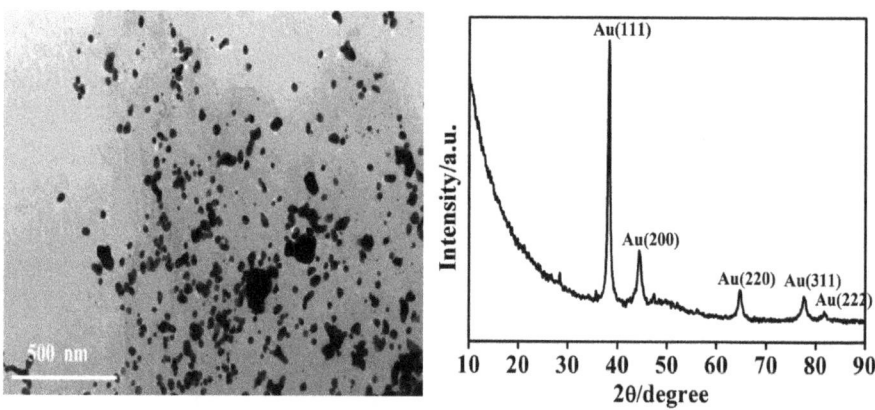

Figure 13. TEM images and X-ray diffraction (XRD) pattern of RGO/polydopamine(PDA)/Au nanohybrids. Taken from Shi et al. [207].

Better sensitivity was attained by Li et al. [208], who developed a 3D nanocomposite based on MoS_2 nanospheres, polyaniline (PANI), and rGO via a one-pot hydrothermal process. Thus, the peak currents obtained from DPV experiments varied linearly in the AA concentration range from 50 µM to 8.0 mM, with a LOD of 22.2 µM. The MoS_2/PANI/rGO-based sensor exhibited high selectivity, reproducibility, good stability, and reliability for the trace determination of these three biocompounds. Another nanocomposite incorporating PANI was prepared by Salahandish et al. [209]. In particular, they synthesized a metal nanoparticle (NP)-grafted N-doped functionalized G (NFG)/PANI nanocomposite on a fluorine-doped tin oxide electrode (FTOE). The synthesis involved the coating of NFG on the FTOE substrate, chronoamperometry of metal NPs on the NFG-coated FTOE, and electropolymerization of PANI on AgNPs modified FTOE (Figure 14). A broad linear range was found between 10–11,460 µM, with a LOD of 8 µM. Results demonstrate that this nanocomposite is a suitable candidate for rapid, reproducible, and selective detection of AA in clinical samples.

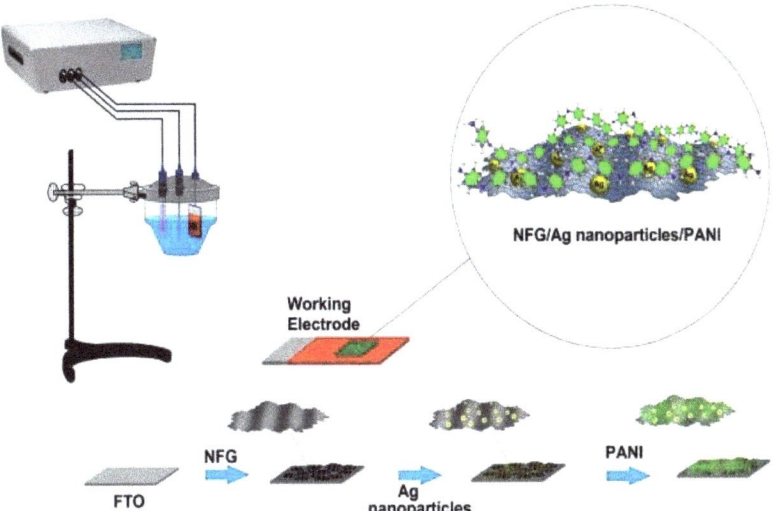

Figure 14. Representation of the synthesis of metal nanoparticles (NPs)-grafted N-doped functionalized graphene (NFG)/polyaniline (PANI) nanocomposites. Taken from *Salahandish* et al. [209].

Abraham et al. [210] developed a rGO/Pd-modified GCE to determine Epinephrine, AA, and uric acid, biomolecules that co-exist in the extracellular fluid of the central nervous system and serum. In this case, GO was synthesized via improved Hummer's method and subsequently subjected to solar exfoliation by exposure to solar rays, which resulted in the formation of rGO. Then, it was suspended in methanol, drop cast on a GCE followed by electrodeposition of Pd. The metal incorporation resulted in improved electrochemical performance in terms of surface area and roughness. CV and DPV experiments were repeated at intervals of one, three, and six days giving reproducible results with an RSD of 2.4%. Thus, a linear range was attained from 300 to 1300 μM, with an LOD of 22 μM. Besides, the influence of pH on the oxidation behaviour was investigated, and it was found that the current increased to up a maximum at pH of 7. The sensor can be effectively used in real systems such as human blood serum and urine.

A more sensitive, inexpensive and reliable sensor was prepared by Kucukkolbasi et al. [211] based on a GO/CdTeQDs/GC electrode prepared via hydrothermal synthesis of the CdTeQDs followed by drop casting. CV and EIS experiments revealed that the modified electrode showed better performance than the bare GC one. The influence of pH buffer concentration, deposition potential, deposition time, and the presence of electroactive interferents on the response of the electrode was investigated. A linear response of the modified electrode was obtained over the concentration range of 32.3–500.0 μM with a LOD of 6.1 μM for AA.

The best sensitivity in the detection of AA has been reported by Chen et al. [212] using three-dimensional holey graphene (3D-HG), a 3D porous network prepared via wet-chemical etching with in-plane nanopores and a very large surface area that favors electrochemically active sites and increases electron-transfer rate. This sensor showed excellent properties for AA, uric acid, and nitrite detection using DPV, with a linear range of 3.2–0.2 μM and a LOD as low as 15 nM. Moreover, the applicability of the 3D-HG modified electrode was tested in real samples, showing very good accuracy and recovery. All the reported results corroborated that G-based materials are great candidates for the individual or simultaneous detection of dopamine, AA, uric acid, or nitrite, with high potential for future diagnosis. However, the real challenge still remains, that is, the development of an economical, reliable, and practical sensor with high sensitivity and selectivity.

On the other hand, several studies regarding the potential of AA as reducing agent for rGO environmentally friendly synthesis have been published. The first study that used AA for the non-toxic and scalable production of rGO was reported by Gao et al. [213], who employed this antioxidant to compete with the six traditional methods to prepare graphene from graphite oxide, in which toxic agents were used. L-AA was used as a reductant together with L-tryptophan, which acted as a stabilizer to avoid the agglomeration and precipitation of the resulting graphene sheets. AA reduces the reactive oxygen species in water, leading to a stable and unreactive process that does not cause cellular damage. Similarly, Fernández-Merino et al. [214] and Zhang et al. [215] developed novel green methods that lead to comparable reduction yields to hydrazine. Stable suspensions of AA-rGO can be prepared not only in water but also in common organic solvents; thus, this bioactive compound represents an ideal substitute for hydrazine in large-scale production. Another potential advantage of using AA as a reductant is that it is only composed of carbon, oxygen, and hydrogen, therefore minimizing the risk of introducing heteroatoms in the reduced products that were not present beforehand.

4.8. Curcumin

Electrochemical sensors have been used as worthy tools for the detection of CM due to their simplicity, accuracy, high sensitivity and selectivity, and reasonable price. Nevertheless, owing to the poor response of this compound, it is difficult to detect it directly at the surfaces of bare electrodes. In this regard, different modifiers have been used to solve this issue and to increase the sensitivity and selectivity of CM detection sensors. Rahimnejad et al. [216] developed a sensitive and accurate sensor based on rGO/CPE prepared via pulverization followed by drop-casting. Measurements via CV and DPV indicated a linear concentration range of 10–6000 µM, with a LOD of 3.183 µM. A similar approach was developed by Zhang et al. [217], who prepared a more sensitive voltammetric method for CM determination using an electrochemically reduced graphene oxide (ERGO)-modified GCE. The modified electrode showed much better electrocatalytic activity towards CM compared with bare GCE and GO/GCE electrodes. A linear voltammetric response was found from 0.2 µM to 60.0 µM, with a LOD of 0.1 µm. Similarly, Li et al. [218] prepared a more sensitive, selective, and accurate G/GCE sensor, which was characterized via CV, EIS, and linear sweep voltammetry (LSV). The currents measured by LSV displayed presented a good linear relationship with CM concentrations in the range of 5.0×10^{-8} to 3.0×10^{-6} µM, with a low detection limit of 0.03 µM.

A comparative assessment of the potential of GO and rGO for electrochemical determination of CM was undertaken by Dey et al. [219]. GCE modified with these two nanomaterials was characterized using SEM, XRD, FTIR, and Raman techniques to understand their morphology and structure. rGO/GCE showed a lower limit detection of 0.9 pM and good signal quality. Further, the repeatability was checked for seven cycles, and interference studies corroborated the selectivity of the method. Even better sensibility was found by Kotan et al. [220], who synthesized L-cysteine functionalized rGO composites were prepared via activation of the carboxylic groups of rGO with ethylcarbodiimidehydrochloride (EDC) (Figure 15) followed by mixing with the Ru@AuNPs and ultrasonication and then drop-cast onto a GCE using an IR heat lamp. The electrochemical determination was studied using SWV with a linearity range of 0.001–0.1 nM and an unprecedented LOD of 0.2 pM.

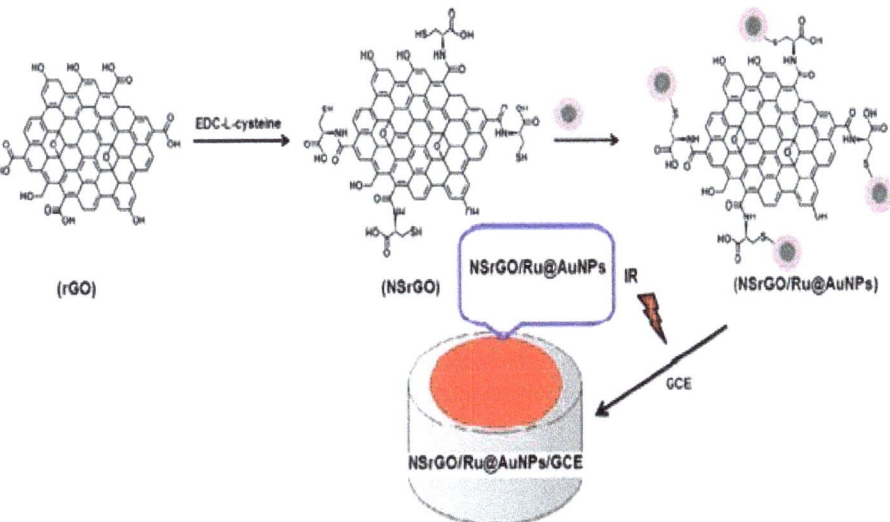

Figure 15. Schematic representation of the synthesis of L-cystein/rGO/Ru@AuNPs. Taken from Kotan et al. [220].

On the other hand, other studies have been reported on the development of G/CM hybrids with antibacterial activity. Marković et al. [221] presented a G/CM nanomesh with antibacterial activity against Gram-positive bacteria like *Staphylococcus aureus*, with a minimum inhibitory concentration (MIC) of 1 mg mL^{-1}. It was found that its cytotoxicity was concentration-dependent. At concentrations higher than 100 mg mL^{-1} some slight cytotoxic effects were observed. In this work, G was exfoliated from highly oriented pyrolytic graphite (HOPG) through an electrochemical exfoliation process with ammonium persulfate as an electrolyte, leading to CM/EHOPG nanomesh hybrids. Similarly, Bugli et al. [222] and Palmieri et al. [223] presented a method using Cm and Go to kill methicillin-resistant *Staphylococcus aureus* (MRSA).

Conversely, Yang et al. [224] developed a β-cyclodextrin (CD) functionalized GO nanocomposite, which displayed excellent antiviral activity and could load curcumin efficiently. Their aim was to find a new strategy to treat respiratory syncytial virus (RSV). Other researchers have employed CM to help in antitumoral treatments. Thus, Hatamie et al. [225] combined CM with rGO sheets, linked by π–π attachment, and studied its effects on human breast cancer cell lines and a normal cell line. Curcumin was utilized for simultaneous reduction of chemically exfoliated GO sheets and functionalization of the rGO ones. The interaction of the rGO sheets and cells resulted in apoptosis as well as a morphological transformation of the cells; thus, it could be used for nanotechnology-based bioapplications against cancer.

The most representative examples reported to date on optical and electrochemical sensors based on graphene and its derivatives for the detection of bioactive compounds are collected in Table 2.

Overall, G-based nanomaterials with high specific surface area, excellent electrical conductivity, good stability, and unique mechanical properties have been found to have enormous potential for the determination of bioactive compounds. Compared to other carbon-based nanomaterials, they can provide more active sites, increase the electrochemical active surface area, improve the mass transport rate, and accelerate the electron transfer rate; hence, better and more reliable results have been obtained.

Table 2. Characteristics of graphene-based sensors for the detection of bioactive compounds.

Bioactive Compound	Carbon Nanomaterial	Processing Method	Detection Method	Linear Range	LOD	Properties	Ref.
Melatonin (MLT)	GO@SiO$_2$ nanocomposite	Modified Hummers´ + Sol-gel with PTEOS and TMOS	dsPE + HPLC with DAD	-	<0.1 µg/mL	Cost-effective, simple, selective and sensitive.	[159]
	G-CSPE	G Sonication + Drop-casting	CV and FPA	-	0.87 µM	Good sensitivity, reversibility, $I_c/I_a \approx 1$.	[160]
	CVD G-CSPE	G Suspension + Drop-casting	DPV	-	15 µg/L	Good sensitivity, reproducibility, versatility, better results than the electrode without G	[161]
	GON-CSPE GRN-CSPE	Longitudinal unzipping + hydrazine reduction + ultrasonication + drop-casting	CV and DPV	-	1.1 µM	Good reproducibility and response time, recovery of 94%–103%.	[162]
	rGO/MIP	Modified Hummers´ + hydrazine reduction + rGO Suspension + Drop-casting + electropolymerization.	CV and SWV	0.05–100 µM	6 nM	Stable and highly sensible.	[163]
	rGO/Fe$_3$O$_4$	Modified Hummers´ + hydrazine reduction + hydrothermal growth	SWV	0.02–5.80 µM	8.4 nM	Good selectivity, repeatability, reproducibility, and biocompatibility.	[164]
	rGO/SnO$_2$-Co$_3$O$_4$ nanocomposite	Modified Hummers´ + SnO$_2$ reduction + hydrothermal growth	CV and SWV	0.02–6.00 µM	4.1 nM	Good sensitivity, selectivity, stability, and repeatability; cost-effective and simple fabrication.	[165]
	N-rGO/CuCo$_2$O$_4$ nanocomposite	Modified Hummers´ + hydrazine reduction + solvothermal method	DPV and SWV	0.01–3.0 µM	4.9 nM	Enhanced selectivity, sensitivity, and biocompatibility.	[166]
	CVD G/CuO-PLL nanocomposite	CVD growth + electrochemical deposition	CV and SWV	0.016–110 µM	12 nM	Good sensitivity and biocompatibility.	[167]
	rGO	Modified Hummers´ + MLT reduction	CV	-	-	Simple, reproducible and biocompatible.	[168]
Gallic acid (GA)	NPGA	hydrothermal reduction of GO with PPD + freeze-drying	DPV and SWV	2.5–1000 µM	67 nM	Large specific surface area and excellent electrical conductivity.	[170]
	G/ZrO$_2$	Hydrothermal growth + physical mixing	DPV and SWV	1 µM–1 mM	124 nM	High surface area, good biocompatibility, and electrical conductivity.	[171]
	rGO/ZrO$_2$/Co$_3$O$_4$	Modified Hummers´ + hydrazine reduction + ultrasonication + drop casting	CV and DPV	6.2–478 nM	1.56 nM	Good sensitivity, selectivity, reproducibility, and stability vs. interferences.	[172]
	MWCNT/rGO nanocomposite	Drop-casting + UV reduction	CV and EIS	29–329 pM	2.57 pM	Excellent sensibility, reproducibility, and long-term stability.	[173]
	CS/Fe$_2$O$_3$/ERGO nanocomposite	Solvothermal synthesis of Fe$_2$O$_3$ + ultrasonicaction + drop casting electrochemical reduction	DPV and EIS	1–100 µM	0.15 µM	Large surface area, excellent electronic conductivity, and high stability.	[174]
	PANI–rGO–TiO$_2$	Solvothermal synthesis of TiO$_2$ + aniline polymerization + mixing + ultrasonication	CV and PC	4.17–250 µM	1.72 µM	Rapid response, high sensitivity, and excellent selectivity.	[175]
	GQDs	Pyrolysis of citric acid	LLE + Fluorescence	5–40 mg/L	1.08 mg/L	Simple, sensitive, and reproducible. Fast response.	[176]
Tannic acid (TA)	GQDs	Pyrolysis of Citric Acid	UV-Vis and Fluorescence	0.1–1 µM	0.26 nM	Good selectivity and applicability.	[179]
	Zn-G	Electrolysis of graphite rods	DPV	2–60 ppb	3.13 ppb	Sustainable and cost-effective.	[180]
Resveratrol (RSV or RES)	CX6@RGO	Ultrasonication + mixing+ freeze drying.	UV-Vis and Fluorescence	2–40 µM	0.47 µM	Fast, simple, sensitive and selective.	[185]
	Porous G	Laser-induced conversion of Kapton/PI tape into 3D porous G	DPV	0.2–50 µM	0.16 µM	Excellent repeatability, stability, reproducibility, and reliability.	[186]
	rGO-GCE	Sonication + electrochemical deposition	CV and DPV	0.8–32 µM	0.2 µM	Long-term stability; low-cost, eco-friendly, and effective.	[187]

Table 2. Cont.

Bioactive Compound	Carbon Nanomaterial	Processing Method	Detection Method	Linear Range	LOD	Properties	Ref.
Oleuropein (OL) and Hydroxytyrosol (HT)	GOPGE	Sonication + drop casting	DPV	0.10–37 µM	30 nM	Good sensitivity and selectivity.	[193]
	TiOx-RGO@GCE	Hummer´s + reduction with AA + sol gel + drop casting	CV and SWV	1–12 µM	18.7 nM	Good sensitivity, simple and accurate.	[194]
	TiO-rGO	Hummer´s + reduction with AA + sol gel + drop casting	CV and SWV	5–30 µM	0.57 nM	Good sensitivity and selectivity.	[195]
	GONs	Ultrasonication + unzipping of MWCNTs + drop casting	CV, EIS, and DPV	-	-	Excellent performance and is fast.	[226]
Tocopherol (TOH)	NF/ERGO/GCE	ultrasonicaction + drop casting electrochemical reduction	DPV	0.5–90 µM	0.06 µM	Excellent selectivity, sensitivity, and reproducibility. Fast and cost-effective.	[199]
	ILs/MIP/GO/QDs	one-step polymerization	Fluorescence	23–92 nM	3.5 nM	Excellent photochemical stability and sensitivity.	[200]
Ascorbic acid (AA)	GLY-GQDs	pyrolysis with EG	Fluorescence	0.03–17.0 µM	25 nM	High sensitivity and selectivity.	[201]
	rGO	Dilution + drop casting	FIA with amperometric detection	65–253 µM	4.7 µM	Simple, sensitive and accurate, and precise.	[202]
	NiO/G	Coprecipitation synthesis of NiO + ultrasonication + drop casting	CV and DPV+ chronoamperometry	-	50 µM	Good selectivity and sensitivity, and cost-effective, easy to handle.	[203]
	GQDs/IL-SPCE	Pyrolysis of Citric Acid + drop casting	CV and EIS	25–400 µM	6.64 µM	High sensitivity and conductivity, good biocompatibility, cost-effective.	[204]
	rGO/AuNPs/SPE	G suspension + mixing electrochemical deposition	CV and DPV	20–375 µM	1.04 µM	High selectivity and sensitivity.	[205]
	Graphene ink coated glass	Water immersion + electrochemical reaction	CV	50–1000 µM	17.8 µM	Simple and cost-effective.	[206]
	rGO/PDA/AuNPs	GO reduction by PDA + mixing	CV + EIS	4.93–9.60 mM	1.64 mM	Good biocompatibility and conductivity.	[207]
	MoS$_2$-PANI/rGO	one-pot hydrothermal synthesis + drop casting	CV and DPV	8 mM–50 µM	22.2 µM	High selectivity, good reproducibility, and stability.	[208]
	NFG/AgNPs/PANI	NFG coating on FTOE + electropolymerization of PANI	CV	10–11460 µM	8 µM	Good reproducibility and excellent selectivity.	[209]
	GCE/Pd/rGO	Sonication + electrodeposition	CV, DPV, and EIS	0.3–1.3 mM	22 µM	Fast response, good selectivity.	[210]
	GCE/GO/CdTeQDs	Hydrothermal synthesis + drop casting	CV + EIS	32.3–500 µM	6.1 µM	Inexpensive, reliable, and sensitive.	[211]
	3D-HG/GCE	Wet-chemical etching + drop casting	DPV	0.2 µM–3.2 mM	15 nM	High sensitivity and selectivity, excellent electrocatalytic activity.	[212]
Curcumin	rGO/CPE	Pulverization + drop casting	CV and DPV	10–6000 µM	3.18 µM	Good replicability catalytic activity, and storage stability.	[216]
	ERGO/GCE	Electrochemical reduction + drop casting	CV	0.2 µM–60 µM	0.1 µM	Good replicability and catalytic activity.	[217]
	G/GCE	Drop casting	CV, EIS	0.05–3.0 µM	0.03 µM	High selectivity and accuracy.	[218]
	rGO/GCE	Drop casting	CV, DPV	0.1 nM–10 nM	0.9 pM	Exceptional sensibility.	[219]
	NSrGO/Ru@AuNPs	L-cysteine functionalization+ Ru@AuNPs grafting	SWV	0.001–0.1 nM	0.2 pM	Exceptional sensibility.	[220]

Phenyl triethoxysilane (PTEOS); Tetramethoxysilane (TMOS); Dispersive solid-phase extraction (dSPE); Graphene-coated carbon screen-printed electrode (G-CSPE); Cyclic voltammetry (CV); Fixed-potential amperometry (FPA); Current of anodic peak (I_a); Current of cathodic peak (I_c); Differential pulse voltammetry (DPV); Molecular imprinted polymer (MIP); Square wave voltammetry (SWV); Nitrogen doped reduced graphene oxide (N-rGO); poly(L-lysine) (PLL); Graphene oxide nanoribbons (GON); Graphene reduced nanoribbons (GRN); Diode Array Detection (DAD); Nitrogen-doped porous graphene aerogel (NPGA); Photocurrent measurements (PC); Graphene Quantum Dots (GQDs); p-sulfonated calix[6]arene (CX6); Nafion (NF); Polydopamine (PDA); Polyaniline (PANI); Nitrogen-doped functionalized graphene (NFG); Three dimensional holey graphene (3D-HG); fluorine-doped tin oxide electrode (FTOE); Flow injection analysis (FIA).

5. Outlook and Future Prospects

Over the last years, G and its derivatives have shown huge potential in the field of optical and electrochemical sensors. Owed to their exceptional electrical, chemical, and mechanical properties, G-based nanomaterials have already been used as sensors for detecting a wide range of analytes, including bioactive compounds, which are essential for human health owing to their multiple biological effects, including antioxidant activity. These sensors display outstanding performance compared to those based on conventional materials in terms of sensitivity, selectivity, response time, and long-term stability. In particular, it is possible to detect picomolar concentrations using electrochemical sensors based on MWCNT/rGO nanocomposites or L-cysteine functionalized-rGO/GCE. Currently, despite the benefits of optical detection, including high selectivity, immunity to electromagnetic interference, and a wide dynamic range, very few optical sensors have been designed for the detection of bioactive compounds. Covalent and non-covalent functionalization of G-based nanomaterials with organic or inorganic systems (i.e., polymers, nanoparticles) offer novel means for the development of the next generation of G-based sensors. However, despite most of the developed sensors represent an important proof-of-concept, the full potential of G-based sensors is far from being reached, and several issues have to be addressed prior to the commercial use of G-based nanocomposites:

(1) New manufacturing/fabrication routed to prepare high-quality G with tailored morphology, and electronic properties are required since the performance of G-based sensors is closely related to the nanomaterial characteristics, namely, purity, defect content, degree of functionalization, and structural morphology. From a practical perspective, novel technologies to produce sensors with reproducible and repeatable characteristics are needed. Thus, improvements in manufacturing to diminish variations among sensors and yield consistent performance, novel sensor designs, and operating modes are required. Likewise, more work on integrating signal conditioning and processing electronics to minimize performance variations, improve selectivity and operating lifetime would be valuable for practical applications.

(2) The functionalization and ultrasonication processes applied to G prior to and during the sensor fabrication may result in a strong reduction in electrical conductivity. Hence, G-based nanocomposites might present electrical properties that do not satisfy the requirements for sensor applications.

(3) Approaches that allow the large-scale synthesis of G at a relatively low cost are highly desirable. Despite significant efforts having been carried out in this direction, current methods are seriously restricted by their low efficiencies, which should be addressed for commercial applications. To the best of our knowledge, a reliable method able to supply the huge demand for pristine GO (or rGO) via an environmentally friendly approach, with a short-sonication time, viable washing steps, and high yield is still lacking.

(4) The actual specific surface area of G-based nanomaterials is considerably lower than the theoretical predictions due to the strong agglomeration tendency of the nanosheets via π-π stacking interactions and the mixing with organic molecules can make it worsen. While some achievements have been attained via the addition of stabilizers, these can have detrimental effects on sensor performance. In this regard, novel approaches to efficiently exfoliate the G sheets and in a green way are pursued.

(5) The toxicity of G-based nanomaterials is not clear yet. Despite considerable efforts in evaluating the potential impact of these nanomaterials on human health and the environment, results are frequently contradictory. They might cause cytotoxicity in humans, and this issue should be clarified. It is important to highlight that "graphene" is not a single nanomaterial but a group of materials, which accounts for the fact that their biological effects may vary depending on their intrinsic properties. A number of parameters, including processing method, lateral dimensions, level of functionalization, defect content, etc., can strongly influence the toxicity of these nanomaterials. Furthermore, their biodegradation mechanisms and extent remain unclear. The number of graphene

layers, the average lateral dimension, and the atomic C/O ratio can play a key role in their biodegradability.

Overall, the development and widespread usage of G-based nanomaterials are largely hindered by the lack of techniques to provide simple, reproducible, and cost-effective sensors at a large scale. The research in this field is still in its infancy. More work is needed to ensure that the sensors are reliable, robust, and have non-toxic and easy manufacturability in a cost-competitive manner. Cost should be reduced in order to attain economic production of these nanomaterials with defined structures and properties at a large scale. In addition, a better understanding of G interactions with the bioactive compounds and the detection (or signal transduction) mechanisms are critical. Moreover, issues related to biodegradation and biocompatibility must be carefully considered, and challenges including device minimization, integration, durability, and lifetime should be addressed. Even so, it is expected that after comprehensive research in the field and continuous innovative efforts, sensors incorporating G-based nanomaterials could offer a new outlook for the detection of a variety of analytes, in particular bioactive compounds. This can be achieved with the collaborations between different disciplines and technologies.

6. Conclusions

In this review, the potential of G-based nanomaterials, namely GO, rGO, and GQDs, for sensing applications, in particular for bioactive compounds with antioxidant properties such as melatonin, gallic acid, tannic acid, resveratrol, hydroxytyrosol, tocopherol, ascorbic acid, and curcumin has been discussed in detail. The synthesis process, functionalization routes, and main properties have been summarized, with particular emphasis on their sensitivity and selectivity. The use of carbon nanomaterials has been demonstrated to be really useful to detect antioxidants through an easy, fast, and green technique, leading to better performance than sensors based on conventional materials. Moreover, the combination between these carbon nanostructures and the antioxidants has opened new properties and applications owed to synergistic effects. Besides, these antioxidants can be used to reduce GO via inexpensive and environmentally friendly methods. Finally, the future outlook for the development of G-based sensors for this type of biocompounds has been outlined. The extensive research progress in nanotechnology for graphene nanomaterials will enable the development of highly sensitive, specific, and green nanosensors at an affordable cost.

Author Contributions: Literature review, C.S.-U.; writing—original draft preparation, C.S.-U.; M.P.S., S.V.-L. and A.M.D.-P.; writing—review and editing, M.P.S., S.V.-L. and A.M.D.-P.; supervision, A.M.D.-P. All authors have read and agreed to the published version of the manuscript.

Funding: Financial support for this work came from the Community of Madrid within the framework of the Multi-year Agreement with the University of Alcalá in the line of action "Stimulus to Excellence for Permanent University Professors", Ref. EPU-INV/2020/012, as well as from the Spanish Ministry of Science, Innovation, and Universities (MICIU) via Project PGC2018-093375-B-I00 co-financed by the EU, are gratefully acknowledged.

Conflicts of Interest: The authors declare no conflict of interest.

References

1. Soldano, C.; Mahmood, A.; Dujardin, E. Production, properties and potential of graphene. *Carbon* **2010**, *48*, 2127–2150. [CrossRef]
2. Novoselov, K.S.; Geim, A.K.; Morozov, S.V.; Jiang, D.; Zhang, Y.; Dubonos, S.V.; Grigorieva, I.V.; Firsov, A.A. Electric field effect in atomically thin carbon films. *Science* **2004**, *306*, 666–669. [CrossRef] [PubMed]
3. Balandin, A.A.; Ghosh, S.; Bao, W.; Calizo, I.; Teweldebrhan, D.; Miao, F.; Lau, C.N. Superior Thermal Conductivity of Single-Layer Graphene. *Nano Lett.* **2008**, *8*, 902–907. [CrossRef]
4. Mayorov, A.S.; Gorbachev, R.V.; Morozov, S.V.; Britnell, L.; Jalil, R.; Ponomarenko, L.A.; Blake, P.; Novoselov, K.S.; Watanabe, K.; Taniguchi, T.; et al. Micrometer-Scale Ballistic Transport in Encapsulated Graphene at Room Temperature. *Nano Lett.* **2011**, *11*, 2396–2399. [CrossRef] [PubMed]
5. Du, X.; Skachko, I.; Barker, A.; Andrei, E.Y. Approaching ballistic transport in suspended graphene. *Nat. Nanotechnol.* **2008**, *3*, 491–495. [CrossRef]

6. Wu, Z.-S.; Ren, W.; Gao, L.; Liu, B.; Jiang, C.; Cheng, H.-M. Synthesis of high-quality graphene with a pre-determined number of layers. *Carbon* **2009**, *47*, 493–499. [CrossRef]
7. Lee, C.; Wei, X.; Kysar, J.W.; Hone, J. Measurement of the elastic properties and intrinsic strength of monolayer graphene. *Science* **2008**, *321*, 385–388. [CrossRef]
8. Díez-Pascual, A.M.; Gómez-Fatou, M.A.; Ania, F.; Flores, A. Nanoindentation in polymer nanocomposites. *Prog. Mater. Sci.* **2015**, *67*, 1–94. [CrossRef]
9. Huang, X.; Yin, Z.; Wu, S.; Qi, X.; He, Q.; Zhang, Q.; Yan, Q.; Boey, F.; Zhang, H. Graphene-Based Materials: Synthesis, Characterization, Properties, and Applications. *Small* **2011**, *7*, 1876–1902. [CrossRef] [PubMed]
10. Weiss, N.O.; Zhou, H.; Liao, L.; Liu, Y.; Jiang, S.; Huang, Y.; Duan, X. Graphene: An Emerging Electronic Material. *Adv. Mater.* **2012**, *24*, 5782–5825. [CrossRef]
11. Huang, X.; Qi, X.; Boey, F.; Zhang, H. Graphene-based composites. *Chem. Soc. Rev.* **2011**, *41*, 666–686. [CrossRef]
12. Díez-Pascual, A.M.; Sánchez, J.A.L.; Capilla, R.P.; Díaz, P.G. Recent Developments in Graphene/Polymer Nanocomposites for Application in Polymer Solar Cells. *Polymers* **2018**, *10*, 217. [CrossRef] [PubMed]
13. Díez-Pascual, A.M.; Díez-Vicente, A.L. Poly(propylene fumarate)/Polyethylene Glycol-Modified Graphene Oxide Nanocomposites for Tissue Engineering. *ACS Appl. Mater. Interfaces* **2016**, *8*, 17902–17914. [CrossRef]
14. Li, X.; Cai, W.; An, J.; Kim, S.; Nah, J.; Yang, D.; Piner, R.; Velamakanni, A.; Jung, I.; Tutuc, E.; et al. Large-Area Synthesis of High-Quality and Uniform Graphene Films on Copper Foils. *Science* **2009**, *324*, 1312–1314. [CrossRef] [PubMed]
15. Charrier, A.; Coati, A.; Argunova, T.; Thibaudau, F.; Garreau, Y.; Pinchaux, R.; Forbeaux, I.; Debever, J.-M.; Sauvage-Simkin, M.; Themlin, J.-M. Solid-state decomposition of silicon carbide for growing ultra-thin heteroepitaxial graphite films. *J. Appl. Phys.* **2002**, *92*, 2479–2484. [CrossRef]
16. Huang, N.M.; Lim, H.N.; Chia, C.H.; Yarmo, M.A.; Muhamad, M.R. Simple Room-Temperature Preparation of High-Yield Large-area Graphene Oxide. *Int. J. Nanomed.* **2011**, *6*, 3443–3448. [CrossRef]
17. Dreyer, D.R.; Ruoff, R.S.; Bielawski, C.W. From Conception to Realization: An Historial Account of Graphene and Some Perspectives for Its Future. *Angew. Chem. Int. Ed.* **2010**, *49*, 9336–9344. [CrossRef] [PubMed]
18. Baraton, L.; He, Z.; Lee, C.S.; Maurice, J.-L.; Cojocaru, C.S.; Gourgues-Lorenzon, A.-F.; Lee, Y.H.; Pribat, D. Synthesis of few-layered graphene by ion implantation of carbon in nickel thin films. *Nanotechnology* **2011**, *22*, 085601. [CrossRef]
19. Su, C.-Y.; Lu, A.-Y.; Xu, Y.; Chen, F.-R.; Khlobystov, A.N.; Li, L.-J. High-Quality Thin Graphene Films from Fast Electrochemical Exfoliation. *ACS Nano* **2011**, *5*, 2332–2339. [CrossRef]
20. Abbasi, E.; Akbarzadeh, A.; Kouhi, M.; Milani, M. Graphene: Synthesis, bio-applications, and properties. *Artif. Cells Nanomed. Biotechnol.* **2014**, *44*, 1–7. [CrossRef]
21. Liu, F.; Wang, C.; Sui, X.; Riaz, M.A.; Xu, M.; Wei, L.; Chen, Y. Synthesis of graphene materials by electrochemical exfoliation: Recent progress and future potential. *Carbon Energy* **2019**, *1*, 173–199. [CrossRef]
22. Mateos, R.; Vera, S.; Valiente, M.; Díez-Pascual, A.M.; Andrés, M.P.S. Comparison of Anionic, Cationic and Nonionic Surfactants as Dispersing Agents for Graphene Based on the Fluorescence of Riboflavin. *Nanomaterials* **2017**, *7*, 403. [CrossRef] [PubMed]
23. Mateos, R.; García-Zafra, A.; Vera-López, S.; Andrés, M.P.S.; Díez-Pascual, A.M. Effect of Graphene Flakes Modified by Dispersion in Surfactant Solutions on the Fluorescence Behaviour of Pyridoxine. *Materials* **2018**, *11*, 888. [CrossRef]
24. Liu, S.; Zeng, T.H.; Hofmann, M.; Burcombe, E.; Wei, J.; Jiang, R.; Kong, J.; Chen, Y. Antibacterial Activity of Graphite, Graphite Oxide, Graphene Oxide, and Reduced Graphene Oxide: Membrane and Oxidative Stress. *ACS Nano* **2011**, *5*, 6971–6980. [CrossRef] [PubMed]
25. Zaaba, N.; Foo, K.; Hashim, U.; Tan, S.; Liu, W.-W.; Voon, C. Synthesis of Graphene Oxide using Modified Hummers Method: Solvent Influence. *Procedia Eng.* **2017**, *184*, 469–477. [CrossRef]
26. Díez-Pascual, A.M.; Urruela, C.S.; Vallés, C.; Vera-López, S.; Andrés, M.P.S. Tailorable Synthesis of Highly Oxidized Graphene Oxides via an Environmentally-Friendly Electrochemical Process. *Nanomaterials* **2020**, *10*, 239. [CrossRef] [PubMed]
27. Sainz-Urruela, C.; Vera-López, S.; Andrés, M.P.S.; Díez-Pascual, A.M. Graphene Oxides Derivatives Prepared by an Electrochemical Approach: Correlation between Structure and Properties. *Nanomaterials* **2020**, *10*, 2532. [CrossRef]
28. Luceño-Sánchez, J.A.; Maties, G.; Gonzalez-Arellano, C.; Diez-Pascual, A.M. Synthesis and Characterization of Graphene Oxide Derivatives via Functionalization Reaction with Hexamethylene Diisocyanate. *Nanomaterials* **2018**, *8*, 870. [CrossRef]
29. Zhu, S.; Tang, S.; Zhang, J.; Yang, B. Control the size and surface chemistry of graphene for the rising fluorescent materials. *Chem. Commun.* **2012**, *48*, 4527–4539. [CrossRef]
30. Zheng, P.; Wu, N. Fluorescence and Sensing Applications of Graphene Oxide and Graphene Quantum Dots: A Review. *Chem. Asian J.* **2017**, *12*, 2343–2353. [CrossRef]
31. Chen, F.; Gao, W.; Qiu, X.; Zhang, H.; Liu, L.; Liao, P.; Fu, W.; Luo, Y. Graphene quantum dots in biomedical applications: Recent advances and future challenges. *Front. Lab. Med.* **2017**, *1*, 192–199. [CrossRef]
32. Wu, J.; Wang, P.; Wang, F.; Fang, Y. Investigation of the Microstructures of Graphene Quantum Dots (GQDs) by Surface-Enhanced Raman Spectroscopy. *Nanomaterials* **2018**, *8*, 864. [CrossRef] [PubMed]
33. Dua, V.; Surwade, S.P.; Ammu, S.; Agnihotra, S.R.; Jain, S.; Roberts, K.E.; Park, S.; Ruoff, R.S.; Manohar, S.K. All-Organic Vapor Sensor Using Inkjet-Printed Reduced Graphene Oxide. *Angew. Chem. Int. Ed.* **2010**, *49*, 2154–2157. [CrossRef] [PubMed]
34. Macnaughton, S.; Ammu, S.; Manohar, S.K.; Sonkusale, S. High-Throughput Heterogeneous Integration of Diverse Nanomaterials on a Single Chip for Sensing Applications. *PLoS ONE* **2014**, *9*, e111377. [CrossRef] [PubMed]

35. Santos, L.; Neto, J.P.; Crespo, A.; Nunes, D.; Costa, N.; Fonseca, I.M.; Barquinha, P.; Pereira, L.; Silva, J.; Martins, R.; et al. WO3 Nanoparticle-Based Conformable pH Sensor. *ACS Appl. Mater. Interfaces* **2014**, *6*, 12226–12234. [CrossRef] [PubMed]
36. Salavagione, H.J.; Díez-Pascual, A.M.; Lázaro, E.; Vera, S.; Gómez-Fatou, M.A. Chemical sensors based on polymer composites with carbon nanotubes and graphene: The role of the polymer. *J. Mater. Chem. A* **2014**, *2*, 14289–14328. [CrossRef]
37. Guaadaoui, A.; Benaicha, S.; Elmajdoub, N.; Bellaoui, M.; Hamal, A. What is a Bioactive Compound? A Combined Definition for a Preliminary Consensus. *Int. J. Nutr. Food Sci.* **2014**, *3*, 174. [CrossRef]
38. IFIS. *Dictionary of Food Science and Technology*; John Wiley & Sons Limited: Hoboken, NJ, USA, 2009; ISBN 0860141861.
39. Hamann, M.T. Bioactive Compounds from Natural Sources: Isolation, Characterisation and Biological Properties. *J. Nat. Prod.* **2001**, *64*, 1382. [CrossRef]
40. Cossy, J.; Arseniyadis, S. *Modern Tools for the Synthesis of Complex Bioactive Molecules*; John Wiley & Sons Limited: Hoboken, NJ, USA, 2012; ISBN 9780470616185.
41. Justino, C.I.L.; Gomes, A.R.; Freitas, A.C.; Duarte, A.C.; Rocha-Santos, T.A.P. Trends in Analytical Chemistry Graphene Based Sensors and Biosensors. *Trends Anal. Chem.* **2017**, *17*, 2161. [CrossRef]
42. Lee, H.; Choi, T.K.; Lee, Y.B.; Cho, H.R.; Ghaffari, R.; Wang, L.; Choi, H.J.; Chung, T.D.; Lu, N.; Hyeon, T.; et al. A graphene-based electrochemical device with thermoresponsive microneedles for diabetes monitoring and therapy. *Nat. Nanotechnol.* **2016**, *11*, 566–572. [CrossRef]
43. Szunerits, S.; Boukherroub, R. Graphene-based biosensors. *Interface Focus* **2018**, *8*, 20160132. [CrossRef] [PubMed]
44. Yu, L.; Yi, Y.; Yao, T.; Song, Y.; Chen, Y.; Li, Q.; Xia, Z.; Wei, N.; Tian, Z.; Nie, B.; et al. All VN-graphene architecture derived self-powered wearable sensors for ultrasensitive health monitoring. *Nano Res.* **2019**, *12*, 331–338. [CrossRef]
45. Teodoro, A.J. Bioactive Compounds of Food: Their Role in the Prevention and Treatment of Diseases. *Oxidative Med. Cell. Longev.* **2019**, *2019*, 1–4. [CrossRef]
46. Del Rio, D.; Rodriguez-Mateos, A.; Spencer, J.P.; Tognolini, M.; Borges, G.; Crozier, A. Dietary (Poly)phenolics in Human Health: Structures, Bioavailability, and Evidence of Protective Effects Against Chronic Diseases. *Antiox. Redox Signal.* **2013**, *18*, 1818–1892. [CrossRef]
47. Hollman, P.C.H.; Cassidy, A.; Comte, B.; Heinonen, M.; Richelle, M.; Richling, E.; Serafini, M.; Scalbert, A.; Sies, H.; Vidry, S. The Biological Relevance of Direct Antioxidant Effects of Polyphenols for Cardiovascular Health in Humans is Not Established. *J. Nutr.* **2011**, *141*, 989S–1009S. [CrossRef]
48. Sang, S.; Lapsley, K.; Jeong, W.S.; Lachance, P.A.; Ho, C.T.; Rosen, R.T. Antioxidative Phenolic Compounds Isolat-ed from Almond Skins (Prunus Amygdalus Batsch). *J. Agric. Food Chem.* **2002**, *50*, 2459–2463. [CrossRef] [PubMed]
49. McClements, D.J.; Decker, E.A. *Fennema's Food Chemistry*, 5th ed.; CRC Press: Boca Raton, FL, USA, 2017; ISBN 9781315372914.
50. Shahidi, F.; Naczk, M. Antioxidant Properties of Food Phenolics. In *Phenolics Food Phenolics*; CRC Press: Boca Raton, FL, USA, 2004; ISBN 9780367395094.
51. Pandi-Perumal, S.R.; Srinivasan, V.; Maestroni, G.J.M.; Cardinali, D.P.; Poeggeler, B.; Hardeland, R. Melatonin: Nature's most versatile biological signal? *FEBS J.* **2006**, *273*, 2813–2838. [CrossRef] [PubMed]
52. Domingos, A.L.G.; Hermsdorff, H.H.M.; Bressan, J. Melatonin intake and potential chronobiological effects on human health. *Crit. Rev. Food Sci. Nutr.* **2017**, *59*, 133–140. [CrossRef]
53. Adamczyk-Sowa, M.; Pierzchala, K.; Sowa, P.; Mucha, S.; Sadowska-Bartosz, I.; Adamczyk, J.; Hartel, M. Melatonin Acts as Antioxidant and Improves Sleep in MS Patients. *Neurochem. Res.* **2014**, *39*, 1585–1593. [CrossRef]
54. Anisimov, V.N.; Popovich, I.G.; Zabezhinski, M.A.; Anisimov, S.V.; Vesnushkin, G.M.; Vinogradova, I.A. Melatonin as antioxidant, geroprotector and anticarcinogen. *Biochim. Biophys. Acta (BBA) Bioenerg.* **2006**, *1757*, 573–589. [CrossRef]
55. Kleszczyński, K.; Fischer, T.W. Melatonin and human skin aging. *Dermato-Endocrinology* **2012**, *4*, 245–252. [CrossRef]
56. Li, Y.; Li, S.; Zhou, Y.; Meng, X.; Zhang, J.-J.; Xu, D.-P.; Li, H.-B. Melatonin for the prevention and treatment of cancer. *Oncotarget* **2017**, *8*, 39896–39921. [CrossRef]
57. Fernandes, F.H.A.; Salgado, H.R.N. Gallic Acid: Review of the Methods of Determination and Quantification. *Crit. Rev. Anal. Chem.* **2016**, *46*, 257–265. [CrossRef]
58. Belmares, R.; Contreras-Esquivel, J.C.; Rodríguez-Herrera, R.; Coronel, A.R.; Aguilar, C.N. Microbial production of tannase: An enzyme with potential use in food industry. *LWT* **2004**, *37*, 857–864. [CrossRef]
59. Dewick, P.M.; Haslam, E. Phenol biosynthesis in higher plants. Gallic acid. *Biochem. J.* **1969**, *113*, 537–542. [CrossRef]
60. Kambourakis, S.; Draths, K.M.; Frost, J.W. Synthesis of Gallic Acid and Pyrogallol from Glucose: Replacing Natural Product Isolation with Microbial Catalysis. *J. Am. Chem. Soc.* **2000**, *122*, 9042–9043. [CrossRef]
61. Lau, S.; Wahn, J.; Schulz, G.; Sommerfeld, C.; Wahn, U. Placebo-controlled study of the mite allergen-reducing effect of tannic acid plus benzyl benzoate on carpets in homes of children with house dust mite sensitization and asthma. *Pediatr. Allergy Immunol.* **2002**, *13*, 31–36. [CrossRef] [PubMed]
62. Singleton, V.L. Naturally Occurring Food Toxicants: Phenolic Substances of Plant Origin Common in Foods. *Adv. Food Res.* **1981**, *27*, 149–242. [CrossRef] [PubMed]
63. Funatogawa, K.; Hayashi, S.; Shimomura, H.; Yoshida, T.; Hatano, T.; Ito, H.; Hirai, Y. Antibacterial Activity of Hydrolyzable Tannins Derived from Medicinal Plants againstHelicobacter pylori. *Microbiol. Immunol.* **2004**, *48*, 251–261. [CrossRef] [PubMed]
64. Lai, J.C.-Y.; Lai, H.-Y.; Rao, N.K.; Ng, S.-F. Treatment for diabetic ulcer wounds using a fern tannin optimized hydrogel formulation with antibacterial and antioxidative properties. *J. Ethnopharmacol.* **2016**, *189*, 277–289. [CrossRef]

65. Soleas, G.J.; Diamandis, E.P.; Goldberg, D.M. Resveratrol: A molecule whose time has come? And gone? *Clin. Biochem.* **1997**, *30*, 91–113. [CrossRef]
66. Sato, M.; Maulik, G.; Bagchi, D.; Das, D.K. Myocardial protection by Protykin, a novel extract oftrans-resveratrol and emodin. *Free. Radic. Res.* **2000**, *32*, 135–144. [CrossRef] [PubMed]
67. Trela, B.C.; Waterhouse, A.L. Resveratrol: Isomeric molar absorptivities and stability. *J. Agric. Food Chem.* **1996**, *44*, 1253–1257. [CrossRef]
68. Soleas, G.J.; Diamandis, E.P.; Goldberg, D.M. Wine as a Biological Fluid: History, Production, and Role in Disease Prevention. *J. Clin. Lab. Anal.* **1997**, *11*, 287–313. [CrossRef]
69. Siemann, E.H.; Creasy, L.L. Concentration of the Phytoalexin Resveratrol in Wine. *Am. J. Enol. Vitic.* **1992**, *43*, 49–52.
70. Athar, M.; Back, J.H.; Tang, X.; Kim, K.H.; Kopelovich, L.; Bickers, D.R.; Kim, A.L. Resveratrol: A review of preclinical studies for human cancer prevention. *Toxicol. Appl. Pharmacol.* **2007**, *224*, 274–283. [CrossRef] [PubMed]
71. Takaoka, M. The Phenolic Substances of White Hellebore (Veratrum Grandiflorum Hoes. Fil). IV. *Nippon. Kagaku Kaishi* **1940**, *61*, 96–98. [CrossRef]
72. Paul, B.; Masih, I.; Deopujari, J.; Charpentier, C. Occurrence of resveratrol and pterostilbene in age-old darakchasava, an ayurvedic medicine from India. *J. Ethnopharmacol.* **1999**, *68*, 71–76. [CrossRef]
73. Jeandet, P.; Douillet-Breuil, A.-C.; Bessis, R.; Debord, S.; Sbaghi, M.; Adrian, M. Phytoalexins from the Vitaceae: Biosynthesis, Phytoalexin Gene Expression in Transgenic Plants, Antifungal Activity, and Metabolism. *J. Agric. Food Chem.* **2002**, *50*, 2731–2741. [CrossRef]
74. Nonomura, S.; Kanagawa, H.; Makimoto, A. Chemical Constituents of Polygonaceous Plants. I. Studies on the Components of ko-j o-kon. (Polygonum Cuspidatum sieb. Et zucc). *Yakugaku Zasshi* **1963**, *83*, 988–990. [CrossRef]
75. Anjaneyulu, A.; Reddy, A.R.; Reddy, D.; Ward, R.; Adhikesavalu, D.; Cameron, T.S. Pacharin: A new dibenzo(2,3-6,7)oxepin derivative from bauhinia racemosa lamk. *Tetrahedron* **1984**, *40*, 4245–4252. [CrossRef]
76. Hanawa, F.; Tahara, S.; Mizutani, J. Antifungal stress compounds from Veratrum grandiflorum leaves treated with cupric chloride. *Phytochemistry* **1992**, *31*, 3005–3007. [CrossRef]
77. Chung, M.-I.; Teng, C.-M.; Cheng, K.-L.; Ko, F.-N.; Lin, C.-N. An Antiplatelet Principle ofVeratrum formosanum. *Planta Med.* **1992**, *58*, 274–276. [CrossRef]
78. Kumar, R.; Jyostna, D.; Krupadanam, G.; Srimannarayana, G. Phenanthrene and stilbenes from Pterolobium hexapetallum. *Phytochemistry* **1988**, *27*, 3625–3626. [CrossRef]
79. Hillis, W.; Hart, J.; Yazaki, Y. Polyphenols of Eucalyptus sideroxylon wood. *Phytochemistry* **1974**, *13*, 1591–1595. [CrossRef]
80. Hathway, D.E.; Seakins, J.W.T. Hydroxystilbenes of Eucalyptus wandoo. *Biochem. J.* **1959**, *72*, 369–374. [CrossRef]
81. Rolfs, C.-H.; Kindl, H. Stilbene Synthase and Chalcone Synthase: Two Different Constitutive Enzymes in Cultured Cells of Picea Excelsa. *Plant Physiol.* **1984**, *75*, 489–492. [CrossRef] [PubMed]
82. Langcake, P.; Pryce, R.J. A new class of phytoalexins from grapevines. *Cell. Mol. Life Sci.* **1977**, *33*, 151–152. [CrossRef] [PubMed]
83. Adrian, M.; Jeandet, P.; Veneau, J.; Weston, L.A.; Bessis, R. Biological Activity of Resveratrol, a Stilbenic Compound from Grapevines, Against Botrytis cinerea, the Causal Agent for Gray Mold. *J. Chem. Ecol.* **1997**, *23*, 1689–1702. [CrossRef]
84. Achmon, Y.; Fishman, A. The antioxidant hydroxytyrosol: Biotechnological production challenges and opportunities. *Appl. Microbiol. Biotechnol.* **2014**, *99*, 1119–1130. [CrossRef] [PubMed]
85. Bernini, R.; Merendino, N.; Romani, A.; Velotti, F. Naturally Occurring Hydroxytyrosol: Synthesis and Anticancer Potential. *Curr. Med. Chem.* **2013**, *20*, 655–670. [CrossRef]
86. Britton, J.; Davis, R.; O'Connor, K.E. Chemical, physical and biotechnological approaches to the production of the potent antioxidant hydroxytyrosol. *Appl. Microbiol. Biotechnol.* **2019**, *103*, 5957–5974. [CrossRef]
87. Stoll, A.; Renz, J.; Brack, A. Antibacterial Materials. VI. Isolation and Constitution of Echinacoside, a Glycoside from the Roots of Echinacea Angustifolia. *DC Helv. Chim. Acta* **1950**, *33*, 1877–1893. [CrossRef]
88. Sun, Y.; Zhou, D.; Shahidi, F. Antioxidant properties of tyrosol and hydroxytyrosol saturated fatty acid esters. *Food Chem.* **2018**, *245*, 1262–1268. [CrossRef] [PubMed]
89. Valenzuela, R.; Illesca, P.; Echeverría, F.; Espinosa, A.; Rincón-Cervera, M.Á.; Ortiz, M.; Hernandez-Rodas, M.C.; Valenzuela, A.; Videla, L.A. Molecular adaptations underlying the beneficial effects of hydroxytyrosol in the pathogenic alterations induced by a high-fat diet in mouse liver: PPAR-α and Nrf2 activation, and NF-κB down-regulation. *Food Funct.* **2017**, *8*, 1526–1537. [CrossRef] [PubMed]
90. Echeverría, F.; Ortiz, M.; Valenzuela, R.; Videla, L.A. Hydroxytyrosol and Cytoprotection: A Projection for Clinical Interventions. *Int. J. Mol. Sci.* **2017**, *18*, 930. [CrossRef] [PubMed]
91. Hazas, M.-C.L.D.L.; Godinho-Pereira, J.; Macià, A.; Almeida, A.F.; Ventura, M.R.; Motilva, M.-J.; Santos, C.N. Brain uptake of hydroxytyrosol and its main circulating metabolites: Protective potential in neuronal cells. *J. Funct. Foods* **2018**, *46*, 110–117. [CrossRef]
92. Wu, L.-X.; Xu, Y.-Y.; Yang, Z.-J.; Feng, Q. Hydroxytyrosol and olive leaf extract exert cardioprotective effects by inhibiting GRP78 and CHOP expression. *J. Biomed. Res.* **2018**, *32*, 371–379. [CrossRef]
93. Fuccelli, R.; Fabiani, R.; Rosignoli, P. Hydroxytyrosol Exerts Anti-Inflammatory and Anti-Oxidant Activities in a Mouse Model of Systemic Inflammation. *Molecules* **2018**, *23*, 3212. [CrossRef]

94. Yamada, K.; Ogawa, H.; Hara, A.; Yoshida, Y.; Yonezawa, Y.; Karibe, K.; Nghia, V.B.; Yoshimura, H.; Yamamoto, Y.; Yamada, M.; et al. Mechanism of the antiviral effect of hydroxytyrosol on influenza virus appears to involve morphological change of the virus. *Antivir. Res.* **2009**, *83*, 35–44. [CrossRef]
95. Kwan, H.Y.; Chao, X.; Su, T.; Fu, X.; Tse, A.K.W.; Fong, W.F.; Yu, Z.-L. The anticancer and antiobesity effects of Mediterranean diet. *Crit. Rev. Food Sci. Nutr.* **2015**, *57*, 82–94. [CrossRef]
96. Tuck, K.L.; Hayball, P.J. Major phenolic compounds in olive oil: Metabolism and health effects. *J. Nutr. Biochem.* **2002**, *13*, 636–644. [CrossRef]
97. Cicerale, S.; Lucas, L.J.; Keast, R.S.J. Antimicrobial, antioxidant and anti-inflammatory phenolic activities in extra virgin olive oil. *Curr. Opin. Biotechnol.* **2012**, *23*, 129–135. [CrossRef] [PubMed]
98. Fabiani, R.; Rosignoli, P.; De Bartolomeo, A.; Fuccelli, R.; Servili, M.; Montedoro, G.F.; Morozzi, G. Oxidative DNA Damage Is Prevented by Extracts of Olive Oil, Hydroxytyrosol, and Other Olive Phenolic Compounds in Human Blood Mononuclear Cells and HL60 Cells. *J. Nutr.* **2008**, *138*, 1411–1416. [CrossRef] [PubMed]
99. Luo, C.; Li, Y.; Wang, H.; Cui, Y.; Feng, Z.; Li, H.; Li, Y.; Wang, Y.; Wurtz, K.; Weber, P.; et al. Hydroxytyrosol promotes superoxide production and defects in autophagy leading to anti-proliferation and apoptosis on human prostate cancer cells. *Curr. Cancer Drug Targets* **2013**, *13*, 625–639. [CrossRef] [PubMed]
100. Schöpf, C.; Göttmann, G.; Meisel, E.-M.; Neuroth†, L. Über β-(3,4-Dioxyphenyl)-äthylalkohol. *Eur. J. Org. Chem.* **1949**, *563*, 86–93. [CrossRef]
101. Servili, M.; Baldioli, M.; Selvaggini, R.; Macchioni, A.; Montedoro, G. Phenolic Compounds of Olive Fruit: One- and Two-Dimensional Nuclear Magnetic Resonance Characterization of Nüzhenide and Its Distribution in the Constitutive Parts of Fruit. *J. Agric. Food Chem.* **1998**, *47*, 12–18. [CrossRef]
102. Hertog, M.G.L.; Feskens, E.J.M.; Kromhout, D.; Hertog, M.G.L.; Hollman, P.C.H.; Hertog, M.G.L.; Katan, M.B. Dietary Antioxidant Flavonoids and Risk of Coronary Heart Disease: The Zutphen Elderly Study. *Lancet* **1993**, *342*, 1007–1011. [CrossRef]
103. Wiseman, H. The Bioavailability of Non-Nutrient Plant Factors: Dietary Flavonoids and Phyto-Oestrogens. *Proc. Nutr. Soc.* **1999**, *58*, 139–146. [CrossRef] [PubMed]
104. Hertog, M.G.L.; Hollman, P.C.H.; Katan, M.B.; Kromhout, D. Intake of potentially anticarcinogenic flavonoids and their determinants in adults in the Netherlands. *Nutr. Cancer* **1993**, *20*, 21–29. [CrossRef]
105. Hertog, M.G.; Sweetnam, P.M.; Fehily, A.M.; Elwood, P.C.; Kromhout, D. Antioxidant flavonols and ischemic heart disease in a Welsh population of men: The Caerphilly Study. *Am. J. Clin. Nutr.* **1997**, *65*, 1489–1494. [CrossRef]
106. Martin-Moreno, J.M.; Willett, W.C.; Gorgojo, L.; Banegas, J.R.; Rodriguez-Artalejo, F.; Fernandez-Rodriguez, J.C.; Maisonneuve, P.; Boyle, P. Dietary fat, olive oil intake and breast cancer risk. *Int. J. Cancer* **1994**, *58*, 774–780. [CrossRef]
107. Keli, S.O.; Hertog, M.G.L.; Feskens, E.J.M.; Kromhout, D. Dietary Flavonoids, Antioxidant Vitamins, and Incidence of Stroke: The Zutphen Study. *Arch. Intern. Med.* **1996**, *156*, 637–642. [CrossRef]
108. Lipworth, L.; Martínez, M.E.; Angell, J.; Hsieh, C.-C.; Trichopoulos, D. Olive Oil and Human Cancer: An Assessment of the Evidence. *Prev. Med.* **1997**, *26*, 181–190. [CrossRef] [PubMed]
109. Wagner, K.-H.; Kamal-Eldin, A.; Elmadfa, I. Gamma-Tocopherol—An Underestimated Vitamin? *Ann. Nutr. Metab.* **2004**, *48*, 169–188. [CrossRef] [PubMed]
110. Jiang, Q.; Christen, S.; Shigenaga, M.K.; Ames, B.N. γ-Tocopherol, the major form of vitamin E in the US diet, deserves more attention. *Am. J. Clin. Nutr.* **2001**, *74*, 714–722. [CrossRef] [PubMed]
111. Rigotti, A. Absorption, transport, and tissue delivery of vitamin E. *Mol. Asp. Med.* **2007**, *28*, 423–436. [CrossRef] [PubMed]
112. David, R.; Lide, E. *CRC Handbook of Chemistry and Physics, Internet Version 2007*, 87th ed.; Lord & Taylor: Boca Raton, FL, USA, 2007; Volume 129, p. 724. [CrossRef]
113. Podszun, M.; Frank, J. Vitamin E–drug interactions: Molecular basis and clinical relevance. *Nutr. Res. Rev.* **2014**, *27*, 215–231. [CrossRef]
114. Mène-Saffrané, L. Vitamin E Biosynthesis and Its Regulation in Plants. *Antioxidants* **2017**, *7*, 2. [CrossRef] [PubMed]
115. Padayatty, S.J.; Katz, A.; Wang, Y.; Eck, P.; Kwon, O.; Lee, J.-H.; Chen, S.; Corpe, C.; Dutta, A.; Dutta, S.K.; et al. Vitamin C as an Antioxidant: Evaluation of Its Role in Disease Prevention. *J. Am. Coll. Nutr.* **2003**, *22*, 18–35. [CrossRef]
116. Institute of Medicine (US) Panel on Dietary Antioxidants and Related Compounds. *Dietary Reference Intakes for Vitamin C, Vitamin E, Selenium, and Carotenoids*; National Academies Press (US): Washington, DC, USA, 2000. [CrossRef]
117. Waugh, W.A.; King, C.G. Isolation and Identification of Vitamin C. *Nutr. Rev.* **2009**, *34*, 81–83. [CrossRef]
118. Gallie, D.R. L-Ascorbic Acid: A Multifunctional Molecule Supporting Plant Growth and Development. *Scientifica* **2013**, *2013*, 795964. [CrossRef] [PubMed]
119. Siendones, E.; González-Reyes, J.A.; Santos-Ocaña, C.; Navas, P.; Córdoba, F. Biosynthesis of Ascorbic Acid in Kidney Bean.l-Galactono-γ-Lactone Dehydrogenase Is an Intrinsic Protein Located at the Mitochondrial Inner Membrane. *Plant Physiol.* **1999**, *120*, 907–912. [CrossRef]
120. Wheeler, G.L.; Jones, M.A.; Smirnoff, N. The biosynthetic pathway of vitamin C in higher plants. *Nat. Cell Biol.* **1998**, *393*, 365–369. [CrossRef]
121. Priyadarsini, K.I. The Chemistry of Curcumin: From Extraction to Therapeutic Agent. *Molecules* **2014**, *19*, 20091–20112. [CrossRef] [PubMed]

122. Aggarwal, B.B.; Kumar, A.; Bharti, A.C. Anticancer potential of curcumin: Preclinical and clinical studies. *Anticancer. Res.* **2003**, *23*, 363–398. [PubMed]
123. Majeed, S. The State of the Curcumin Market. Available online: https://www.grandviewresearch.com/industry-analysis/turmeric-extract-curcumin-market (accessed on 22 March 2021).
124. Vera-Ramirez, L.; Pérez-Lopez, P.; Varela-Lopez, A.; Ramirez-Tortosa, M.; Battino, M.; Quiles, J.L. Curcumin and liver disease. *BioFactors* **2013**, *39*, 88–100. [CrossRef]
125. Wright, L.; Frye, J.; Gorti, B.; Timmermann, B.; Funk, J. Bioactivity of Turmeric-derived Curcuminoids and Related Metabolites in Breast Cancer. *Curr. Pharm. Des.* **2013**, *19*, 6218–6225. [CrossRef]
126. Lestari, M.L.; Indrayanto, G. Curcumin. In *Profiles of Drug Substances, Excipients and Related Methodology*; Elsevier: Amsterdam, The Netherlands, 2014; Volume 39, pp. 113–204. [CrossRef]
127. Pulido-Moran, M.; Moreno-Fernandez, J.; Ramirez-Tortosa, C.; Ramirez-Tortosa, M. Curcumin and Health. *Molecules* **2016**, *21*, 264. [CrossRef]
128. Hewlings, S.J.; Kalman, D.S. Curcumin: A Review of Its Effects on Human Health. *Foods* **2017**, *6*, 92. [CrossRef]
129. Nelson, K.M.; Dahlin, J.L.; Bisson, J.; Graham, J.; Pauli, G.F.; Walters, M.A. The Essential Medicinal Chemistry of Curcumin: Miniperspective. *J. Med. Chem.* **2017**, *60*, 1620–1637. [CrossRef]
130. Nelson, K.M.; Dahlin, J.L.; Bisson, J.; Graham, J.; Pauli, G.F.; Walters, M.A. Curcumin May (Not) Defy Science. *ACS Med. Chem. Lett.* **2017**, *8*, 467–470. [CrossRef]
131. Schiborr, C.; Kocher, A.; Behnam, D.; Jandasek, J.; Toelstede, S.; Frank, J. The oral bioavailability of curcumin from micronized powder and liquid micelles is significantly increased in healthy humans and differs between sexes. *Mol. Nutr. Food Res.* **2014**, *58*, 516–527. [CrossRef]
132. Araiza-Calahorra, A.; Akhtar, M.; Sarkar, A. Recent advances in emulsion-based delivery approaches for curcumin: From encapsulation to bioaccessibility. *Trends Food Sci. Technol.* **2018**, *71*, 155–169. [CrossRef]
133. Sanidad, K.Z.; Sukamtoh, E.; Xiao, H.; McClements, D.J.; Zhang, G. Curcumin: Recent Advances in the Development of Strategies to Improve Oral Bioavailability. *Annu. Rev. Food Sci. Technol.* **2019**, *10*, 597–617. [CrossRef] [PubMed]
134. Kita, T.; Imai, S.; Sawada, H.; Kumagai, H.; Seto, H. The Biosynthetic Pathway of Curcuminoid in Turmeric (*Curcuma longa*) as Revealed by13C-Labeled Precursors. *Biosci. Biotechnol. Biochem.* **2008**, *72*, 1789–1798. [CrossRef] [PubMed]
135. Chung, K.-T.; Wong, T.Y.; Wei, C.-I.; Huang, Y.-W.; Lin, Y. Tannins and Human Health: A Review. *Crit. Rev. Food Sci. Nutr.* **1998**, *38*, 421–464. [CrossRef] [PubMed]
136. Gambini, J.; López-Grueso, R.; Olaso-González, G.; Inglés, M.; Abdelazid, K.; El Alami, M.; Bonet-Costa, V.; Borrás, C.; Viña, J. Resveratrol: Distribución, propiedades y perspectivas. *Rev. Española Geriatr. Gerontol.* **2013**, *48*, 79–88. [CrossRef]
137. Alsemeh, A.E.; Samak, M.A.; El-Fatah, S.S.A. Therapeutic prospects of hydroxytyrosol on experimentally induced diabetic testicular damage: Potential interplay with AMPK expression. *Cell Tissue Res.* **2019**, *380*, 173–189. [CrossRef]
138. Reiter, E.; Jiang, Q.; Christen, S. Anti-inflammatory properties of α- and γ-tocopherol. *Mol. Asp. Med.* **2007**, *28*, 668–691. [CrossRef]
139. Wang, X.; Li, Q.; Xu, J.; Wu, S.; Xiao, T.; Hao, J.; Yu, P.; Mao, L. Rational Design of Bioelectrochemically Multifunctional Film with Oxidase, Ferrocene, and Graphene Oxide for Development of in Vivo Electrochemical Biosensors. *Anal. Chem.* **2016**, *88*, 5885–5891. [CrossRef] [PubMed]
140. Rezaei, A.; Akhavan, O.; Hashemi, E.; Shamsara, M. Ugi Four-Component Assembly Process: An Efficient Approach for One-Pot Multifunctionalization of Nanographene Oxide in Water and Its Application in Lipase Immobilization. *Chem. Mater.* **2016**, *28*, 3004–3016. [CrossRef]
141. Nulakani, N.V.R.; Subramanian, V. A Theoretical Study on the Design, Structure, and Electronic Properties of Novel Forms of Graphynes. *J. Phys. Chem. C* **2016**, *120*, 15153–15161. [CrossRef]
142. Bai, H.; Li, C.; Shi, G. Functional Composite Materials Based on Chemically Converted Graphene. *Adv. Mater.* **2011**, *23*, 1089–1115. [CrossRef]
143. Strom, T.A.; Dillon, E.P.; Hamilton, C.E.; Barron, A.R. Nitrene addition to exfoliated graphene: A one-step route to highly functionalized graphene. *Chem. Commun.* **2010**, *46*, 4097–4099. [CrossRef] [PubMed]
144. Xu, Y.; Liu, Z.; Zhang, X.; Wang, Y.; Tian, J.; Huang, Y.; Ma, Y.; Zhang, X.; Chen, Y. A Graphene Hybrid Material Covalently Functionalized with Porphyrin: Synthesis and Optical Limiting Property. *Adv. Mater.* **2009**, *21*, 1275–1279. [CrossRef]
145. Eskandari, P.; Abousalman-Rezvani, Z.; Roghani-Mamaqani, H.; Salami-Kalajahi, M.; Mardani, H. Polymer grafting on graphene layers by controlled radical polymerization. *Adv. Colloid Interface Sci.* **2019**, *273*, 102021. [CrossRef]
146. Díez-Pascual, A.M.; García-García, D.; Andrés, M.P.S.; Vera, S. Determination of riboflavin based on fluorescence quenching by graphene dispersions in polyethylene glycol. *RSC Adv.* **2016**, *6*, 19686–19699. [CrossRef]
147. Krishnan, S.K.; Singh, E.; Singh, P.; Meyyappan, M.; Nalwa, H.S. A review on graphene-based nanocomposites for electrochemical and fluorescent biosensors. *RSC Adv.* **2019**, *9*, 8778–8881. [CrossRef]
148. Zhou, M.; Zhai, Y.; Dong, S. Electrochemical Sensing and Biosensing Platform Based on Chemically Reduced Graphene Oxide. *Anal. Chem.* **2009**, *81*, 5603–5613. [CrossRef]
149. Coroş, M.; Pruneanu, S.; Staden, R.-I.S.-V. Review—Recent Progress in the Graphene-Based Electrochemical Sensors and Biosensors. *J. Electrochem. Soc.* **2020**, *167*, 037528. [CrossRef]
150. Li, D.; Zhang, W.; Yu, X.; Wang, Z.; Su, Z.; Wei, G. When biomolecules meet graphene: From molecular level interactions to material design and applications. *Nanoscale* **2016**, *8*, 19491–19509. [CrossRef]

151. Domi, B.; Rumbo, C.; García-Tojal, J.; Elena Sima, L.; Negroiu, G.; Tamayo-Ramos, J.A. Interaction Analysis of Commercial Graphene Oxide Nanoparticles with Unicellular Systems and Biomolecules. *Int. J. Mol. Sci.* **2019**, *21*, 205. [CrossRef] [PubMed]
152. Carbone, M.; Gorton, L.; Antiochia, R. An Overview of the Latest Graphene-Based Sensors for Glucose Detection: The Effects of Graphene Defects. *Electroanalysis* **2015**, *27*, 16–31. [CrossRef]
153. Bitounis, D.; Ali-Boucetta, H.; Hong, B.H.; Min, D.-H.; Kostarelos, K. Prospects and Challenges of Graphene in Biomedical Applications. *Adv. Mater.* **2013**, *25*, 2258–2268. [CrossRef]
154. Karunanithi, D.; Radhakrishna, A.; Sivaraman, K.P.; Biju, V.M.N. Quantitative determination of melatonin in milk by LC-MS/MS. *J. Food Sci. Technol.* **2014**, *51*, 805–812. [CrossRef]
155. Vitale, A.A.; Ferrari, C.C.; Aldana, H.; Affanni, J.M. Highly sensitive method for the determination of melatonin by normal-phase high-performance liquid chromatography with fluorometric detection. *J. Chromatogr. B Biomed. Sci. Appl.* **1996**, *681*, 381–384. [CrossRef]
156. Simonin, G.; Bru, L.; Lelièvre, E.; Jeanniot, J.-P.; Bromet, N.; Walther, B.; Boursier-Neyret, C. Determination of melatonin in biological fluids in the presence of the melatonin agonist S 20098: Comparison of immunological techniques and GC-MS methods. *J. Pharm. Biomed. Anal.* **1999**, *21*, 591–601. [CrossRef]
157. Pucci, V.; Ferranti, A.; Mandrioli, R.; Raggi, M.A. Determination of melatonin in commercial preparations by micellar electrokinetic chromatography and spectrofluorimetry. *Anal. Chim. Acta* **2003**, *488*, 97–105. [CrossRef]
158. Lu, J.; Lau, C.; Lee, M.K.; Kai, M. Simple and convenient chemiluminescence method for the determination of melatonin. *Anal. Chim. Acta* **2002**, *455*, 193–198. [CrossRef]
159. Niu, J.; Zhang, X.; Qin, P.; Yang, Y.; Tian, S.; Yang, H.; Lu, M. Simultaneous Determination of Melatonin, l-Tryptophan, and two l-Tryptophan-Derived Esters in Food by HPLC with Graphene Oxide/SiO2 Nanocomposite as the Adsorbent. *Food Anal. Methods* **2018**, *11*, 2438–2446. [CrossRef]
160. Apetrei, I.M.; Apetrei, C. Voltammetric determination of melatonin using a graphene-based sensor in pharmaceutical products. *Int. J. Nanomed.* **2016**, *11*, 1859–1866. [CrossRef] [PubMed]
161. Miccoli, A.; Restani, P.; Floroian, L.; Taus, N.; Badea, M.; Cioca, G.; Bungau, S. Sensitive Electrochemical Detection Method of Melatonin in Food Supplements. *Rev. Chim.* **2018**, *69*, 854–859. [CrossRef]
162. Gomez, F.J.V.; Martín, A.; Silva, M.F.; Escarpa, A. Screen-printed electrodes modified with carbon nanotubes or graphene for simultaneous determination of melatonin and serotonin. *Microchim. Acta* **2015**, *182*, 1925–1931. [CrossRef]
163. Gupta, P.; Goyal, R.N. Graphene and Co-polymer composite based molecularly imprinted sensor for ultratrace determination of melatonin in human biological fluids. *RSC Adv.* **2015**, *5*, 40444–40454. [CrossRef]
164. Bagheri, H.; Afkhami, A.; Hashemi, P.; Ghanei, M. Simultaneous and sensitive determination of melatonin and dopamine with Fe3O4 nanoparticle-decorated reduced graphene oxide modified electrode. *RSC Adv.* **2015**, *5*, 21659–21669. [CrossRef]
165. Zeinali, H.; Bagheri, H.; Monsef-Khoshhesab, Z.; Khoshsafar, H.; Hajian, A. Nanomolar simultaneous determination of tryptophan and melatonin by a new ionic liquid carbon paste electrode modified with SnO2-Co3O4@rGO nanocomposite. *Mater. Sci. Eng. C* **2017**, *71*, 386–394. [CrossRef]
166. Tadayon, F.; Sepehri, Z. A new electrochemical sensor based on a nitrogen-doped graphene/CuCo2O4 nanocomposite for simultaneous determination of dopamine, melatonin and tryptophan. *RSC Adv.* **2015**, *5*, 65560–65568. [CrossRef]
167. Liu, Y.; Li, M.; Li, H.; Wang, G.; Long, Y.; Li, A.; Yang, B. In Situ Detection of Melatonin and Pyridoxine in Plants Using a CuO–Poly(l-lysine)/Graphene-Based Electrochemical Sensor. *ACS Sustain. Chem. Eng.* **2019**, *7*, 19537–19545. [CrossRef]
168. Esfandiar, A.; Akhavan, O.; Irajizad, A. Melatonin as a powerful bio-antioxidant for reduction of graphene oxide. *J. Mater. Chem.* **2011**, *21*, 10907–10914. [CrossRef]
169. Akhavan, O.; Ghaderi, E.; Esfandiar, A. Wrapping Bacteria by Graphene Nanosheets for Isolation from Environment, Reactivation by Sonication, and Inactivation by Near-Infrared Irradiation. *J. Phys. Chem. B* **2011**, *115*, 6279–6288. [CrossRef] [PubMed]
170. Al-Ansi, N.; Salah, A.; Bawa, M.; Adlat, S.; Yasmin, I.; Abdallah, A.; Qi, B. 3D nitrogen-doped porous graphene aerogel as high-performance electrocatalyst for determination of gallic acid. *Microchem. J.* **2020**, *155*, 104706. [CrossRef]
171. Chikere, C.O.; Faisal, N.H.; Kong-Thoo-Lin, P.; Fernandez, C. Interaction between Amorphous Zirconia Nanoparticles and Graphite: Electrochemical Applications for Gallic Acid Sensing Using Carbon Paste Electrodes in Wine. *Nanomaterials* **2020**, *10*, 537. [CrossRef] [PubMed]
172. Puangjan, A.; Chaiyasith, S. An efficient ZrO2/Co3O4/reduced graphene oxide nanocomposite electrochemical sensor for simultaneous determination of gallic acid, caffeic acid and protocatechuic acid natural antioxidants. *Electrochim. Acta* **2016**, *211*, 273–288. [CrossRef]
173. Ganesh, H.V.S.; Patel, B.R.; Fini, H.; Chow, A.M.; Kerman, K. Electrochemical Detection of Gallic Acid-Capped Gold Nanoparticles Using a Multiwalled Carbon Nanotube-Reduced Graphene Oxide Nanocomposite Electrode. *Anal. Chem.* **2019**, *91*, 10116–10124. [CrossRef]
174. Gao, F.; Zheng, D.; Tanaka, H.; Zhan, F.; Yuan, X.; Gao, F.; Wang, Q. An Electrochemical Sensor for Gallic Acid Based on Fe2O3/Electro-Reduced Graphene Oxide Composite: Estimation for the Antioxidant Capacity Index of Wines. *Mater. Sci. Eng. C Mater. Biol. Appl.* **2015**, *57*, 279–287. [CrossRef]
175. Ma, W.; Han, D.; Gan, S.; Zhang, N.; Liu, S.; Wu, T.; Zhang, Q.; Dong, X.; Niu, L. Rapid and specific sensing of gallic acid with a photoelectrochemical platform based on polyaniline–reduced graphene oxide–TiO2. *Chem. Commun.* **2013**, *49*, 7842. [CrossRef] [PubMed]

176. Benítez-Martínez, S.; Valcárcel, M. Graphene quantum dots as sensor for phenols in olive oil. *Sens. Actuators B Chem.* **2014**, *197*, 350–357. [CrossRef]
177. Liu, G.; Yu, R.; Lan, T.; Liu, Z.; Zhang, P.; Liang, R. Gallic acid-functionalized graphene hydrogel as adsorbent for removal of chromium (iii) and organic dye pollutants from tannery wastewater. *RSC Adv.* **2019**, *9*, 27060–27068. [CrossRef]
178. Croitoru, A.; Oprea, O.; Nicoara, A.; Trusca, R.; Radu, M.; Neacsu, I.; Ficai, D.; Ficai, A.; Andronescu, E. Multifunctional Platforms Based on Graphene Oxide and Natural Products. *Medicines* **2019**, *55*, 230. [CrossRef]
179. Sinduja, B.; John, S.A. Sensitive determination of tannic acid using blue luminescent graphene quantum dots as fluorophore. *RSC Adv.* **2016**, *6*, 59900–59906. [CrossRef]
180. Palisoc, S.T.; Cansino, E.J.F.; Dy, I.M.O.; Razal, C.F.A.; Reyes, K.C.N.; Racines, L.R.; Natividad, M.T. Electrochemical determination of tannic acid using graphite electrodes sourced from waste zinc-carbon batteries. *Sens. BioSens. Res.* **2020**, *28*, 100326. [CrossRef]
181. Zhao, S.; Xie, S.; Zhao, Z.; Zhang, J.; Li, L.; Xin, Z. Green and High-Efficiency Production of Graphene by Tannic Acid-Assisted Exfoliation of Graphite in Water. *ACS Sustain. Chem. Eng.* **2018**, *6*, 7652–7661. [CrossRef]
182. Luo, J.; Lai, J.; Zhang, N.; Liu, Y.; Liu, R.; Liu, X. Tannic Acid Induced Self-Assembly of Three-Dimensional Graphene with Good Adsorption and Antibacterial Properties. *ACS Sustain. Chem. Eng.* **2016**, *4*, 1404–1413. [CrossRef]
183. Lim, M.-Y.; Shin, H.; Shin, D.M.; Lee, S.-S.; Lee, J.-C. Poly(vinyl alcohol) nanocomposites containing reduced graphene oxide coated with tannic acid for humidity sensor. *Polymer* **2016**, *84*, 89–98. [CrossRef]
184. Yoo, S.; Li, X.; Wu, Y.; Liu, W.; Wang, X.; Yi, W. Ammonia Gas Detection by Tannic Acid Functionalized and Reduced Graphene Oxide at Room Temperature. *J. Nanomater.* **2014**, *2014*, 1–6. [CrossRef]
185. Li, C.-P.; Tan, S.; Ye, H.; Cao, J.; Zhao, H. A novel fluorescence assay for resveratrol determination in red wine based on competitive host-guest recognition. *Food Chem.* **2019**, *283*, 191–198. [CrossRef] [PubMed]
186. Zhang, C.; Ping, J.; Ying, Y. Evaluation of trans-resveratrol level in grape wine using laser-induced porous graphene-based electrochemical sensor. *Sci. Total. Environ.* **2020**, *714*, 136687. [CrossRef]
187. Liu, L.; Zhou, Y.; Kang, Y.; Huang, H.; Li, C.; Xu, M.; Ye, B. Electrochemical Evaluation of Trans -Resveratrol Levels in Red Wine Based on the Interaction between Resveratrol and Graphene. *J. Anal. Methods Chem.* **2017**, *2017*, 5749025. [CrossRef] [PubMed]
188. Gurunathan, S.; Han, J.W.; Kim, E.S.; Park, J.H.; Kim, J.-H. Reduction of graphene oxide by resveratrol: A novel and simple biological method for the synthesis of an effective anticancer nanotherapeutic molecule. *Int. J. Nanomed.* **2015**, *10*, 2951–2969. [CrossRef] [PubMed]
189. Salas, E.C.; Sun, Z.; Lüttge, A.; Tour, J.M. Reduction of Graphene Oxide via Bacterial Respiration. *ACS Nano* **2010**, *4*, 4852–4856. [CrossRef]
190. Akhavan, O.; Ghaderi, E.; Akhavan, A. Size-dependent genotoxicity of graphene nanoplatelets in human stem cells. *Biomaterials* **2012**, *33*, 8017–8025. [CrossRef]
191. Akhavan, O.; Ghaderi, E.; Emamy, H.; Akhavan, F. Genotoxicity of graphene nanoribbons in human mesenchymal stem cells. *Carbon* **2013**, *54*, 419–431. [CrossRef]
192. He, X.-P.; Deng, Q.; Cai, L.; Wang, C.-Z.; Zang, Y.; Li, J.; Chen, G.-R.; Tian, H. Fluorogenic Resveratrol-Confined Graphene Oxide for Economic and Rapid Detection of Alzheimer's Disease. *ACS Appl. Mater. Interfaces* **2014**, *6*, 5379–5382. [CrossRef] [PubMed]
193. Gomez, F.J.V.; Spisso, A.; Silva, M.F. Pencil graphite electrodes for improved electrochemical detection of oleuropein by the combination of Natural Deep Eutectic Solvents and graphene oxide. *Electrophoresis* **2017**, *38*, 2704–2711. [CrossRef]
194. Kurtulbaş, E.; Yazar, S.; Ortaboy, S.; Atun, G.; Şahin, S. Evaluation of the phenolic antioxidants of olive (*Olea europaea*) leaf extract obtained by a green approach: Use of reduced graphene oxide for electrochemical analysis. *Chem. Eng. Commun.* **2019**, *207*, 920–932. [CrossRef]
195. Yazar, S.; Kurtulbaş, E.; Ortaboy, S.; Atun, G.; Şahin, S. Screening of the antioxidant properties of olive (*Olea europaea*) leaf extract by titanium based reduced graphene oxide electrode. *Korean J. Chem. Eng.* **2019**, *36*, 1184–1192. [CrossRef]
196. Şahin, S.; Ciğeroğlu, Z.; Özdemir, O.K.; Bilgin, M.; Elhussein, E.; Gülmez, Ö. Recovery of hydroxytyrosol onto graphene oxide nanosheets: Equilibrium and kinetic models. *J. Mol. Liq.* **2019**, *285*, 213–222. [CrossRef]
197. Şahin, S.; Elhussein, E.A.A.; Salam, M.A.; Bayazit, Ş.S. Recovery of polyphenols from water using Zr-based metal-organic frameworks and their nanocomposites with graphene nanoplatelets. *J. Ind. Eng. Chem.* **2019**, *78*, 164–171. [CrossRef]
198. Baioun, A.; Kellawi, H.; Falah, A. A Modified Electrode by a Facile Green Preparation of Reduced Graphene Oxide Utilizing Olive Leaves Extract. *Carbon Lett.* **2017**, *24*, 47–54. [CrossRef]
199. Filik, H.; Avan, A.A.; Aydar, S.; Çakar, Ş. Determination of Tocopherol Using Reduced Graphene Oxide-Nafion Hybrid-Modified Electrode in Pharmaceutical Capsules and Vegetable Oil Samples. *Food Anal. Methods* **2016**, *9*, 1745–1753. [CrossRef]
200. Liu, H.; Fang, G.; Zhu, H.; Li, C.; Liu, C.; Wang, S. A novel ionic liquid stabilized molecularly imprinted optosensing material based on quantum dots and graphene oxide for specific recognition of vitamin E. *Biosens. Bioelectron.* **2013**, *47*, 127–132. [CrossRef]
201. Liu, R.; Yang, R.; Qu, C.; Mao, H.; Hu, Y.; Li, J.; Qu, L. Synthesis of glycine-functionalized graphene quantum dots as highly sensitive and selective fluorescent sensor of ascorbic acid in human serum. *Sens. Actuators B Chem.* **2017**, *241*, 644–651. [CrossRef]
202. De Faria, L.V.; Lisboa, T.P.; De Farias, D.M.; Araujo, F.M.; Machado, M.M.; De Sousa, R.A.; Matos, M.A.C.; Muñoz, R.A.A.; Matos, R.C. Direct analysis of ascorbic acid in food beverage samples by flow injection analysis using reduced graphene oxide sensor. *Food Chem.* **2020**, *319*, 126509. [CrossRef]

203. Swamy, K.; Gunnam, R.; Kesamsetty, V.; Manjunatha, H.; Janardan, S.; Chandra, K.; Naidu, K.C.B.; Ramesh, S.; Kothamas; Babu, S.; et al. Simultaneous Detection of Dopamine, Tyrosine and Ascorbic Acid Using NiO/Graphene Modified Graphite Electrode. *Biointerface Res. Appl. Chem.* **2020**, *10*. [CrossRef]
204. Kunpatee, K.; Traipop, S.; Chailapakul, O.; Chuanuwatanakul, S. Simultaneous determination of ascorbic acid, dopamine, and uric acid using graphene quantum dots/ionic liquid modified screen-printed carbon electrode. *Sens. Actuators B Chem.* **2020**, *314*, 128059. [CrossRef]
205. Ji, D.; Liu, Z.; Liu, L.; Low, S.S.; Lu, Y.; Yu, X.; Zhu, L.; Li, C.; Liu, Q. Smartphone-based integrated voltammetry system for simultaneous detection of ascorbic acid, dopamine, and uric acid with graphene and gold nanoparticles modified screen-printed electrodes. *Biosens. Bioelectron.* **2018**, *119*, 55–62. [CrossRef] [PubMed]
206. Fu, L.; Wang, A.; Lai, G.; Su, W.; Malherbe, F.; Yu, J.; Lin, C.-T.; Yu, A. Defects regulating of graphene ink for electrochemical determination of ascorbic acid, dopamine and uric acid. *Talanta* **2018**, *180*, 248–253. [CrossRef] [PubMed]
207. Shi, L.; Wang, Z.; Chen, X.; Yang, G.; Liu, W. Reduced Graphene Oxide/Polydopamine/Gold Electrode as Elecrochemical Sensor for Simultaneous Determination of Ascorbic Acid, Dopamine, and Uric Acid. *Int. J. Electrochem. Sci.* **2019**, *14*, 8882–8891. [CrossRef]
208. Li, S.; Ma, Y.; Liu, Y.; Xin, G.; Wang, M.; Zhang, Z.; Liu, Z. Electrochemical sensor based on a three dimensional nanostructured MoS2 nanosphere-PANI/reduced graphene oxide composite for simultaneous detection of ascorbic acid, dopamine, and uric acid. *RSC Adv.* **2019**, *9*, 2997–3003. [CrossRef]
209. Salahandish, R.; Ghaffarinejad, A.; Naghib, S.M.; Niyazi, A.; Majidzadeh, K.; Janmaleki, M.; Sanati-Nezhad, A. Sandwich-structured nanoparticles-grafted functionalized graphene based 3D nanocomposites for high-performance biosensors to detect ascorbic acid biomolecule. *Sci. Rep.* **2019**, *9*, 1226. [CrossRef] [PubMed]
210. Renjini, S.; Pinky, A.; Aparna, S.; Anitha, K.V. Graphene-Palladium Composite for the Simultaneous Electrochemical Determination of Epinephrine, Ascorbic acid and Uric Acid. *J. Electrochem. Soc.* **2019**, *166*, B1321–B1329. [CrossRef]
211. Kucukkolbasi, S.; Erdogan, Z.O.; Baslak, C.; Sogut, D.; Kus, M. A Highly Sensitive Ascorbic Acid Sensor Based on Graphene Oxide/CdTe Quantum Dots-Modified Glassy Carbon Electrode. *Russ. J. Electrochem.* **2019**, *55*, 107–114. [CrossRef]
212. Chen, Z.; Zhang, Y.; Zhang, J.; Zhou, J. Electrochemical Sensing Platform Based on Three-Dimensional Holey Graphene for Highly Selective and Ultra-Sensitive Detection of Ascorbic Acid, Uric Acid, and Nitrite. *J. Electrochem. Soc.* **2019**, *166*, B787–B792. [CrossRef]
213. Gao, J.; Liu, F.; Liu, Y.; Ma, N.; Wang, Z.; Zhang, X. Environment-Friendly Method To Produce Graphene That Employs Vitamin C and Amino Acid. *Chem. Mater.* **2010**, *22*, 2213–2218. [CrossRef]
214. Fernández-Merino, M.J.; Guardia, L.; Paredes, J.I.; Villar-Rodil, S.; Solís-Fernández, P.; Martínez-Alonso, A.; Tascón, J.M.D. Vitamin C Is an Ideal Substitute for Hydrazine in the Reduction of Graphene Oxide Suspensions. *J. Phys. Chem. C* **2010**, *114*, 6426–6432. [CrossRef]
215. Zhang, J.; Yang, H.; Shen, G.; Cheng, P.; Zhang, J.; Guo, S. Reduction of graphene oxide vial-ascorbic acid. *Chem. Commun.* **2010**, *46*, 1112–1114. [CrossRef] [PubMed]
216. Rahimnejad, M.; Zokhtareh, R.; Moghadamnia, A.A.; Asghary, M. An Electrochemical Sensor Based on Reduced Graphene Oxide Modified Carbon Paste Electrode for Curcumin Determination in Human Blood Serum. *Port. Electrochim. Acta* **2020**, *38*, 29–42. [CrossRef]
217. Zhang, D.; Ouyang, X.; Ma, J.; Li, L.; Zhang, Y. Electrochemical Behavior and Voltammetric Determination of Curcumin at Electrochemically Reduced Graphene Oxide Modified Glassy Carbon Electrode. *Electroanalysis* **2016**, *28*, 749–756. [CrossRef]
218. Li, K.; Li, Y.; Yang, L.; Wang, L.; Ye, B. The electrochemical characterization of curcumin and its selective detection in Curcuma using a graphene-modified electrode. *Anal. Methods* **2014**, *6*, 7801–7808. [CrossRef]
219. Dey, N.; Devasena, T.; Sivalingam, T. A Comparative evaluation of Graphene oxide based materials for Electrochemical non-enzymatic sensing of Curcumin. *Mater. Res. Express* **2018**, *5*, 025406. [CrossRef]
220. Kotan, G.; Kardaş, F.; Yokuş, Ö.A.; Akyıldırım, O.; Saral, H.; Eren, T.; Yola, M.L.; Atar, N. A novel determination of curcumin via Ru@Au nanoparticle decorated nitrogen and sulfur-functionalized reduced graphene oxide nanomaterials. *Anal. Methods* **2016**, *8*, 401–408. [CrossRef]
221. Marković, Z.M.; Kepić, D.P.; Matijašević, D.M.; Pavlović, V.B.; Jovanović, S.P.; Stanković, N.K.; Milivojević, D.D.; Spitalsky, Z.; Holclajtner-Antunović, I.D.; Bajuk-Bogdanović, D.V.; et al. Ambient light induced antibacterial action of curcumin/graphene nanomesh hybrids. *RSC Adv.* **2017**, *7*, 36081–36092. [CrossRef]
222. Bugli, F.; Cacaci, M.; Palmieri, V.; Di Santo, R.; Torelli, R.; Ciasca, G.; Di Vito, M.; Vitali, A.; Conti, C.; Sanguinetti, M.; et al. Curcumin-loaded graphene oxide flakes as an effective antibacterial system against methicillin-resistant Staphylococcus aureus. *Interface Focus* **2018**, *8*, 20170059. [CrossRef]
223. Palmieri, V.; Bugli, F.; Cacaci, M.; Di Santo, R.; Vitali, A.; Torelli, R.; Di Vito, M.; Conti, C.; Sanguinetti, M.; De Spirito, M.; et al. Antibacterial Properties of Curcumin Loaded Graphene Oxide Flakes. *Biophys. J.* **2018**, *114*, 362. [CrossRef]
224. Yang, X.X.; Li, C.M.; Li, Y.F.; Wang, J.; Huang, C.Z. Synergistic antiviral effect of curcumin functionalized graphene oxide against respiratory syncytial virus infection. *Nanoscale* **2017**, *9*, 16086–16092. [CrossRef] [PubMed]

225. Hatamie, S.; Akhavan, O.; Sadrnezhaad, S.K.; Ahadian, M.M.; Shirolkar, M.M.; Wang, H.Q. Curcumin-reduced graphene oxide sheets and their effects on human breast cancer cells. *Mater. Sci. Eng. C* **2015**, *55*, 482–489. [CrossRef] [PubMed]
226. Moreno-Guzman, M.; Martín, A.; del Marín, M.C.; Sierra, T.; Ansón-Casaos, A.; Martínez, M.T.; Escarpa, A. Electrochemical Behavior of Hybrid Carbon Nanomaterials: The Chemistry behind Electrochemistry. *Electrochim. Acta* **2016**, *214*, 286–294. [CrossRef]

Article

On the Consistency of the Exfoliation Free Energy of Graphenes by Molecular Simulations

Anastasios Gotzias [1,*], Elena Tocci [2] and Andreas Sapalidis [1]

1. National Centre for Scientific Research "Demokritos", Institute of Nanoscience and Nanotechnology INN, 15310 Athens, Greece; a.sapalidis@inn.demokritos.gr
2. Institute on Membrane Technology ITM–CNR, National Research Council, 87036 Rende, Italy; e.tocci@itm.cnr.it
* Correspondence: a.gotzias@inn.demokritos.gr; Tel.: +30-210-6503408

Abstract: Monolayer graphene is now produced at significant yields, by liquid phase exfoliation of graphites in solvents. This has increased the interest in molecular simulation studies to give new insights in the field. We use decoupling simulations to compute the exfoliation free energy of graphenes in a liquid environment. Starting from a bilayer graphene configuration, we decouple the Van der Waals interactions of a graphene monolayer in the presence of saline water. Then, we introduce the monolayer back into water by coupling its interactions with water molecules and ions. A different approach to compute the graphene exfoliation free energy is to use umbrella sampling. We apply umbrella sampling after pulling the graphene monolayer on the shear direction up to a distance from a bilayer. We show that the decoupling and umbrella methods give highly consistent free energy results for three bilayer graphene samples with different size. This strongly suggests that the systems in both methods remain closely in equilibrium as we move between the states before and after the exfoliation. Therefore, the amount of nonequilibrium work needed to peel the two layers apart is minimized efficiently.

Keywords: liquid exfoliation; layered materials; decoupling simulations; umbrella sampling

1. Introduction

Graphenes are two-dimensional, single-layer carbon nanosheets with unprecedented physical, mechanical, optical and electronic properties [1–4]. They are classified with different layered materials like boron nitrides, metal oxides, dichalcogenides and the recently introduced class of metal organic framework nanosheets (MONs) [5–7]. Most of their fascinating properties are attributed to the atomic-scale thickness, the continuous 2D connectivity and the ultra-large specific surface area. Graphene nanosheets can be produced by liquid-phase exfoliation and dispersion of graphites [8–12]. This is achieved by chemical functionalization and sonication of graphitized matrices in the presence of certain solvents. Highly polar solvents can destroy the graphitic structure or leave defects and functional moieties on the dispersed layers. By definition, the defects and functional groups change the intrinsic atomic structure and the electrical and mechanical properties of graphene. Graphene oxide, for example, is an insulator rather than a semi metal, and therefore, it is conceptually different than graphene. The special properties of graphenes are also expected to change as the number of layers increases. This is important, because partial exfoliation may release multilayers and clusters instead of large scale of graphene monolayers. In addition, the exfoliated monolayers tend to aggregate into multilayer configurations within the solvent. Nowadays, several high-yield exfoliation methods have been reported that produce unoxidized and defect-free graphene. The material is used to produce graphene-based composites or films, a key requirement for applications such as thin-film transistors, conductive transparent electrodes, photovoltaics and biomedical implants [13–15]. The availability of high-quality graphene samples has

increased the interest for explicit molecular simulations of the exfoliation processes and relevant applications [16–19].

The energy required to exfoliate graphene is balanced by the solvent-graphene interactions [9]. Solvent molecules enter the inner core of the graphite and break the Van der Waals forces between the layers. In order to simulate exfoliation, we configure paths either by pulling one layer at a distance from the remaining structure or by breaking the interactions of the layer in small changing steps [20,21]. If the steps are effectively small, the changes on the system are considered reversible, and the reversible work needed to transform one state into the other is equal to the free energy difference of the states before and after exfoliation. The free energy difference between two states, labeled A and B, is given by $\Delta F = F_A - F_B$. ΔF is related to the ratio of the partition functions of the states [22]. If the states lie far apart in phase space, the estimation of this ratio is intractable [23,24]. The reversible path serves to overcome this hurdle. It provides a connection between the two states so that we can evaluate the changes in F using thermodynamic integration [25–27]. That is, we split the overall perturbation into small perturbation steps for which the phase spaces continuously overlap. In this respect, the ratio of the partition functions for the consecutive perturbation steps is calculated more easily [28,29].

We may configure various paths to transform state A into state B [30,31]. A common approach is to use a reaction coordinate to move reversibly from the A-like state to a B-like state. Another approach is to modify the system's Hamiltonian. In this case, we mix the energies or the parameters of states A and B, according to a decoupling parameter, λ, which varies from 0 to 1 [32–34]. In many circumstances, it is convenient to introduce an intermediate state (or states) labeled C and evaluate the free energy difference by subtracting the free energies with respect to the state C: $\Delta F = (F_A - F_C) - (F_B - F_C)$ [35–39].

In a previous work, we employed one method to compute the exfoliation free energy of graphenes [40]. In brief, we considered a normal and a shear reaction coordinate to exfoliate (dissociate) a graphene layer from a bilayer configuration in an aqueous solution. We computed the free energy differences using umbrella sampling in small steps along the exfoliation paths. We reported that the free energy difference was greater when the exfoliation was coordinated on the shear than the normal direction. This was attributed to the awareness that the shear exfoliation of graphenes is reversible whereas the normal exfoliation is not. Notably, this outcome is in accordance with experiment. For instance, ball milling utilizes high shear force to delaminate layered materials and generate 2D nanosheets [41,42]. In this regard, molecular simulations succeed to realize a proof of principle. That is, the slip between the layers takes place in the in-plane direction, under the effect of shear force to yield free-standing graphene monolayers.

In this work, we compute the exfoliation free energy of graphenes using decoupling molecular simulations. The simulations are performed in two stages. In the first stage, we decouple the Van der Waals interactions of a single layer starting from a bilayer graphene configuration, in the presence of saline water. In the second stage we decouple the interactions of the same layer starting from a single layer configuration. The exfoliation free energy is computed by subtracting the free energy differences between the two stages. We consider three bilayer graphene samples with small, medium and large size. Then, we compare the free energy estimates of the decoupling simulations with those of the umbrella sampling in which, the same graphene nanosheets dissociate on the shear direction in respect to the bilayer plane.

2. Materials and Methods

Solvation free energies were computed by simulation by decoupling a nanoparticle from the solvent by thermodynamic integration, using the identity,

$$\Delta F = \int_0^1 d\lambda \left\langle \frac{\partial H(\lambda)}{\partial \lambda} \right\rangle \tag{1}$$

where H is parameterized Hamiltonian, and λ is the decoupling parameter, which parameterizes the atomistic interactions between the solvent and the nanoparticle [43,44]. The coupled state ($\lambda = 0$) corresponds to a simulation where the nanoparticle is interacting fully with the solvent, whereas the uncoupled state ($\lambda = 1$) corresponds to a simulation where the nanoparticle is not interacting with the solvent. Considering the van der Waals interaction between two atoms i and j, the λ-dependent Lennard Jones expression takes the form,

$$U_{ij} = (1-\lambda)^n 4\epsilon_{ij} \left(\frac{1}{\left[\alpha\lambda^n + \left(\frac{r_{ij}}{\sigma_{ij}}\right)^6\right]^2} - \frac{1}{\alpha\lambda^n + \left(\frac{r_{ij}}{\sigma_{ij}}\right)^6} \right) \quad (2)$$

where $r_{ij} = |r_i - r_j|$ is the interatomic distance, ϵ_{ij} and σ_{ij} are the Lorentz-Berthelot combination parameters. We set $\alpha = 0.5$ and $n = 1$. The term, $\alpha\lambda^n$, at the denominators makes the potential convergent as $r \to 0$, for $\lambda < 1$. This is useful especially for advanced steps of the decoupling ($0.7 < \lambda < 1$) as it improves the sampling at distances close to σ_{ij} from the atoms of the solute [32,45,46].

To configure the bilayer graphenes we considered two copies of a single layer graphene. The single layers were all-carbon sheets, having a honeycomb pattern of carbons and peripheral hydrogens. The layers were planar and square. We generated the coordinate files of the graphene layers using the BuildCstruct script (http://chembytes.wikidot.com/buildcstruct, accessed on 28 July 2021) [47]. We built three graphene layers with edges 1.0 nm, 1.5 nm and 2.5 nm. Using these structures, we configured three bilayer graphenes with small, medium and large size, respectively. We placed the copied layers in a parallel orientation, at a distance 0.3 nm from each other. We modeled the sp^2 carbon atoms on the basis of the OPLSAA references of naphthalene and of aliphatic carbons [48–51]. We modeled hydrogen interactions using the interaction parameters of benzene hydrogens. The structures were uncharged. We used the GROMACS package, version 2018 to build the simulations [52]. All simulations were submitted to the high performance computing services of the Greek National Infrastructure for research and technology, GRNET-ARIS.

We placed the bilayer graphene at the center of a cubic box. We set the nearest edge of the cube at 1.5 nm from the outermost atom of the bilayer. We set the cut off radius to 1.4 nm to be consistent with the minimum image convention. We used periodic boundary conditions in all directions. We solvated the simulation box with simple point charge (SPC) water and we added 100 mM NaCl. Ions were interacting fully with the SPC molecules. They were not interacting electrostatically with the bilayer, because the graphenes were uncharged. Complementary, we performed equivalent simulations for single layer graphene samples with small, medium and large size. We followed the same simulation protocol as for the bilayer graphenes, only that we used a single layer instead of two. The dimensions of the simulation cubes for all the studied systems and the number of molecules described therein, are listed in Table 1.

Before the decoupling simulations, the system energy was relaxed using a steepest descent minimization, followed by an equilibration in the NPT ensemble over 1 ns. We used the Berendsen coupling to regulate the temperature at 310 K and the pressure at 1 bar. The final configuration of the last NPT run, was used as starting configuration for the decoupling simulations. The bilayer graphene and the solvent (water and ions) were coupled to separate coupling paths. We used the Nose-Hoover method for the temperature coupling and the Parinello-Rahman for the pressure. We labeled the two graphene layers using a different index. One layer served as an immobile reference on which, we applied position restraints on the atoms. The other layer was interacting with the reference layer and the ambient molecules on the basis of Equation (2). We used 31 λ steps evenly distributed in $[0,1]$, with $d\lambda = 0.032$. Simulations for each of the 31 total λ steps ran over 10 nanoseconds. The thermodynamic integration was computed using the Bennett acceptance ratio (BAR) method [25].

We considered a two-stage process to compute the exfoliation free energy of graphenes. First we deleted a single graphene sheet from the bilayer configuration, then we solvated the graphene sheet back into the solvent. The free energy difference on the first stage was computed by the decoupling simulations of bilayer graphene samples. Likewise, the free energy difference on the second stage was computed by the decoupling simulations of single layer graphenes. The exfoliation free energy was estimated by subtracting the free energy difference on the second stage from that on the first stage.

A different technique to compute the exfoliation free energy of bilayer graphenes is to use umbrella sampling simulations. The umbrella sampling method was discussed in detail in a previous work [40]. There, we employed umbrella sampling to compute the binding free energies of the same, as in this work, three bilayer graphene samples. Although we used the term "binding" instead of "exfoliation" in the definition of the free energy, we refer to the same thermodynamic quantity. In brief, we configured a path in which a single graphene layer dissociated from a bilayer configuration (state A) to an arbitrary far distance (state B) in the solvent. This was achieved, by setting a force to pull the graphene along the coordinate of the path. The other layer (reference) remained at a fixed position. We performed umbrella sampling on a sequence of configuration points along the dissociation path. Each umbrella simulation output a probability distribution function. It was critical, that the neighboring umbrella distributions along the sequence of configurations were overlapping. If two consecutive umbrella distributions did not overlap, we sampled more configuration points until the new distributions bridged the non-overlapping gaps. In this respect, the spacings between the sampled configurations on the path from state A to state B, could be uneven or arbitrarily small. After completing a sufficient amount of umbrella sampling, we gathered the corresponding probability distributions and computed the free energies using the weighted histogram analysis method (WHAM) [53].

Table 1. Model properties of graphene layers with small, medium and large size, edge lengths of the simulation cubes and the number of solvent molecules (water and ions) contained in the cubes for the single and bilayer graphene configurations.

Graphene layer	Small			Medium			Large
Side Length (nm)	1.0			1.5			2.5
Carbon atoms	59			111			263
Hydrogen atoms	23			31			47
	Single Layer Graphene			Bilayer Graphene		
Simulation Cube	Small	Medium	Large	Small	Medium	Large
Cube edge (nm)	4.72	5.36	6.67	4.74	5.38	6.68
Water molecules (SPC)	3369	4929	9651	3411	4924	9608
Ions Na^+, Cl^-	6	9	18	6	9	18

3. Results

In Figure 1, we present configurations from the decoupling simulations performed on three bilayer graphene samples with small, medium and large size. We place the samples at the centre of a cubic box and solvate with saline water. In Figure 1, we decouple only the red graphene layer. The black layer interacts regularly with the environment and serves as a reference. We scale down the Van der Waals interactions of the decoupled graphene by increasing the parameter λ on the basis of Equation (2). At small λ values, the bilayer configurations are stable due to the adequately strong interlayer interactions. The layers preserve a parallel orientation and they can only twist at small angles against each other. With increasing λ, the Van der Waals interactions of the decoupled graphene decrease. This makes the decoupled graphene disconnect from the reference layer. As $\lambda \to 1$, the decoupled graphene moves freely inside the box. The decoupling interactions become small enough, so that we may to observe configurations in which the two layers intersect. This is due to the term $\alpha\lambda^n$ on the denominators of Equation (2), which makes the Van der Waals potential go to zero in a well-behaved manner with λ. The reference layer remains at a fixed position through the simulations, by imposing position restraints on the atoms. However, we can see in the snapshots of Figure 1 that when the two layers are apart, the reference layer is displaced from the box centre to the opposite direction of the decoupled graphene. This is because we change the periodicity on the representation of the simulated trajectories, setting the center of mass (COM) of both layers at the center of the simulation box.

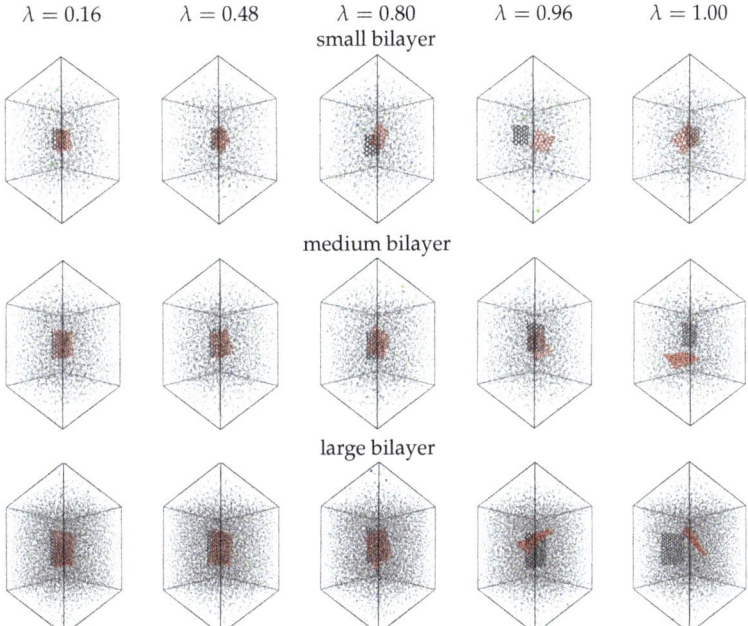

Figure 1. Configurations of the decoupling simulations of small, medium and large bilayer graphene samples at different values of the decoupling parameter, λ. As λ increases, the van der Waals interactions of the decoupled graphene layer decrease. The decoupled graphene layer is colored in red and the reference in black. Water molecules are shown with lines. Na^+ and Cl^- ions are shown with red and blue spheres. The time-frame of the presented configurations is random.

Figure 2, shows the derivatives of the free energy as a function of the decoupling parameter, λ. Increasing λ from zero to one represents a gradual change from a system with full Van der Waals interactions, to a system where the decoupled graphene is not interacting with the environment. We use 31 λ steps evenly distributed in [0,1]. We consider both single layer and bilayer configurations. In the bilayer configurations only one layer is being decoupled. The free energy values are normalized by the area of the graphenes in order to highlight the finite-size effects of the calculations. Such effects are expected to be appreciable for small interfaces. As a result, small graphenes obtain greater free energy derivatives than the large. We also observe greater energy derivatives for the bilayer than the single layer configurations. This is due to the additional interactions of the second layer. At the low λ range, the energy difference is positive indicating the stability of the starting configurations. At advanced steps of the decoupling, $\lambda > 0.7$, the free energy difference is negative. At high λ range, the decoupling interactions are near zero so that the water molecules are allowed to sample the volumes occupied by the atomic sites of the decoupled graphene. This is the same reason why we may observe the layer intersections in Figure 1. The small interaction energy implies a rearrangement of the solvent phase that creates new, lower in energy configurations, making the energy decrease. The bilayer graphenes obtain a first negative energy difference at higher λ, than the single layer graphenes. This is because of the interactions of the reference layer that do not change upon decoupling.

Likewise, the negative energy differences of the bilayers are smaller than those of the single layer configurations.

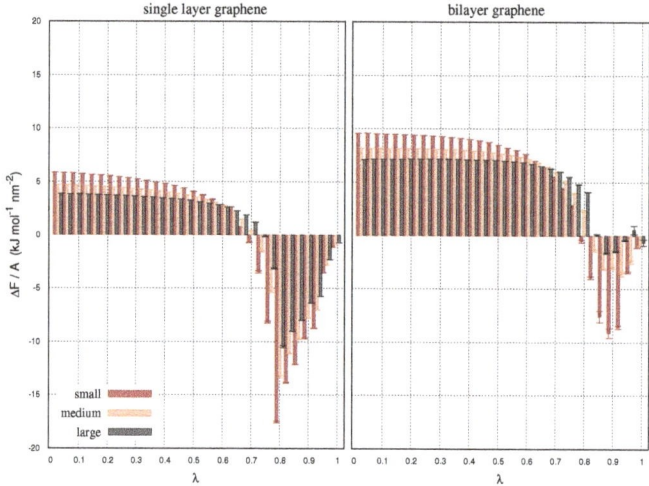

Figure 2. Free energy derivatives as a function of the decoupling parameter, λ, for the single and bilayer graphene samples with small, medium and large size. The free energy values are normalized by the area size of the graphenes.

In Figure 3 we show the relative free energies of the decoupling simulations as a function of λ. The energies result from the summation of the energy derivatives shown in Figure 2, up to the value of λ. Using decoupling simulations we simulate the transformation of a single layer (state B) or of a bilayer (state A) into a system where one layer is removed (state C). At full decoupling (i.e., $\lambda = 1$) the relative free energy of the single and the bilayer graphenes is expressed by $F_B - F_C$ and $F_A - F_C$, respectively. In the panels of Figure 3, the free energies converge nearly on the same value, regardless of the size of the layers. This is important as it approves that we used appropriate spacings $d\lambda$ in the thermodynamic integration. The convergence of the free energy differences also confirms the statistical consistency of our calculations, since the contributing integrals have resulted from individual simulations.

The exfoliation free energy of the bilayer graphenes is given by the difference between the relative energies in the two panels in Figure 3, i.e., $F_{AB} = (F_A - F_C) - (F_B - F_C)$. The free energies are plotted in the left-hand panel in Figure 4, as a function of the decoupling parameter, λ. The free energy is proportional to λ up to a range, where the energy presents a step-like increase. This step is attributed to the peak of the bilayer energy curves at $0.7 < \lambda < 0.9$, in Figure 3. At higher values of λ, the energy values reach a plateau. At $\lambda = 1$ we compute comparable exfoliation free energies for the three systems. In general, it is difficult to obtain converged free energy estimates by subtracting the energies of different Hamiltonians. The descrepancies of the free energy are attributed to the different contributions of the solvent-solvent interactions to the overall energy. Solvent-solvent interactions can be negligible compared to the magnitude of the solvent-graphene or the graphene-graphene interactions. However, when the graphene layers are small, the contributions of the solvent-solvent interactions may affect the result. The exfoliation free energies along with the values and statistical variances of $F_B - F_C$ and $F_A - F_C$ for the single and bilayer configurations are listed in Table 2.

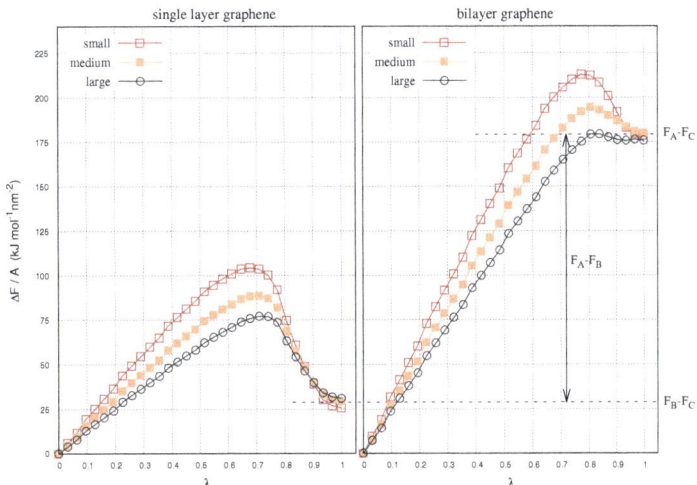

Figure 3. Cumulative free energies as a function of the decoupling parameter, λ, for the single and bilayer graphene samples with small, medium and large size. The free energy values are normalized by the area size of the graphenes.

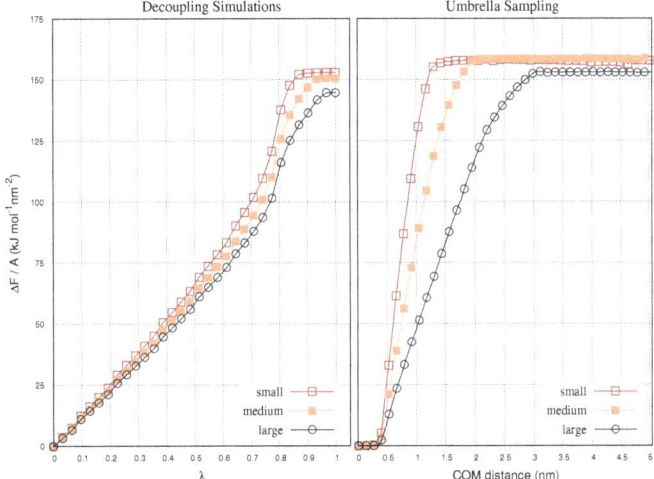

Figure 4. (**Left**) Exfoliation free energies of bilayer graphene samples with small, medium and large size, computed by decoupling simulations. The energies are shown as a function of the decoupling parameter λ, and computed by subtracting the relevant curves shown in the panels of Figure 3, based on the expression $\Delta F_{AB} = (F_B - F_C) - (F_C - F_A)$. (**Right**) Exfoliation free energy of the same bilayer graphene samples computed by umbrella sampling. The energies are shown as a function of the interlayer distance as the one layer is pulled away from the other on the shear direction. The umbrella sampling simulations are performed in a previous work [40].

Table 2. Differences of free energy by decoupling bilayer, $F_A - F_C$, and single layer, $F_B - F_C$, graphene samples with small, medium and large size. The free energy variance ($\pm \sigma$) is given by the average of the variances over the 31 λ steps. Correction is computed by $kTln(V_{\lambda=1}/V_{\lambda=0})$. F_{AB} is the exfoliation free energy expressed by $(F_A - F_C) - (F_B - F_C)$. F_{AB}^u is the exfoliation free energy for the corresponding bilayer graphene samples computed using umbrella sampling in a previous work [40]. The energies are normalized by the area of the graphene layers, expressed in kJ mol^{-1} nm^{-2}.

	Bilayer Graphene		Single Layer Graphene			
	$\Delta F_{sim} = F_A - F_C$	Correction	$\Delta F_{sim} = F_B - F_C$	Correction	F_{AB}	F_{AB}^u
small	$178.71 \pm 5.28 \times 10^{-2}$	1.93×10^{-2}	$25.77 \pm 1.84 \times 10^{-2}$	1.77×10^{-2}	152.94	157.41
medium	$180.12 \pm 3.02 \times 10^{-2}$	8.76×10^{-3}	$29.55 \pm 1.56 \times 10^{-2}$	8.89×10^{-3}	150.57	158.59
large	$175.73 \pm 3.29 \times 10^{-2}$	3.89×10^{-3}	$31.08 \pm 8.70 \times 10^{-3}$	4.00×10^{-3}	144.65	152.88

For the sake of comparison, we plot the exfoliation free energies of the bilayer graphenes computed with umbrella sampling simulations, in the right-hand panel of Figure 4. The umbrella simulations are detailed in a previous work [40]. The exfoliated graphene is pulled on the shear direction relevant to the initial plane of the bilayer. It is pulled over 0.5 ns using a pull rate 10 nm ns^{-1}, so that a final COM distance of 5 nm between the layers is achieved. At this distance, the layers do not interact with each other, so that we may assume the layer exfoliated. The free energies are expressed in Figure 4 as a function of the center of mass (COM) distance between the layers. The free energies obtain a plateau when the interlayer distance becomes greater than the lateral size of the graphenes. Pulling further the graphene, does not change the free energy of the system. The umbrella simulations give comparable free energies with the decoupling simulations for the same graphene samples. This means that the two methods, i.e., shear pulling umbrella sampling and decoupling of van der Waals interactions of the graphene layers, are similar in terms of reversibility.

4. Discussion

Free energy calculations are independent of the simulation protocol that is used. However, the level of precision does depend very much on the choice of the transformation path [54,55]. Different configurations having an absolute free energy F_A should on application of an adiabatic transformation all end up with the same free energy F_B. If this condition is not satisfied the transformation has been carried out too rapidly and the transformation strictly speaking is not adiabatic. The states, A and B, before and after exfoliation are separated by a free energy barrier. With decoupling simulations, we cross the barrier by introducing an intermediate state C. With umbrella simulations, we cross the barrier with more umbrella sampling. Nevertheless, there are two distinct types of barriers. The first type is is due to interactions of the perturbed degrees of freedom i.e., those of the exfoliated layer with itself and the environment. The other type of energy barrier is attributed to the interactions of the non perturbed degrees of freedom, i.e., those of the solvent (water and ions). These interactions cannot be smoothed by adding intermediate states or with additional sampling. In this respect, the rearrangement of the solvent molecules can increase the system's entropy affecting the free energy calculations.

Apart from the solvent phase perturbations, discrepancies of the free energy are often attributed to the statistical variance of the molecular simulations. We can see in Table 2 that the variances are small, however, they do not reflect the level of accuracy of the estimates. The error of a free energy computed by molecular simulation, ΔF_{sim}, in respect to that of an experiment, ΔF_{exp}, is given by

$$\Delta F_{exp} = \Delta F_{sim} - kTln(V_{\lambda=1}/V_{\lambda=0}) \quad (3)$$

This equation involves only the average volume $V_{\lambda=0}$ of the fully coupled system, and average volume $V_{\lambda=1}$ of the box of pure solvent (or of the solvent and the reference layer) containing the same number of solvent molecules as the coupled system [32,56]. The term with the ratio of the volumes, i.e., the correction, for the single and bilayer decoupling simulations, is tabulated in Table 2. In each case, the correction is between 10% and 40% of the reported variance. Because of this magnitude we can safely neglect this term in our calculations.

5. Conclusions

The availability of high-yield exfoliation methods to produce isolated, unoxidized and defect-free graphene has increased the interest in fundamental studies on the exfoliation processes by molecular simulations. We compute the liquid-phase exfoliation free energy of three graphene layers with different sizes using two simulation protocols, namely decoupling molecular dynamics and umbrella sampling. Using the two simulations, we estimate very similar free energies regardless of the size of the layers. This reflects the thermodynamic consistency of the methodologies and confirms that we designed the corresponding exfoliation paths to be equally reversible. The modeled systems, containing graphenes and water and ions, are reasonably simple, to be able to obtain good statistical sampling, allowing us to realize a lower bound on the amount of the sampling necessary to simulate more complex structures like functionalized graphenes and other 2D layered materials and metal organic nanosheets.

Author Contributions: Conceptualisation, A.G., A.S., E.T.; methodology, A.G., E.T.; visualisation, A.G.; writing, A.G. All authors have read and agreed to the published version of the manuscript.

Funding: This research has received funding by the project IDEA, ERANETMED2-72-357.

Data Availability Statement: Upon request.

Conflicts of Interest: The authors declare no conflict of interest.

References

1. Backes, C.; Abdelkader, A.; Alonso, C.; Andrieux-Ledier, A.; Arenal, R.; Azpeitia, J.; Balakrishnan, N.; Banszerus, L.; Barjon, J.; Bartali, R.; et al. Production and processing of graphene and related materials. *2D Mater.* **2020**, *7*. [CrossRef]
2. Sekhon, S.S.; Kaur, P.; Kim, Y.H.; Sekhon, S.S. 2D graphene oxide–aptamer conjugate materials for cancer diagnosis. *NPJ 2D Mater. Appl.* **2021**, *5*, 21. [CrossRef]
3. SI, A.; Kyzas, G.Z.; Pal, K.; de Souza, F.G., Jr. Graphene functionalized hybrid nanomaterials for industrial-scale applications: A systematic review. *J. Mol. Struct.* **2021**, *1239*, 130518. [CrossRef]
4. Pereira, L.F.C. Investigating mechanical properties and thermal conductivity of 2D carbon-based materials by computational experiments. *Comput. Mater. Sci.* **2021**, *196*, 110493. [CrossRef]
5. Sarkar, A.S.; Stratakis, E. Dispersion behaviour of two dimensional monochalcogenides. *J. Colloid Interface Sci.* **2021**, *594*, 334–341. [CrossRef]
6. Zhang, C.J. Interfacial assembly of two-dimensional MXenes. *J. Energy Chem.* **2021**, *60*, 417–434. [CrossRef]
7. Ashworth, D.J.; Foster, J.A. Metal–organic framework nanosheets (MONs): A new dimension in materials chemistry. *J. Mater. Chem. A* **2018**, *6*, 16292–16307. [CrossRef]
8. Hernandez, Y.; Nicolosi, V.; Lotya, M.; Blighe, F.M.; Sun, Z.; De, S.; McGovern, I.T.; Holland, B.; Byrne, M.; Gun'Ko, Y.K.; et al. High-yield production of graphene by liquid-phase exfoliation of graphite. *Nat. Nanotechnol.* **2008**, *3*, 563–568. [CrossRef]
9. Coleman, J.N.; Lotya, M.; O'Neill, A.; Bergin, S.D.; King, P.J.; Khan, U.; Young, K.; Gaucher, A.; De, S.; Smith, R.J.; et al. Two-Dimensional Nanosheets Produced by Liquid Exfoliation of Layered Materials. *Science* **2011**, *331*, 568–571. [CrossRef]
10. Cui, X.; Shi, W.; Lu, C. Large-scale visualization of the dispersion of liquid-exfoliated two-dimensional nanosheets. *Chem. Commun.* **2021**, *57*, 4303–4306. [CrossRef]
11. Vacacela Gomez, C.; Guevara, M.; Tene, T.; Villamagua, L.; Usca, G.T.; Maldonado, F.; Tapia, C.; Cataldo, A.; Bellucci, S.; Caputi, L.S. The liquid exfoliation of graphene in polar solvents. *Appl. Surf. Sci.* **2021**, *546*, 149046. [CrossRef]
12. Chen, X.; Dubois, M.; Radescu, S.; Rawal, A.; Zhao, C. Liquid-phase exfoliation of F-diamane-like nanosheets. *Carbon* **2021**, *175*, 124–130. [CrossRef]
13. Natter, N.; Kostoglou, N.; Koczwara, C.; Tampaxis, C.; Steriotis, T.; Gupta, R.; Paris, O.; Rebholz, C.; Mitterer, C. Plasma-Derived Graphene-Based Materials for Water Purification and Energy Storage. *C* **2019**, *5*, 16. [CrossRef]
14. Naeem, M.; Kuan, H.C.; Michelmore, A.; Yu, S.; Mouritz, A.P.; Chelliah, S.S.; Ma, J. Epoxy/graphene nanocomposites prepared by in-situ microwaving. *Carbon* **2021**, *177*, 271–281. [CrossRef]
15. Gotzias, A. Injecting Carbon Nanostructures in Living Cells. In Proceedings of the Workshops of the 11th EETN Conference on Artificial Intelligence 2020 (SETN2020 Workshops), Athens, Greece, 2–4 September 2020.

16. Kong, X.; Zhuang, J.; Zhu, L.; Ding, F. The complementary graphene growth and etching revealed by large-scale kinetic Monte Carlo simulation. *NPJ Comput. Mater.* **2021**, *7*, 14. [CrossRef]
17. Stevens, K.; Tran-Duc, T.; Thamwattana, N.; Hill, J.M. Modeling Interactions between Graphene and Heterogeneous Molecules. *Computation* **2020**, *8*, 107. [CrossRef]
18. Folorunso, O.; Hamam, Y.; Sadiku, R.; Sinha Ray, S.; Adekoya, G. Comparative study of graphene-polypyrrole and borophene-polypyrrole composites: Molecular dynamics modeling approach. *Eng. Solid Mech.* **2021**, *9*, 311–322. [CrossRef]
19. Parab, A.D.; Budi, A.; Slocik, J.M.; Rao, R.; Naik, R.R.; Walsh, T.R.; Knecht, M.R. Molecular-Level Insights into Biologically Driven Graphite Exfoliation for the Generation of Graphene in Aqueous Media. *J. Phys. Chem. C* **2020**, *124*, 2219–2228. [CrossRef]
20. Liang, L.; Chen, E.Y.; Shen, J.W.; Wang, Q. Molecular modelling of translocation of biomolecules in carbon nanotubes: Method, mechanism and application. *Mol. Simul.* **2016**, *42*, 827–835. [CrossRef]
21. Cai, L.; Lv, W.; Zhu, H.; Xu, Q. Molecular dynamics simulation on adsorption of pyrene-polyethylene onto ultrathin single-walled carbon nanotube. *Phys. E Low-Dimens. Syst. Nanostruct.* **2016**, *81*, 226–234. [CrossRef]
22. Han, K.K. A new Monte Carlo method for estimating free energy and chemical potential. *Phys. Lett. A* **1992**, *165*, 28–32. [CrossRef]
23. Christ, C.D.; van Gunsteren, W.F. Enveloping distribution sampling: A method to calculate free energy differences from a single simulation. *J. Chem. Phys.* **2007**, *126*, 184110. [CrossRef] [PubMed]
24. Wu, D.; Kofke, D.A. Phase-space overlap measures. II. Design and implementation of staging methods for free-energy calculations. *J. Chem. Phys.* **2005**, *123*, 084109. [CrossRef]
25. Bennett, C.H. Efficient estimation of free energy differences from Monte Carlo data. *J. Comput. Phys.* **1976**, *22*, 245–268. [CrossRef]
26. Frenkel, D.; Smit, B. *Understanding Molecular Simulation: From Algorithms to Applications*, 2nd ed.; Academic Press, Inc.: Orlando, FL, USA, 1996; Volume 50. [CrossRef]
27. Schultz, A.J.; Kofke, D.A. Identifying and estimating bias in overlap-sampling free-energy calculations. *Mol. Simul.* **2021**, *47*, 379–389. [CrossRef]
28. Wu, D. Understanding free-energy perturbation calculations through a model of harmonic oscillators: Theory and implications to improve the sampling efficiency by molecular simulation. *J. Chem. Phys.* **2010**, *133*, 244116. [CrossRef]
29. Sidler, D.; Schwaninger, A.; Riniker, S. Replica exchange enveloping distribution sampling (RE-EDS): A robust method to estimate multiple free-energy differences from a single simulation. *J. Chem. Phys.* **2016**, *145*, 154114. [CrossRef]
30. Perthold, J.W.; Petrov, D.; Oostenbrink, C. Toward Automated Free Energy Calculation with Accelerated Enveloping Distribution Sampling (A-EDS). *J. Chem. Inf. Model.* **2020**, *60*, 5395–5406. [CrossRef]
31. Wu, J.Z.; Azimi, S.; Khuttan, S.; Deng, N.; Gallicchio, E. Alchemical Transfer Approach to Absolute Binding Free Energy Estimation. *J. Chem. Theory Comput.* **2021**, *17*, 3309–3319. [CrossRef] [PubMed]
32. Shirts, M.R.; Pitera, J.W.; Swope, W.C.; Pande, V.S. Extremely precise free energy calculations of amino acid side chain analogs: Comparison of common molecular mechanics force fields for proteins. *J. Chem. Phys.* **2003**, *119*, 5740–5761. [CrossRef]
33. Mobley, D.L.; Chodera, J.D.; Dill, K.A. On the use of orientational restraints and symmetry corrections in alchemical free energy calculations. *J. Chem. Phys.* **2006**, *125*, 084902–084902. [CrossRef] [PubMed]
34. Fathizadeh, A.; Elber, R. A mixed alchemical and equilibrium dynamics to simulate heterogeneous dense fluids: Illustrations for Lennard-Jones mixtures and phospholipid membranes. *J. Chem. Phys.* **2018**, *149*, 072325. [CrossRef] [PubMed]
35. Wu, D.; Kofke, D.A. Phase-space overlap measures. I. Fail-safe bias detection in free energies calculated by molecular simulation. *J. Chem. Phys.* **2005**, *123*, 054103. [CrossRef] [PubMed]
36. Cournia, Z.; Allen, B.; Sherman, W. Relative Binding Free Energy Calculations in Drug Discovery: Recent Advances and Practical Considerations. *J. Chem. Inf. Model.* **2017**, *57*, 2911–2937. [CrossRef] [PubMed]
37. Heinzelmann, G.; Gilson, M.K. Automation of absolute protein-ligand binding free energy calculations for docking refinement and compound evaluation. *Sci. Rep.* **2021**, *11*, 1116. [CrossRef] [PubMed]
38. Lee, T.S.; Allen, B.K.; Giese, T.J.; Guo, Z.; Li, P.; Lin, C.; McGee, T.D.; Pearlman, D.A.; Radak, B.K.; Tao, Y.; et al. Alchemical Binding Free Energy Calculations in AMBER20: Advances and Best Practices for Drug Discovery. *J. Chem. Inf. Model.* **2020**, *60*, 5595–5623. [CrossRef]
39. Hahn, D.F.; Hünenberger, P.H. Alchemical Free-Energy Calculations by Multiple-Replica λ-Dynamics: The Conveyor Belt Thermodynamic Integration Scheme. *J. Chem. Theory Comput.* **2019**, *15*, 2392–2419. [CrossRef] [PubMed]
40. Gotzias, A. Binding Free Energy Calculations of Bilayer Graphenes Using Molecular Dynamics. *J. Chem. Inf. Model.* **2021**, *61*, 1164–1171. [CrossRef]
41. Ranjan, R.; Bajpai, V. Graphene-based metal matrix nanocomposites: Recent development and challenges. *J. Compos. Mater.* **2021**, *55*, 2369–2413. [CrossRef]
42. Zhou, D.; Zhao, L.; Li, B. Recent progress in solution assembly of 2D materials for wearable energy storage applications. *J. Energy Chem.* **2021**, *62*, 27–42. [CrossRef]
43. Noroozi, J.; Ghotbi, C.; Sardroodi, J.J.; Karimi-Sabet, J.; Robert, M.A. Solvation free energy and solubility of acetaminophen and ibuprofen in supercritical carbon dioxide: Impact of the solvent model. *J. Supercrit. Fluids* **2016**, *109*, 166–176. [CrossRef]
44. Bux, K.; Moin, S.T. Solvation of cholesterol in different solvents: A molecular dynamics simulation study. *Phys. Chem. Chem. Phys.* **2020**, *22*, 1154–1167. [CrossRef]

45. König, G.; Glaser, N.; Schroeder, B.; Kubincová, A.; Hünenberger, P.H.; Riniker, S. An Alternative to Conventional λ-Intermediate States in Alchemical Free Energy Calculations: λ-Enveloping Distribution Sampling. *J. Chem. Inf. Model.* **2020**, *60*, 5407–5423. [CrossRef] [PubMed]
46. Mecklenfeld, A.; Raabe, G. Efficient solvation free energy simulations: Impact of soft-core potential and a new adaptive λ-spacing method. *Mol. Phys.* **2017**, *115*, 1322–1334. [CrossRef]
47. Shkolin, A.V.; Fomkin, A.A.; Yakovlev, V.Y.; Men'shchikov, I.E. Model Nanoporous Supramolecular Structures Based on Carbon Nanotubes and Hydrocarbons for Methane and Hydrogen Adsorption. *Colloid J.* **2018**, *80*, 739–750. [CrossRef]
48. Jorgensen, W.L.; Maxwell, D.S.; Tirado-Rives, J. Development and Testing of the OPLS All-Atom Force Field on Conformational Energetics and Properties of Organic Liquids. *J. Am. Chem. Soc.* **1996**, *118*, 11225–11236. [CrossRef]
49. Karataraki, G.; Sapalidis, A.; Tocci, E.; Gotzias, A. Molecular Dynamics of Water Embedded Carbon Nanocones: Surface Waves Observation. *Computation* **2019**, *7*, 50. [CrossRef]
50. Gotzias, A.; Sapalidis, A. Pulling Simulations and Hydrogen Sorption Modelling on Carbon Nanotube Bundles. *C* **2020**, *6*, 11. [CrossRef]
51. Tieleman, D.P.; MacCallum, J.L.; Ash, W.L.; Kandt, C.; Xu, Z.; Monticelli, L. Membrane protein simulations with a united-atom lipid and all-atom protein model: Lipid–protein interactions, side chain transfer free energies and model proteins. *J. Phys. Condens. Matter* **2006**, *18*, S1221–S1234. [CrossRef] [PubMed]
52. Abraham, M.J.; Murtola, T.; Schulz, R.; Páll, S.; Smith, J.C.; Hess, B.; Lindahl, E. GROMACS: High performance molecular simulations through multi-level parallelism from laptops to supercomputers. *SoftwareX* **2015**, *1–2*, 19–25. [CrossRef]
53. Kumar, S.; Rosenberg, J.M.; Bouzida, D.; Swendsen, R.H.; Kollman, P.A. The weighted histogram analysis method for free-energy calculations on biomolecules. I. The method. *J. Comput. Chem.* **1992**, *13*, 1011–1021. [CrossRef]
54. Lemkul, J.A.; Allen, W.J.; Bevan, D.R. Practical Considerations for Building GROMOS-Compatible Small-Molecule Topologies. *J. Chem. Inf. Model.* **2010**, *50*, 2221–2235. [CrossRef]
55. Chelli, R.; Gellini, C.; Pietraperzia, G.; Giovannelli, E.; Cardini, G. Path-breaking schemes for nonequilibrium free energy calculations. *J. Chem. Phys.* **2013**, *138*, 214109. [CrossRef] [PubMed]
56. Hinkle, K.R.; Phelan, F.R. Solvation of Carbon Nanoparticles in Water/Alcohol Mixtures: Using Molecular Simulation To Probe Energetics, Structure, and Dynamics. *J. Phys. Chem. C* **2017**, *121*, 22926–22938. [CrossRef] [PubMed]

Article

Fluorescence Study of Riboflavin Interactions with Graphene Dispersed in Bioactive Tannic Acid

María Paz San Andrés [1,2,*], Marina Baños-Cabrera [1], Lucía Gutiérrez-Fernández [1], Ana María Díez-Pascual [1,2] and Soledad Vera-López [1,2]

[1] Universidad de Alcalá, Facultad de Ciencias, Departamento de Química Analítica, Química Física e Ingeniería Química, Ctra. Madrid-Barcelona Km. 33.6, 28805 Alcalá de Henares, Madrid, España (Spain); marineta.bc@gmail.com (M.B.-C.); lucia.gutierrezf@edu.uah.es (L.G.-F.); am.diez@uah.es (A.M.D.-P.); soledad.vera@uah.es (S.V.-L.)

[2] Universidad de Alcalá, Instituto de Investigación Química Andrés M. del Río (IQAR), Ctra. Madrid-Barcelona Km. 33.6, 28805 Alcalá de Henares, Madrid, España (Spain)

* Correspondence: mpaz.sanandres@uah.es

Abstract: The potential of tannic acid (TA) as a dispersing agent for graphene (G) in aqueous solutions and its interaction with riboflavin have been studied under different experimental conditions. TA induces quenching of riboflavin fluorescence, and the effect is stronger with increasing TA concentration, due to π-π interactions through the aromatic rings, and hydrogen bonding interactions between the hydroxyl moieties of both compounds. The influence of TA concentration, the pH, and the G/TA weight ratio on the quenching magnitude, have been studied. At a pH of 4.1, G dispersed in TA hardly influences the riboflavin fluorescence, while at a pH of 7.1, the nanomaterial interacts with riboflavin, causing an additional quenching to that produced by TA. When TA concentration is kept constant, quenching of G on riboflavin fluorescence depends on both the G/TA weight ratio and the TA concentration. The fluorescence attenuation is stronger for dispersions with the lowest G/TA ratios, since TA is the main contributor to the quenching effect. Data obey the Stern–Volmer relationship up to TA 2.0 g L^{-1} and G 20 mg L^{-1}. Results demonstrate that TA is an effective dispersant for graphene-based nanomaterials in liquid medium and a green alternative to conventional surfactants and synthetic polymers for the determination of biomolecules.

Keywords: tannic acid; graphene; fluorescence; quenching; riboflavin

1. Introduction

Graphene (G) is a carbon nanomaterial that comprises a two-dimensional honeycomb lattice of sp^2 hybridized carbon atoms with one-atom thickness, a high electron density, and a large surface area [1]. It has excellent electrical, optical, and thermal properties, combined with high mechanical resistance, transparency, low density and flexibility [2]. The G structure with delocalized π bonds accounts for its high electrical and thermal conductivity [3]. Its properties make it an ideal material for a wide range of applications, ranging from structural nanocomposites [4], to electronics, optics, sensors, and biodevices.

G can interact covalently and non-covalently with other molecules or polymers and is able to incorporate them on both sides of its sheets. Covalent interactions occur via formation of a chemical bond, and two main approaches can be discerned: grafting-from and grafting-to [5]. In the former, G is used as a growing platform of the molecules or polymer chains, whereas the later consists of the direct coupling of G or pre-functionalized G with the polymer chains, which should incorporate reactive functional groups. Nonetheless, these strategies can alter the conjugated π-system of G and bring together structural defects that led to worse performance. On the other hand, the non-covalent approach is based on the physical adsorption and/or wrapping of molecules or polymers via weak interactions such as H-bonding, hydrophobic (van der Waals) π-π, H-π, cation-π, and anion-π, which preserve the intrinsic electronic properties of this nanomaterial. G has high surface adsorption capacity, and is able to retain molecules with different structures, especially those with aromatic rings [6]. Further, it is hydrophobic and has a strong tendency to agglomerate, so it must be dispersed in liquids (such as organic solvents or water) frequently with the aid of ultrasounds and/or dispersing agents [7,8]. Different dispersants have been used to improve their solubility and other properties. Good aqueous dispersions of G sheets have also been achieved using a derivative of pyrene, 1-pyrenebutyrate [9]. Surfactants are amongst the most widely used dispersing agents for G in aqueous media. Sodium cholate with mild sonication for a long time disperses G up to concentrations of 0.3 mg mL^{-1} [10]. Cationic, anionic, and nonionic surfactants have been used [11], and while ionic surfactants adsorbed on G provide electrostatic repulsion between sheets that prevent their aggregation, nonionic ones lead to stabilization by steric interactions [12]. Nonionic surfactants (such as Polyoxyethylene-100-stearyl ether (Brij 700) or Poly(ethylene glycol)-block-poly(propylene glycol)-block-poly(ethylene glycol) (Pluronic P-123) can exfoliate and disperse G in water at high concentrations, while ionic surfactants such as sodium taurodeoxycholate (TDOC) or sodium dodecyl sulfate (SDS) are less effective [13]. In our research group, the ability of surfactants of different natures (cationic hexadecyltrimethylammonium bromide (CTAB), zwitterionic laurylsulfobetaine (LSB), and nonionic polyoxiethylen-23-lauryl ether (Brij L23) to disperse G in an aqueous medium has been thoroughly investigated. Further, the interaction of G with α-tocopherol [14], with the anionic surfactant SDS, or with pyrene [15], as well as their interaction with vitamins, such as riboflavin [16] and pyridoxine [17], have been investigated in detail. The ability of surfactants to disperse G depends on a number of parameters, including their charge, concentration, and length of the hydrocarbon chain. Polymers, such as polyethylenglicol (PEG) and copolymers such as polyethylene glycol-polypropylene glycol-polyethylene glycol (poloxamer P-407), have also been studied to disperse G and graphene oxide (GO) [18,19].

Tannic acid (TA) is a water-soluble polyphenolic compound that belongs to the group of hydrolysable tannins, found naturally in fruits, seeds, plants, and tree bark. It is present in foods such as wine, coffee, tea, and beer. It has antibacterial, antienzymatic, and astringent properties, although it inhibits the absorption of iron in the body by complex formation. It is also used to treat skin ulcers, wounds, and toothaches. Further, it is employed as a food additive (E-181), and it is a clarifying agent and flavor enhancer [20]. Its chemical structure (Figure 1) consists of a central glucose molecule, which hydroxyl groups are linked by ester bonds with gallic acid moieties. One molecule contains 8 to 10 moles of gallic acid per mole of glucose [21]. In the literature, different pK_a values for TA have been reported, most recently, 6.3, 7.4, and 8.6 [22]. In a previous study [23], TA was used as an additive to modify the structure of a graphene derivative (GO) for the development of GO/carbonized paper/TA ternary composites in the solid state to be used as flexible electrodes in energy storage applications. TA molecules adsorbed onto the GO via π-π stacking interactions and increased the interlayer spacing between GO sheets, hence the contact area with the carbonized paper. On the other hand, due to its structure and biocompatibility, TA is a good candidate to replace synthetic surfactants in order to obtain low-cost, environmentally friendly G dispersions in water. TA has been used to obtain G by exfoliating graphite with ultrasound, and the influence of TA concentration, pH, and sonication time have been studied, among other parameters [24]. Other polyphenols such as resveratrol [25], and even raw tea [26] or coffee [27], have also been used for the dispersion of G and its derivatives.

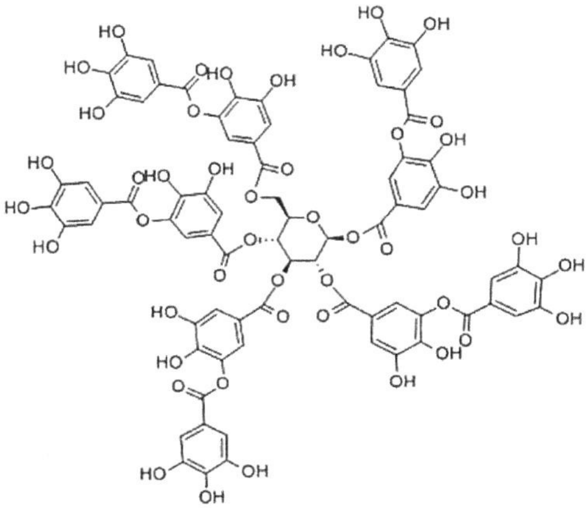

Figure 1. Chemical structure of tannic acid.

Riboflavin, known as vitamin B_2, is a molecule of biological interest, a water-soluble vitamin excreted in the urine and an essential nutrient, since its daily consumption is necessary. It is present in the body principally as a basic constituent of the coenzymes flavin adenine dinucleotide (FAD) and flavin mononucleotide (FMN), and it participates in cell development and human growth [28]. Further, it contributes to the metabolism of other vitamins, carbohydrates and proteins, it participates in multiple metabolic reactions, such as the formation of red blood cells, functioning of the nervous system, and DNA synthesis [29], it is essential for energy production, and acts as an antioxidant. Riboflavin scarcity can induce hematologic, cardiovascular, and gastrointestinal diseases and has been identified as a risk factor for cancer [30]. Henceforth, considering the great nutritional significance of this water-soluble vitamin, it has been taken as model in this study.

Riboflavin has native fluorescence, which has been studied in different media. Some amino acids, such as cysteine and methionine, produce the attenuation of the fluorescence of riboflavin, known as "quenching" [31], which can be produced by energy transfer (dynamic quenching) or by formation of a ground state complex (static quenching) [32]. In both cases, a linear relationship can be obtained between the ratio of the fluorescence intensity in the absence (F_0) and in the presence of the quencher molecule (F) and the quencher concentration ([Q]). The intercept of the plot should be equal to one, and the slope is the Stern–Volmer constant (K_{SV}) in the dynamic quenching Equation (1), being $K_{SV} = \tau K_q$ (τ = fluorescence lifetime, K_q = bimolecular quenching constant) while in the static quenching K_{SV} is replaced by K_S Equation (2), the constant of formation of the complex:

$$F_0/F = 1 + k_q \tau_0 [Q] = 1 + K_{SV} [Q] \quad (1)$$

$$F_0/F = 1 + K_S [Q] \quad (2)$$

Steady-state fluorescence measurements do not allow distinguishing between the two types of quenching, hence lifetime measurements are necessary. Nevertheless, the quenching constants (either dynamic (K_{SV}) or static (K_S)), can be calculated from the plot of F_0/F versus [Q], which provides information about the interactions between the analyte and the quencher molecule.

The main aim of this work is to investigate the possibilities of tannic acid as a dispersant agent for G in water, a biocompatible medium. Fluorescent measurements have been carried out to study the interaction between G dispersed in tannic acid and a fluorescent molecule of high biological interest, such as riboflavin. Thus, the interaction of this vitamin with tannic acid or G/tannic acid system has been evaluated and the results have been compared with those previously obtained for G dispersed in synthetic surfactants and polymers, in order to assess the effectiveness of tannic acid as dispersing agent.

2. Results and Discussion

2.1. Riboflavin Fluorescence in Water and in TA Aqueous Solutions without G

The fluorescence three-dimensional spectrum of riboflavin 0.6 mg L^{-1} in water is shown in Figure 2a. Two fluorescence intensity maxima are observed at excitation wavelengths of 380 nm and 455 nm. The emission maxima are found at 520 nm for the two excitation wavelengths. The excitation and emission maximum wavelengths hardly shift upon changing the medium from water to TA aqueous solutions at pH 4.1. Figure 2b shows the emission spectra at l_{exc} = 455 nm of riboflavin in water and in the presence of 0.5 and 2.0 g L^{-1} TA aqueous solutions. A decrease in the fluorescence intensity of riboflavin is observed with increasing TA concentration due to the interaction between both compounds. A fluorescence quenching phenomenon occurs owing to the interaction between TA and riboflavin that can take place via π-π stacking through the aromatic rings, via hydrogen bonding between the hydroxyl moieties of both molecules, or even via hydrophobic or electrostatic forces. TA presents the ability of multiple hydrogen bonding and crosslinking due to its numerous hydroxyl groups, and the pH and temperature dependence of its interactions with biomolecules [33] or polymers [34] have been reported.

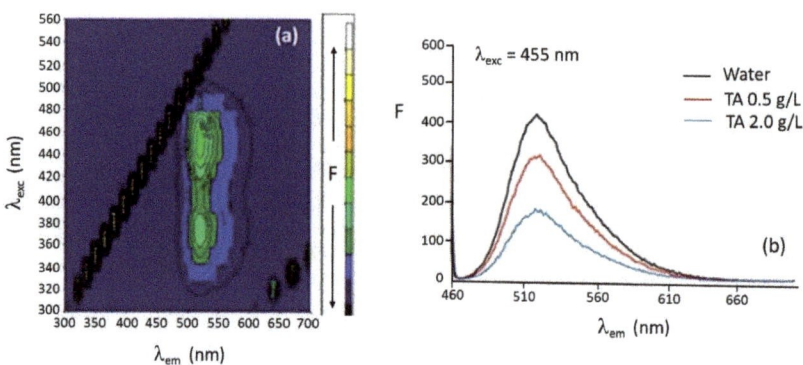

Figure 2. 3D fluorescence spectrum of riboflavin 0.6 mg L^{-1} in water (**a**) and emission spectra at λ_{exc} 455 nm in water and tannic acid 0.5 y 2.0 g L^{-1} (**b**).

The fluorescence spectra obtained in water, as well as in 0.5 and 2.0 g L^{-1} TA solutions, are similar at pHs of 4.1 and 7.1. A clear decrease in the intensity is observed with increasing TA concentration, which causes fluorescence quenching of riboflavin. The ratio between the fluorescence intensity of riboflavin in water and in the presence of TA solutions at the maximum emission wavelength (520 nm) for the excitation wavelength of 455 nm are calculated, and the results are shown in Table 1.

Table 1. Ratio of the fluorescence intensity of riboflavin in water (F_W) and in tannic acid 0.5 and 2.0 g L^{-1} (F) at pHs of 4.1 and 7.1 for a λ_{exc} = 455 nm and a λ_{em} = 520 nm.

pH	F/F$_W$	
	[TA] = 0.5 g L^{-1}	[TA] = 2.0 g L^{-1}
4.1	0.67	0.39
7.1	0.74	0.36

The decrease in fluorescence intensity in the presence of TA is significantly stronger for 2.0 g L^{-1} than for 0.5 g L^{-1} at both pHs. Therefore, it can be concluded that the interaction between both molecules occurs at both pHs, either by energy transfer from riboflavin in the excited state to TA (dynamic quenching), or by formation of a complex in the ground state between both molecules (static quenching). As indicated above, both molecules can interact mainly via π-π and H-bonding interactions. The fact that there is hardly any difference between the fluorescence quenching found at pH 4.1 and 7.1 suggests that π-π interactions predominate. Nonetheless, hydrogen bonds should also play a key role in the interaction between both molecules. Our results are in agreement with former works that studied the interactions between molecules with hydroxyl moieties and polyphenols; these weaken with increasing temperature, but depend only slightly on the solution pH [34,35].

To assess the influence of TA concentration on the quenching of riboflavin fluorescence intensity, solutions with TA concentrations ranging from 0 to 2.0 g L^{-1} were prepared, and the Stern–Volmer equation was plotted. Figure 3a shows the decrease in fluorescence intensity as a function of TA concentration and Figure 3b shows the F_0/F ratio for increasing TA concentrations, where F is the intensity of riboflavin in TA and F_0 is the fluorescence intensity of riboflavin in water in the absence of TA at pHs of 4.1 and 7.1.

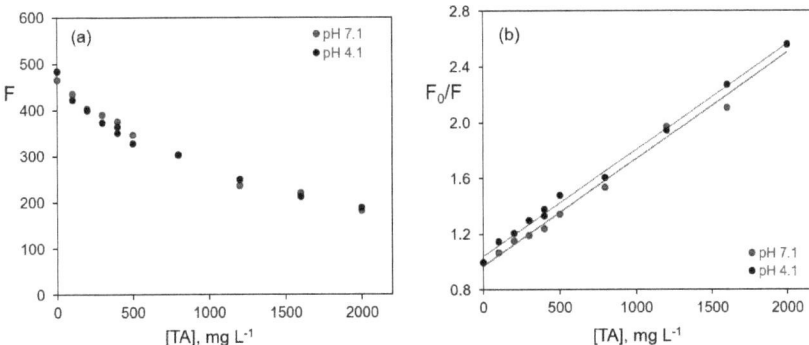

Figure 3. Riboflavin (0.6 mg L^{-1}) fluorescence intensity versus TA concentration (**a**) and Stern–Volmer plot (**b**). Each data point is the average of at least 2 independent measurements.

As can be observed in Figure 3b, the fluorescence quenching produced by TA on riboflavin follows the Stern–Volmer equation and the F_0/F ratio is linear up to at least 2.0 g L^{-1} of TA. The behavior is similar for both pHs, indicating that the dissociation of some TA hydroxyls does not influence the interaction between the two molecules.

The calculated values for the intercept, the slope with its standard deviations (SD), and the correlation coefficient (r), as well as the values of the quenching constants (K), either dynamic or static (K_{SV} or K_S) for both pHs, are collected in Table 2.

Table 2. Quenching constants (K) for riboflavin in TA solutions 2.0 g L^{-1} at pHs of 4.1 and 7.1.

	Intercept ± SD	Slope ± SD	r	K, L mg^{-1}
pH 4.1	1.05 ± 0.02	0.00075 ± 0.00002	0.9977	0.00075
pH 7.1	0.94 ± 0.06	0.00072 ± 0.00004	0.9920	0.00072

The quenching constants expressed considering the molar concentrations of TA are 1404 M^{-1} and 1348 M^{-1} for pHs of 4.1 and 7.1, respectively.

2.2. Riboflavin Fluorescence in the Presence of Graphene Dispersions in Tannic Acid 2.0 g L^{-1}

The dispersions of G in TA were prepared at pH 4.1, which is the pH of the aqueous solution of 2.0 g L^{-1} TA. Appropriate amounts of G were added to obtain G/TA mass ratios of 1.0% and 0.5%. The dispersions were prepared by keeping the G/TA mass ratio constant, which implies that the concentrations of G and TA vary when making the corresponding dilutions with water, and with a variable mass ratio G/TA, using TA 2.0 g L^{-1} as solvent. All the obtained dispersions were found to be stable, that is, they did not settle even after a few days (see photographs in Figure S1). Figure 4 shows the variation of the fluorescence intensity of riboflavin, for the two G percentages tested, versus G concentration (Figure 4a), the change in the F_0/F ratio versus TA concentration (Figure 4b), and as a function of G concentration (Figure 4c). In all cases, a distinction is made between a G/TA mass ratio constant or variable.

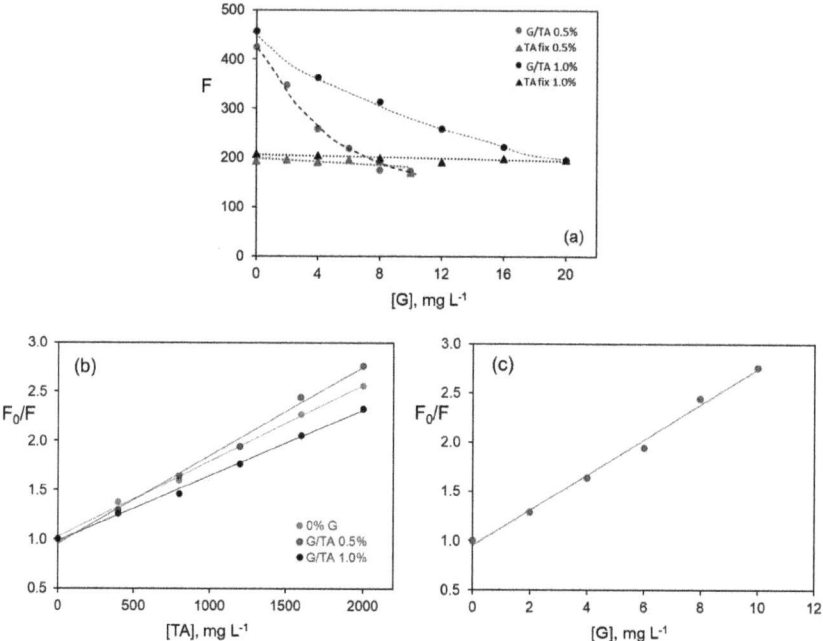

Figure 4. Change in F as a function of G concentration (**a**); change in F_0/F versus TA concentration (**b**); F_0/F versus G concentration (**c**) for G dispersions in TA 2.0 g L^{-1}, pH of 4.1 ([Riboflavin] = 0.6 mg L^{-1}; λ_{ex} = 455; λ_{em} = 520 nm).

From the results in Figure 4a, for a variable G/TA mass ratio that is fixed TA concentration (data marked with black and red triangles), the presence of G does not influence the decrease in riboflavin fluorescence. For both data series, the fluorescence values are almost the same to that obtained in the presence of an aqueous solution of TA 2.0 g L^{-1}.

However, when the G/TA mass ratio is constant (data marked with red and black circles), two sets of results are observed. For a G concentration of 1.0 wt%, the fluorescence values of riboflavin are the same as those obtained in the absence of G (Figure 3a), indicating again that G does not produce noticeable additional quenching to that found in TA solutions. A more pronounced quenching is observed for a G concentration of 0.5 wt%, probably due to the sum of contributions of G and TA. Assuming that for a constant G concentration of 1.0 wt% the predominant quenching is due to TA, Figure 4b shows the Stern–Volmer graphs, where slight modifications can be observed in the values of the quenching constants with the percentage of G. In any case, this modification is insignificant, and it can be considered that the constants are almost the same.

In the presence of a G concentration of 0.5 wt% there seems to be an additional quenching to that produced by TA, as shown in Figure 4a. If so, Stern–Volmer can be applied assuming that G is responsible for the process. Figure 4c shows the Stern–Volmer graph, the data are adjusted with a correlation coefficient of 0.997 and yield a quenching constant value of 0.0009 L mg^{-1} in G 0.5 wt%, about 29% higher than that obtained if the TA were the only quencher (0.0007 L mg^{-1} in the absence of G).

From the results obtained, the role of G in the quenching process is not clear. This may be due to an inadequete dispersion of G in the TA solutions, to the existence of notable differences in how TA interacts with G that causes poor accessibility of riboflavin to the dispersed G sheets, to the influence of pH, the TA concentration, or the G weight percentage. To try to elucidate these questions, a series of experiments introducing various variables by performing G dispersion in aqueous TA solutions are carried out, as detailed below.

The fluorescence of riboflavin decreases compared to that obtained in water by a factor $F/F_W = 0.43$ in the presence of TA for the two G percentages studied, but remains constant as the G concentration is increased, with $F/F_0 = 1.0$. However, for dispersions obtained via dilution with TA of the same concentration as the dispersion, no decrease in the fluorescence of riboflavin (0.6 mg L^{-1}) is observed, which should be related to the presence of G. This fact is corroborated by comparing with the results obtained for dispersions prepared by diluting with water. In this series, the fluorescence intensity decreases with increasing G and TA concentration up to the same value obtained for the G dispersion in 2.0 g L^{-1} TA.

Results suggest that at pH of 4.1, G dispersed in TA scarcely interacts with riboflavin, hence no change in its fluorescence is observed, either because G is not properly dispersed, or because the amount dispersed in this medium is very low. The interaction of riboflavin with TA is much stronger than that of riboflavin with G dispersed under these conditions. Therefore, it can be concluded that the quenching of TA on riboflavin occurs through the hydroxyl moieties of ribose and not via the aromatic rings, since G has a large number of aromatic rings all throughout its structure. A chemical model illustrating the potential interactions among G, TA, and riboflavin is shown in Scheme 1.

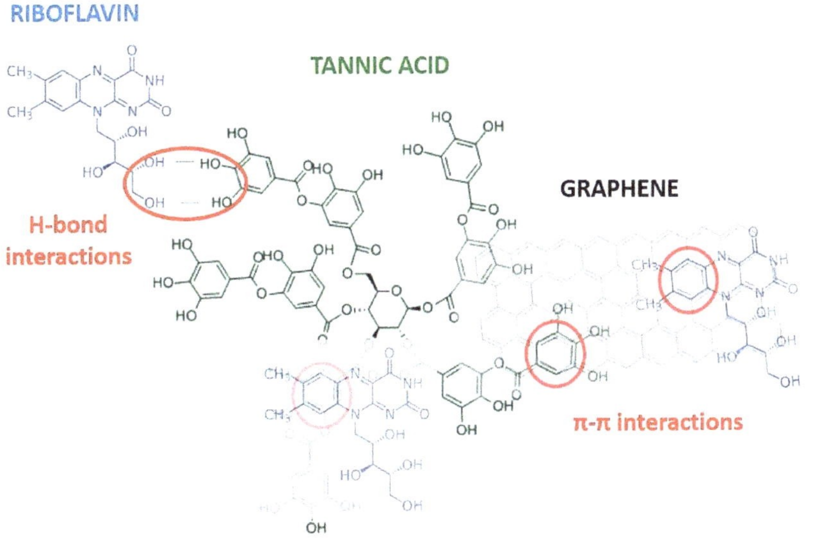

Scheme 1. Illustration of the interactions among G, TA, and riboflavin.

It can be inferred that the quenching phenomenon of riboflavin fluorescence in the G dispersions is due to the presence of TA, hence the quenching constants obtained decrease with increasing G concentration. Thus, as the G/TA weight ratio increases from 0.5% to 1.0%, the higher the G concentration in the dispersion, the smaller the slopes of the Stern–Volmer plots, that is, the lower the quenching constants. Therefore, it is found that the quenching magnitude is inversely proportional to the G concentration, while it is directly proportional to the TA concentration in each solution. This is due to the fact that the lower the concentration of G, the higher the TA concentration, and TA is a more efficient quencher than G.

The values of the intercepts, slopes along with their standard deviations, correlation coefficients, and quenching constants obtained from the Stern–Volmer plot for G dispersions

(0.5 wt% and 1.0 wt%) in TA at a concentration of 2.0 g L^{-1} and at pH of 4.1, are collected in Table 3. Values at the top and bottom of the table correspond to dispersions with a constant TA concentration and a constant G/TA weight ratio, respectively.

Table 3. Quenching constants (K) of riboflavin fluorescence by G dispersions (0.5% and 1.0 wt%) in TA 2.0 g L^{-1} at pH of 4.1 with constant G/AT ratio. Values at the top correspond to TA as quencher, and at the bottom, to G as quencher.

%G	a	b (TA)	r	K, L mg^{-1}
0	1.05 ± 0.02	0.00075 ± 0.00002	0.9977	0.00075
0.5	0.96 ± 0.04	0.00090 ± 0.00003	0.9970	0.00090
1.0	0.99 ± 0.01	0.00066 ± 0.00008	0.9998	0.00066
		b (G)		
0.5	0.96 ± 0.04	0.179 ± 0.007	0.9970	0.179
1.0	0.99 ± 0.01	0.0662 ± 0.0008	0.9998	0.0662

It can be observed that the values of the quenching constants do not change for the series with a constant TA concentration, regardless of the percentage of G in the dispersion. However, when the G/TA weight ratio is maintained, the quenching constants decrease since G concentration is double in the dispersion with 0.5%, compared to that with 1.0% for the same TA concentration (2.0 g L^{-1}). Therefore, it is confirmed that the quenching observed is caused by the TA, since the quenching constants decrease when increasing the amount of G in the dispersion (i.e., K decreases by a factor near 2 as G concentration is doubled). To make it clearer, the parameters obtained upon increasing G concentration are highlighted in blue, since they do not correspond to the quenching produced by G on riboflavin, but to the quenching induced by TA.

2.3. Study on the Improvement of G Dispersions in Aqueous Solutions of TA

2.3.1. Influence of the Sonication Time on the Preparation of 0.5 wt% G Dispersion in TA 2.0 g L^{-1} at pH 4.1

The optimization of the parameters influencing the preparation of the G dispersions is crucial to attain dispersions with optimal properties. Among them, the time of application of the ultrasonic probe to the mixture can be crucial. With the aim to elucidate whether the increase in the sonication time, during the preparation stage, can improve the state of dispersion of G in TA at pH 4.1, as has been previously reported for the exfoliation of graphite in TA [23], a similar experimental procedure to that previously described was used, albeit increasing the time the probe was applied to the G/TA mixture. Thus, the sonication time was increased from 5 to 15 min, and the fluorescence intensity obtained for both cases is shown in Figure S2.

No significant change in riboflavin fluorescence is observed with increasing probe sonication time; therefore, henceforth all dispersions have been prepared with a probe time of 5 min.

2.3.2. Influence of Solution pH on the Fluorescence of Riboflavin in the Presence of G Dispersions in TA 2.0 g L^{-1}

With the aim to improve the dispersion of G in TA solutions, the solution pH was increased to 7.1, at which, part of the hydroxyl groups of TA are dissociated. Dispersions are prepared following the same protocol. Upon preparation of the dispersions, two series of measurements are carried out, similarly to those made at pH 4.1, keeping the mass ratio G/TA constant, which implies that the concentrations of G and TA vary when making the corresponding dilutions with water, and with a variable mass ratio G/TA, using TA 2.0 g L^{-1} as solvent.

Figure 5 shows the comparison of the results previously obtained at pH 4.1 and those obtained at pH 7.1.

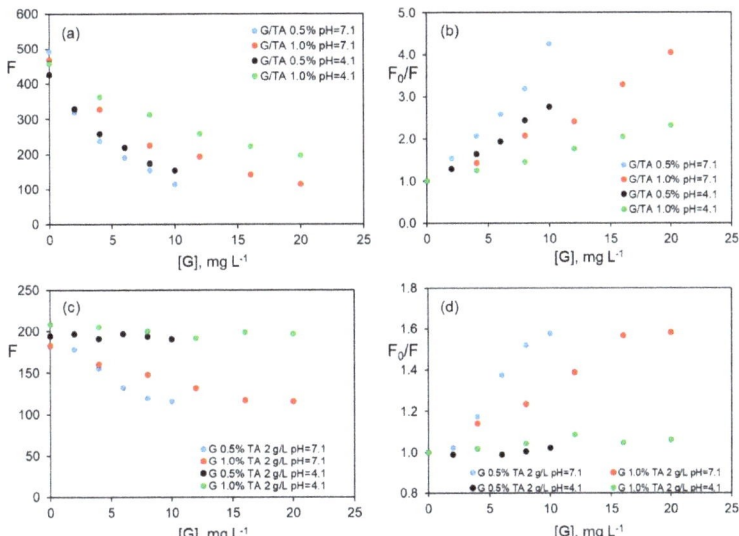

Figure 5. Variation of F (**a**) and F_0/F (**b**) versus [G] at a constant G/AT ratio and F (**c**) and F_0/F (**d**) versus [G] at constant [TA] for G dispersions in TA 2 g L^{-1}.

The change in the fluorescence intensity of riboflavin, $\lambda_{exc}/\lambda_{em} = 455/520$ nm, versus G concentration for dispersions prepared in TA 2.0 g L^{-1}, weight ratio G/TA constant (G 0.5% and 1.0% wt%) at the two working pH, is displayed in Figure 5a.

As can be observed, a stronger decrease in the fluorescence of riboflavin at pH 7.1, for the constant G/TA mass ratio and a 1.0 wt% of G, is observed. However, when the percentage of G is 0.5, no significant differences for both pH values are observed.

Figure 5b shows the fit to the Stern–Volmer equation for both pH values. Contrary to the behaviour found at pH 4.1, the data do not fit the model at pH 7.1, with an intercept value different from unity.

Figure 5c shows the results obtained for the change in fluorescence when the G/TA mass ratio is variable and the TA concentration is constant, for pH values of 4.1 and 7.1. Figure 5d shows the fit to the Stern–Volmer equation. In this case, higher quenching also occurs at pH 7.1.

Although the results are not conclusive, it is possible to assume that at pH 7.1, the interaction of a more dissociated TA with G allows for dispersions more accessible to the riboflavin molecule. Quenching is a result of two types of interactions, through the hydroxyl moieties of ribose of TA and via the aromatic rings of G. Therefore, there is a double contribution to the quenching phenomenon of riboflavin fluorescence.

2.3.3. Influence of TA Concentration in the G Dispersions on the Fluorescence of Riboflavin at pH 7.1

As can be observed in Figure 6, at pH 7.1 a fluorescence quenching phenomenon caused by the presence of G takes place, in addition to that caused by TA in the absence of the nanomaterial.

In solutions with a constant TA concentration (2.0 g L^{-1}), the decrease in the fluorescence of riboflavin is the same for the two G percentages studied, although the G concentration for the dispersion with 1.0% (20 mg L^{-1}) is twice that for 0.5% (10 mg L^{-1}).

In order to study the effect of TA concentration in the G/TA dispersions on the fluorescence of riboflavin, two sets of dispersions were prepared with a TA concentration of 0.5 g L^{-1} and G weight ratios of 2.0% and 4.0%. These percentages correspond to the same G concentrations added in TA dispersions 2.0 g L^{-1}, for G weight ratios of 0.5 and

1.0% w/w, setting the pH at 7.1. The results are plotted in Figure 7. By comparing the change in the fluorescence at a variable and constant TA concentration, it is possible to determine whether the interaction of riboflavin with G depends on the concentration of G and tannic acid or only on the weight ratio between both compounds.

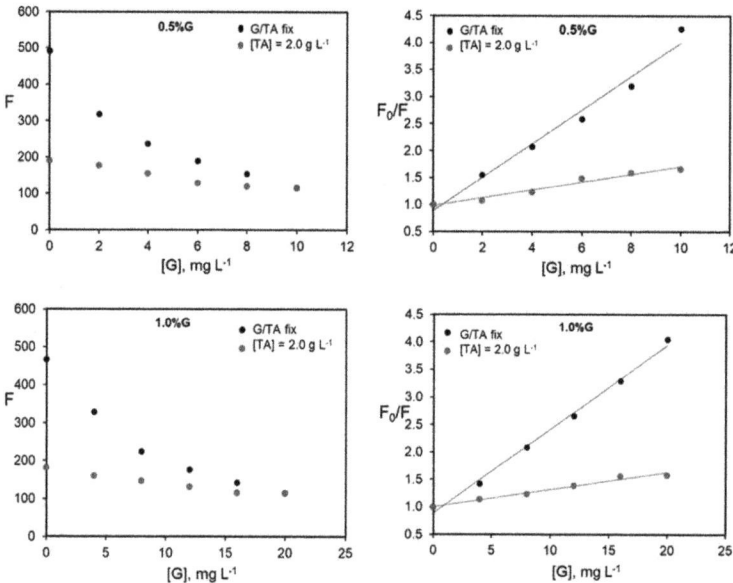

Figure 6. Change in the fluorescence intensity of riboflavin (0.6 mg L^{-1}) and F_0/F ratio as a function of G concentration for G dispersions 0.5 y 1.0 wt% in TA 2 g L^{-1} ($\lambda_{ex}/\lambda_{em}$ = 455/520 nm, pH = 7.1).

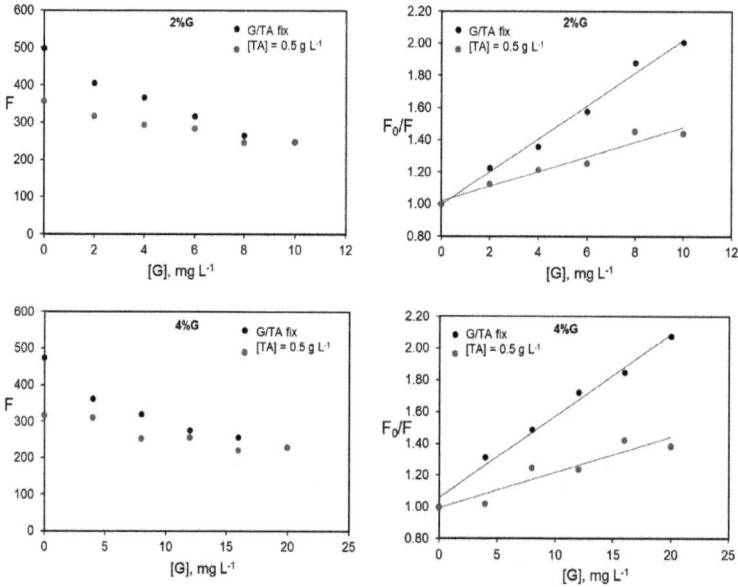

Figure 7. Change in the fluorescence intensity of riboflavin (0.6 mg L^{-1}) and F_0/F ratio as a function of G concentration for G dispersions 2.0 wt% and 4.0 wt% in TA 0.5 g L^{-1} ($\lambda_{ex}/\lambda_{em}$ = 455/520 nm; pH = 7.1).

The comparison of Figures 6 and 7 reveals that for the same G concentrations, dispersions in TA 0.5 g L^{-1} lead to a smaller decrease in riboflavin fluorescence than those prepared in TA 2.0 g L^{-1}. This behaviour was expected, since TA is the main factor responsible for the decrease in riboflavin fluorescence intensity.

For the two G percentages studied, smaller slopes are found for a TA concentration of 0.5 g L^{-1} and, as mentioned previously, for a TA concentration of 0.5 g L^{-1}, the slope found for dispersions with a G/TA weight ratio of 4.0% is approximately half that found for dispersions with a weight ratio of 2.0%.

The decrease in the intensity of riboflavin fluorescence is considerably stronger when TA concentration changes than when it remains constant. While TA interacts strongly with riboflavin, the interaction of G with the vitamin is much weaker. This behavior is the same for G dispersions in TA 0.5 and 2.0 g L^{-1}. When TA concentration (0.5 g L^{-1}) is constant, a straight line is obtained with a small slope, lower than that in TA 2.0 g L^{-1}. In addition, for this TA concentration, data variability is greater, especially for the dispersion with a G/TA weight ratio of 4.0%, which was tested three times to check for reproducibility and in all cases the data lead to a poor linear fit.

Figure 8 shows F_0/F ratio (Stern–Volmer plot) as a function of G and TA concentration for G dispersions in TA 2.0 g L^{-1} and TA 0.5 g L^{-1} and the four percentages of G (0.5, 1.0, 2.0 and 4.0%) for a constant G/TA ratio. The values in the absence of G have also been included for comparison.

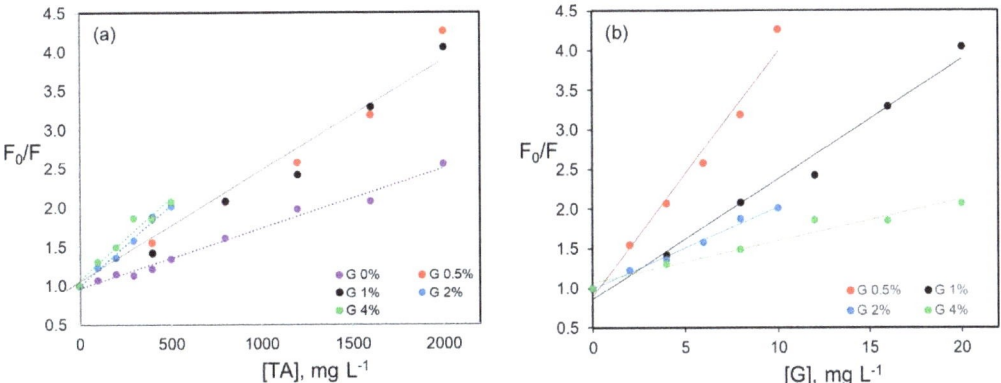

Figure 8. F_0/F ratio as a function of TA concentration (**a**) and G concentration (**b**). [Riboflavin] = 0.6 mg L^{-1}, $\lambda_{ex}/\lambda_{em}$ = 455/520 nm, pH = 7.1.

It can be observed from Figure 8a that the presence of G in the medium increases the quenching effect, and as the concentration of TA in the medium increases, this effect is less pronounced than that found for the lowest concentration, 0.5 g L^{-1}. Further, for the same TA concentration, the F_0/F ratio does not depend on the G percentage. On the other hand, when F_0/F ratio is plotted against the G concentration, the slope decreases as the percentage of G in the dispersion increases, regardless of the TA concentration.

The fluorescence intensity ratios (F_o/F_w, F/F_o and F/F_w), obtained for the two TA concentrations and the different G weight percentages studied in this work, are compared in Figure 9.

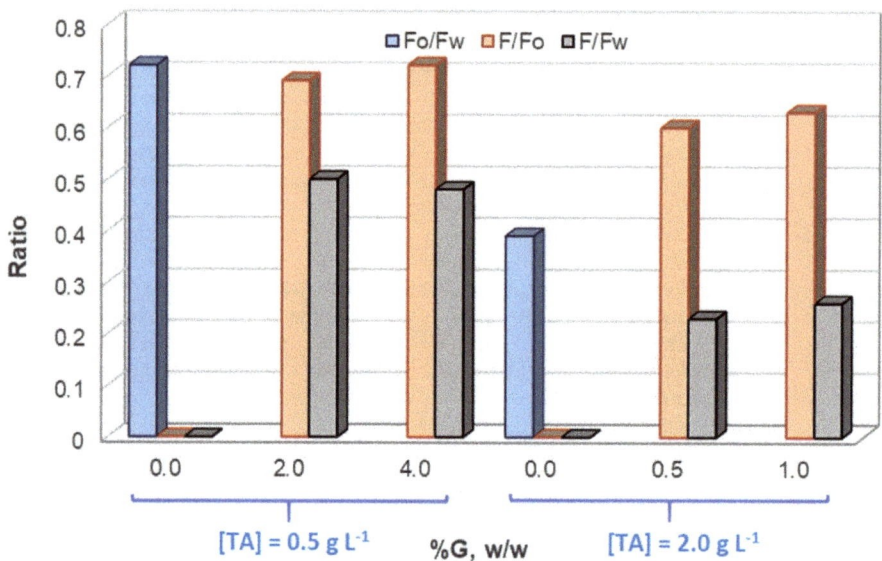

Figure 9. Plot of the fluorescence intensity ratios of riboflavin in the different media. F_W: water; F_0: Tannic Acid; F: G/TA dispersions.

It can be observed that the quenching effect caused by TA is directly proportional to its concentration (28% and 62% in TA 0.5 and 2.0 g L^{-1}, respectively). An additional decrease in riboflavin fluorescence is found in the presence of G, and the effect is independent to the G percentage, while proportional to the TA concentration.

The same behaviour is found when the quenching efficiency in TA solutions is compared with that obtained for G dispersions in TA, regardless of the percentage of G. Thus, the quenching is weaker for dispersions in TA 0.5 g L^{-1} (efficiency of 50% and 52% for G percentages of 2.0% and 4.0%, respectively) than in 2.0 g L^{-1} (efficiency of 77% and 74% for the same G percentages). This suggests that the interaction with riboflavin is stronger in the dispersions with the highest TA concentration. Finally, for dispersions with a constant G/TA ratio, the fluorescence quenching (i.e., F/F_W) is again independent to the G percentage in the dispersion with increasing TA concentration. In this case, the decrease in fluorescence intensity corresponds to a synergistic effect of both G and TA in the dispersion.

2.4. Quenching Constants of Riboflavin in the Presence of G Dispersions in TA for Variable TA Concentration

Figure 9 shows that the quenching effect induced by TA considerably increases in the presence of G, albeit does not depend on the G/TA weight ratio in the dispersion.

Table 4 lists the fluorescence quenching constants of riboflavin for G dispersions in TA 2.0 and 0.5 g L^{-1} for different G/TA weight ratios, for variable G and TA concentrations. The coefficients of the linear fits along with their standard deviations (SD) are also included. In all cases, points that deviated from the linearity were removed.

Table 4. Quenching constants of riboflavin for G dispersions (0.5 wt% and 1.0 wt%) in TA 2.0 g L^{-1} at pH = 7.1 as well as G dispersions (2.0% and 4.0% w/w) in TA 0.5 g L^{-1} for a constant G/TA ratio.

%G	a ± SD	b ± SD (TA)	r	K, L mg^{-1}
0	1.02 ± 0.02	0.00070 ± 0.00002	0.9982	0.00070
0.5	1.00 ± 0.02	0.00135 ± 0.00002	0.9995	0.00135
1.0	0.89 ± 0.08	0.00154 ± 0.00006	0.9976	0.00154
		b ± SD (G)		
0.5 (n = 4)	0.97 ± 0.02	0.2705 ± 0.005	0.9995	0.2705
1.0 (n = 5)	0.8925 ± 0.072	0.154 ± 0.006	0.9976	0.154
%G	a ± SD	b ± SD (TA)	r	K, L mg^{-1}
0	1.02 ± 0.02	0.00070 ± 0.00002	0.9982	0.00070
2.0	1.00 ± 0.02	0.00199 ± 0.00008	0.9975	0.00199
4.0	1.05 ± 0.03	0.0020 ± 0.0001	0.9953	0.0020
		b ± SD (G)		
2.0	1.00 ± 0.02	0.100 ± 0.004	0.9975	0.100
4.0	1.05 ± 0.03	0.051 ± 0.003	0.9953	0.051

The parameters obtained upon increasing G concentration are again highlighted in blue, since these also include the influence of TA on the vitamin fluorescence intensity. The effect of both G and TA concentration has been simultaneously assessed using multiple regression for a 95% confidence level, and the regression coefficients obtained along with their confidence interval (CI), are displayed in Table 5.

Table 5. Multiple linear regression analysis (95% confidence) for the fluorescence of riboflavin in the presence of G dispersions in TA for a constant G/TA ratio ($n = 32$; G wt% = 0, 0.5, 1.0, 2.0, 4.0).

F/F$_0$ = a + b [G] + c [TA]					
a ± C.I	b ± C.I.	c ± C.I.	R^2, %	K (G), L mg^{-1}	K (TA), L mg^{-1}
0.95 ± 0.05	0.050 ± 0.005	0.00091 ± 0.00005	95.21	0.05	0.0009

CI: 95.0% confidence interval for the coefficients; n: sample size.

The experimental values of F$_0$/F as a function of the predicted values by the multiple linear regression are plotted in Figure S3, showing a correlation of 95%. As can be observed in Table 5, this equation allows for the calculation of the simultaneous effect of TA and G upon the fluorescence of riboflavin, evaluating the quenching contribution of each variable.

When F$_0$/F values are plotted independently versus TA concentration, quenching constants of 0.020 and 0.014 L mg^{-1} are found for TA concentrations of 0.5 and 2.0 g L^{-1}, respectively, irrespective the percentage of G in both cases. For G, K values of 0.27 and 0.15 L mg^{-1} are obtained for dispersions with 0.5% and 1.0% G in TA 2.0 g L^{-1}, respectively. However, those with 2.0% and 4.0% G in TA 0.5 g L^{-1} show smaller K values, 0.10 and 0.05 L mg^{-1}, respectively. It is difficult in this case to separate the effect of TA and G, since the concentrations of both compounds varies simultaneously. In the multiple regression analysis, the influence of both concentrations on the F$_0$/F ratio is taken into account, and a linear relationship is obtained with an intercept that does not statistically differ from zero. Further, the coefficients are 0.05 for G, which is the lowest value of those obtained with the two concentrations of TA, and 0.0009 for TA, which is very close to that obtained for this compound in the absence of G. It can be concluded that the presence of G enhances the riboflavin-TA interaction, albeit it appears that TA is the main contributor to the quenching effect. This is consistent with the fact that the slope of F_0/F versus G concentration is considerably higher than that versus TA.

2.5. Quenching Constants of Riboflavin in the Presence of G Dispersions in TA for a Constant TA Concentration

The change in F_0/F ratio versus G concentration for both variable and constant TA concentrations were plotted in Figures 6 and 7. For a constant TA concentration, the observed changes can be only attributed to the variation of G concentration. In Figure 10, the variation of the F_0/F ratio with G concentration for all the dispersions, when TA concentration is kept constant, can be observed.

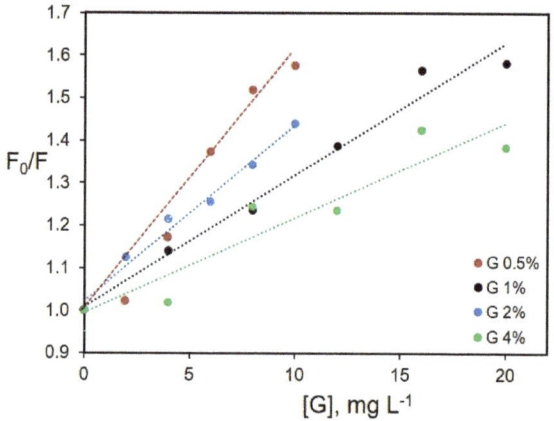

Figure 10. F_0/F ratio as a function of G concentration for a constant TA concentration. [Riboflavin] = 0.6 mg L^{-1}, $\lambda_{ex}/\lambda_{em}$ = 455/520 nm, pH = 7.1.

Table 6 lists the fluorescence quenching constants of riboflavin for G dispersions in TA 2.0 and 0.5 g L^{-1} for a constant TA concentration. The intercept and slope values for the linear fits along with their standard deviation (SD) and the correlation constant are also included. In these cases, K values correspond only to the presence of G because TA decreases the fluorescence of riboflavin, but in the same way in all solutions.

Table 6. Quenching constants of riboflavin for G dispersions 0.5 wt% and 1.0 wt% in TA 2.0 g L^{-1} and G dispersions 2.0 wt% and 4.0 wt% in TA 0.5 g L^{-1} measured for a constant TA concentration at pH 7.1. Multiple linear regression analysis (95% confidence) is also included (F/F$_0$ = a + b [G]).

%G; [TA]	a ± CI	b ± CI	r	K, L mg^{-1}
0.5; 2.0 g L^{-1}	1.01 ± 0.03	0.060 ± 0.004	0.9953	0.06
1.0; 2.0 g L^{-1}	1.01 ± 0.02	0.029 ± 0.001	0.9967	0.03
2.0; 0.5 g L^{-1}	1.01 ± 0.01	0.042 ± 0.002	0.9965	0.04
4.0; 0.5 g L^{-1}	1.01 ± 0.05	0.024 ± 0.004	0.9683	0.02
Multiple Linear Regression (n = 19)	1.02 ± 0.04	0.031 ± 0.005	90.5%	0.03

CI: 95.0% confidence interval for the coefficients; n: sample size.

The multiple linear regression is also included in the table, although TA concentration has only two values (0.5 y 2.0 g L^{-1}), and therefore, the influence of this variable is not significant (*p*-value = 0.54).

The F_0/F ratio shows a linear fit with a good correlation coefficient (90.5%), nonetheless five data had to be removed, since the variability was considerably higher than the one found when plotted versus TA, especially for the dispersion with 4.0% G.

The values found for the quenching constants are higher for the lowest G percentage of each TA concentration studied, similar to the behaviour described above, for variable TA concentrations. For the dispersion with 2.0 g L^{-1} of TA and 0.5 wt% G, a K value of 0.06 was obtained when the two points corresponding to the lowest concentrations were removed (open circles in Figure 10), albeit the Stern–Volmer plot is not linear, hence data corresponding to this dispersion are highlighted in colour. For the dispersion with 1.0 wt% G, a K value of 0.029 was obtained, and for those with 2.0 and 4.0% G in 0.5 g L^{-1} TA, the values were 0.042 and 0.024 L mg^{-1} respectively. However, in these dispersions, and especially for that with 4.0%G, a high data variability was found that was repeated when different series of dispersions were prepared under the same conditions. Therefore, the relationship between the experimental data and the Stern–Volmer equation is not as good as required, and some values must be removed.

The fit obtained for the multiple regression taking, into account all the G and TA concentrations studied in this work, is worse than that calculated for variable TA concentration, and 5 data had to be removed to obtain a good correlation coefficient (90.5%). Figure S4 shows the relationship between the experimental F_0/F values and those predicted by the mathematical model.

The different variabilities in the measurements found for the two series studied in this work may be due to the fact that when the initial dispersion is diluted with water, the G/TA ratio remains constant, hence the system formed between both compounds is stationary. However, by diluting the dispersion with tannic acid, this G/TA ratio no longer remains constant and therefore the G/AT system is being altered. In such cases, not only does the TA present in the dispersion interact with riboflavin, but also, the free TA used for the dilution interacts with it, destabilizing the system and thus altering the quenching process, causing a high variability in the data. The quenching constants obtained for variable G concentration are much lower than those obtained with variable TA concentration, which corroborates the strong influence of TA on the value of such constants due to its stronger interaction with riboflavin than G.

2.6. Morphology of G Dispersions in TA

The structure and morphology of G dispersions in TA at pHs of 4.1 and 7.1 have been investigated by Transmission Electron Microscopy (TEM), and typical micrographs obtained for both pHs at different G concentrations are compared in Figure 11.

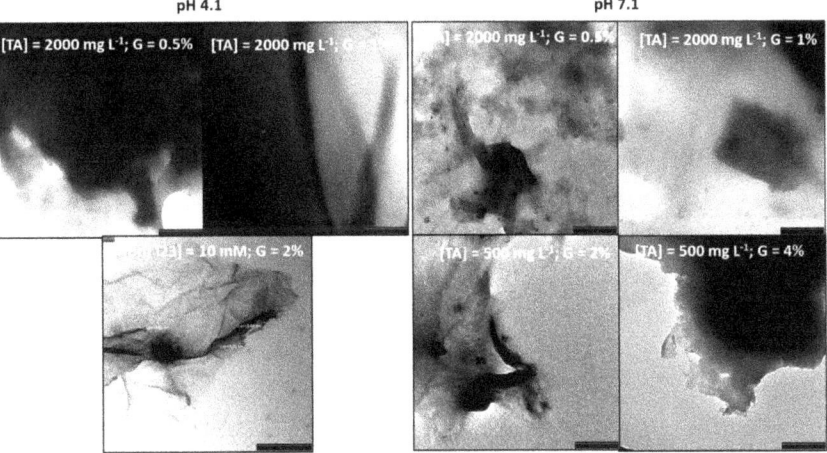

Figure 11. Representative TEM micrographs of G dispersions (0.5 wt% and 1.0 wt%) in TA 2.0 g L^{-1} and dispersions (2.0 wt% and 4.0 wt%) in TA 0.5 g L^{-1} at pHs of 4.1 and 7.1. 40,000× magnification. For comparative purposes, a G dispersion (2.0 wt%) in 10 mM Brij L23 is also included.

For the two pHs, the G sheets seem to be highly covered by TA, especially at low G percentages, and display a soapy aspect. Further, small black dots can be observed in the images with low G content, which could arise from TA aggregates, as reported previously for TA coatings on polymeric nanoparticles [36].

At pH 4.1, the average thicknesses obtained are 6.32 ± 0.7, 4.93 ± 0.3 and 3.58 ± 0.2 for G weight percentages of 0.5, 1.0 and 2.0 wt%, respectively. At pH 7.1, the mean values are 4.39 ± 0.6, 2.67 ± 0.5, 2.25 ± 0.4 and 2.14 ± 0.4 for G weight percentages of 0.5, 1.0, 2.0 and 4.0 wt%. Accordingly, at pH 7.1, the G sheets appear to be better dispersed and are thinner than at pH 4.1. The improved exfoliation found at a pH of 7.1 is likely related to the higher dissociation degree of the hydroxyl groups of TA. Thus, at the higher pH, TA should be negatively charged, and when adsorbed onto the G sheets, it would provoke a repulsion between the sheets. G layers wrapped in TA are stabilized against re-aggregation by the repulsive electrostatic interactions between nearby TA-coated flakes, producing quenching on the riboflavin fluorescence.

For both pHs, the thickness decreases with the increasing G percentage. This can be rationalized considering that the higher the G concentration, the lower the amount of TA in the dispersion, hence the nanomaterial is less covered by TA, and the thickness diminishes. These observations are in agreement with the results obtained from fluorescence measurements (Figure 4).

In order to compare with the morphology obtained upon dispersion of G in synthetic surfactants, a TEM micrograph of G 2.0 wt% dispersed in 10 mM Brij L23 is also included in Figure 11. In this case, the mean thickness found is 3.30 ± 0.4, comparable to those found for dispersions in TA. Indeed, the exfoliation attained herein may be as effective as that obtained for G dispersions in synthetic surfactant solutions [16], which led to flake thicknesses in the range of 1–8 nm. Further, it was found that the exfoliation level depends on the surfactant nature: the thinnest layers were observed for dispersions in the non-ionic surfactant, in which the stabilization mechanism seems to be based on steric and polar effects. Overall, TEM images obtained in this work confirm the good exfoliation of G upon ultrasonication in TA aqueous solutions and indicate that the degree of exfoliation depends on the G/TA weight ratio.

2.7. Comparison of the Quenching effect of TA, Synthetic Surfactants and Polymers on the Fluorescence of Riboflavin

Ionic surfactants induce fluorescence quenching on riboflavin, the highest effect caused by the anionic surfactant SDS [16]. Both cationic surfactants CTAB and DTAB attenuate the fluorescence to a lesser extent than SDS, indicating that the quenching magnitude does not depend on the length of the hydrocarbon chain, which is albeit influenced by the charge (i.e., between 0.03 and 0.09 L mg^{-1}, depending on the percentage of G with respect to the surfactant). The quenching constants obtained are lower than those obtained with TA. The nonionic Brij L23 interacts strongly with riboflavin and disperses G very well, giving constants of 0.05 and 0.01 L mg^{-1} for surfactant concentrations of 0.002 and 0.010 M, respectively.

Other dispersants with a polymeric nature, such as PEG [18], provide even lower constants and the relationship between F_0/F and G concentration is not linear like Poloxamer 407 [19]. Therefore, it can be concluded that TA is an effective dispersing agent for G, as already observed in relation to graphite exfoliation [24], and its dispersion efficiency is better at pH 7.1 than at 4.1, since the interaction of G with riboflavin is weak at this low pH. However, at pH 7.1 the exfoliation is better given that TA is negatively charged, causing the repulsion between neighbouring G coated sheets. Overall, results corroborate that this antioxidant biocompatible compound is an effective dispersant for graphene-based nanomaterials and can be used as a green alternative to conventional surfactants and synthetic polymers for the determination of biomolecules.

3. Materials and Methods

3.1. Reagents

All the reagents were of analytical grade. Graphene, made up of less than 6 sheets of thickness less than 2 nm, was supplied by Avanzare Innovación Tecnológica SL. (Logroño, Spain). Tannic acid and riboflavin were purchased from Sigma Aldrich (Madrid, Spain). Sodium hydroxide was from Panreac (Barcelona, Spain), phosphoric acid was from Sigma Aldrich (Madrid, Spain), and the ultra-pure water was obtained in a Milli-Q system from Millipore (Milford, CT, USA).

3.2. Instrumentation

Fluorescence spectra were recorded on a Perkin-Elmer LS-50B fluorimeter (Walthman, MA, USA) at 25 ± 1 °C equipped with a Thermomix BU bath from Braun. The excitation and emission slits were set at 5 nm. The cuvettes were quartz with 1 cm light path. The software used to record the spectra was FL WinLab from Perkin-Elmer (Walthman, MA, USA).

G dispersions were prepared with an Elmasonic S40 ultrasound bath, a Hielscher UP400S ultrasound probe (Teltow, Germany), and an Orto Alresa Digicen centrifuge (Madrid, Spain).

Transmission electron microscopy (TEM) measurements were performed on a Zeiss EM-10C/CR microscope (Oberkochen, Germany) with a voltage of 60 kV.

3.3. Procedure

3.3.1. Preparation of the Solutions

Tannic acid solutions of 0.5 g L^{-1} and 2.0 g L^{-1} in ultrapure water were prepared at two different pHs, 4.1 and 7.1, at which some tannic acid hydroxyl groups are ionized. The pH was adjusted with a 0.01 M sodium hydroxide (NaOH) solution. A stock solution of riboflavin was prepared with 28% v/v H_3PO_4 buffer and stored at 4 °C in glass beakers in the dark. Working solutions were prepared by dilution in the buffer, and the riboflavin concentration for studying the interaction with TA and with G dispersions in TA was of 0.6 mg L^{-1}.

3.3.2. Preparation of G Dispersions in TA

Dispersions of G in TA were prepared by weighing the appropriate amount of G and adding the necessary volume of a TA aqueous solution, 0.5 g L^{-1} or 2.0 g L^{-1}, until obtaining the desired G/TA mass ratio, in a final volume of 100 mL. The mixture was then placed in an ultrasonic bath for 30 min followed by sonication with the ultrasonic probe for 5 min at a power of 160 W, and then centrifuged for 1 h at 4000 rpm. The supernatant was collected, and the solid remaining at the bottom of the tube was separated.

From the supernatant, the final dispersions were prepared in two ways. First, by diluting with ultrapure water in order to obtain variable concentrations of G and TA while maintaining the G/TA weight ratio constant. The other way was performed via dilution with TA of the same concentration used to prepare the dispersion, in order to vary the G concentration while keeping the TA concentration constant, that is, the G/TA weight ratio variable. For each type of dispersion, four G weight ratios were prepared: 0.5, 1.0, 2.0, and 4.0 wt%.

3.3.3. Riboflavin Fluorescence Spectra

Firstly, the fluorescence spectra of riboflavin, 0.6 mg L^{-1}, were recorded, at T = 25.0 ± 0.1 °C, as three-dimensional contour graphs to choose the optimal excitation and emission wavelengths. The recorded spectra were obtained in water, in the 0.5 and 2.0 g L^{-1} TA aqueous solutions, at the pHs of 4.1 and 7.1, and in presence of different G/TA dispersions. The initial excitation wavelength was set at 220 nm, and 25 spectra were registered with an increment of 10 nm.

Statistical calculations were performed using the Statgraphics Centurion XVII program.

3.3.4. Transmission Electron Microscopy (TEM)

G dispersions in TA solutions at pHs of 4.1 and 7.1 were observed with a transmission electron microscope in order to assess the influence of the G/TA weight ratio and TA concentration on the state of dispersion of the nanomaterial. At least 20 measurements at different locations of the sample surface were carried out, and the average thickness along with the standard deviation are provided.

4. Conclusions

The effectiveness of TA as a dispersing agent for G in aqueous solutions has been carefully examined under different experimental conditions. TA provoked quenching of riboflavin fluorescence, and its magnitude depended on a number of parameters, including the TA concentration, the solution pH, and the G/TA weight ratio in the dispersion. Results indicate similar quenching effects for solution with pHs of 4.1 and 7.1, while it became stronger with increasing TA concentration. The interaction between both molecules was a result of hydrogen bonds between their hydroxyl groups and π-π stacking between their aromatic rings.

At pH 4.1, the fluorescence intensity was about the same in the presence and the absence of G, indicating that the nanomaterial dispersed in TA hardly alters the fluorescence of riboflavin, since the interaction between both molecules should be very weak. The sonication time applied during the preparation of the dispersions at pH 4.1 did not change the interaction of riboflavin with G.

At pH 7.1, G dispersed in TA interacted with riboflavin, hence a synergistic effect of both on attenuating the fluorescence of the vitamin was detected. The decrease in fluorescence was stronger for dispersions with the lowest G percentages given that their TA concentration is higher, and this compound is the mayor contributor to the quenching effect. For dispersions with a constant TA concentration, the quenching magnitude depends on both the G/TA weight ratio and the TA concentration. In these measurements, a high data variability was found, due to the distribution of riboflavin between the G dispersed in TA at a constant G/TA weight ratio in equilibrium and the alteration produced by the added TA with the same concentration that changes the G/TA ratio, which is difficult to attain an equilibrium position.

The quenching of riboflavin caused by TA follows the Stern–Volmer relationship up to concentrations of at least 2.0 g L^{-1} (1.2 mM), and up to G contents of 20 mg L^{-1}. This linear relationship between F_0/F and TA concentration can be used to determine the concentration of this antioxidant compound in the absence of other molecules that induce fluorescence quenching. Overall, it is demonstrated that this biocompatible molecule is an effective and environmentally friendly substitute for synthetic surfactants and polymers as dispersant for graphene-based nanomaterials and would aid in suppressing agglomerates and improving material processability and properties.

Supplementary Materials: The following are available online at https://www.mdpi.com/article/10.3390/ijms22105270/s1, Figure S1: Representative photographs of a G 0.5% dispersion in TA 2.0 g L^{-1} at pH 4.1 (a) and pH 7.1 (b) after one week. Figure S2: Comparison of the effect of time sonication (5 and 15 min) for G 0.5 wt% dispersions in tannic acid 2.0 g L^{-1} for the two series studied in this work ($\lambda_{ex}/\lambda_{em}$= 455/520 nm; pH = 4.1. Figure S3: Multiple linear regression for F_0/F obtained with all concentrations of G and TA in solutions with variable TA concentration. [Riboflavin] = 0.6 mg L^{-1}, $\lambda_{exc}/\lambda_{em}$= 455/520 nm, pH = 7.1. Figure S4: Multiple linear regression for F_0/F obtained with all G concentrations and TA 2.0 g L^{-1} and 0.5 g L^{-1} for a constant TA concentration 0.5 mg L^{-1} and 2.0 mg L^{-1}. [Riboflavin] = 0.6 mg L^{-1} 455/520 nm, pH = 7.1.

Author Contributions: M.B.-C. and L.G.-F. performed the experiments; M.P.S. supervision and writing—original draft preparation; S.V.-L., A.M.D.-P. and M.P.S. analyzed and discussed the data, writing—review and editing. All authors have read and agreed to the published version of the manuscript.

Funding: This research was funded by Spanish Ministry of Science and Innovation and Universities (MICIU), grant number PGC2018-093375-B-I00.

Data Availability Statement: The authors confirm that the data supporting the findings of this study are available within the article and/or its supplementary materials.

Acknowledgments: Financial support from the Community of Madrid within the framework of the Multi-year Agreement with the University of Alcalá in the line of action "Stimulus to Excellence for Permanent University Professors", Ref. EPU-INV/2020/012, is also gratefully acknowledged.

Conflicts of Interest: The authors declare no conflict of interest.

References

1. Novoselov, K.S.; Geim, A.K.; Morozov, S.V.; Jiang, D.; Zhang, Y.; Dubonos, S.V.; Grigorieva, I.V.; Firsov, A.A. Electric field effect in atomically thin carbon films. *Science* **2004**, *306*, 666–669. [CrossRef] [PubMed]
2. Geim, A.K.; Novoselov, K.S. The rise of graphene. *Nat. Mater.* **2007**, *6*, 183–191. [CrossRef]
3. Balandin, A.A.; Ghosh, S.; Bao, W.; Calizo, I.; Teweldebrhan, D.; Miao, F.; Lau, C.N. Superior thermal conductivity of single-layer graphene. *Nano Lett.* **2008**, *8*, 902–907. [CrossRef]
4. Díez-Pascual, A.M.; Gómez-Fatou, M.A.; Ania, F.; Flores, A. Nanoindentation in polymer nanocomposites. *Prog. Mater. Sci.* **2015**, *67*, 1–94. [CrossRef]
5. Salavagione, H.J.; Díez-Pascual, A.M.; Lázaro, E.; Vera, S.; Gómez-Fatou, M.A. Chemical sensors based on polymer composites with carbon nanotubes and graphene: The role of the polymer. *J. Mater. Chem. A* **2014**, *2*, 14289–14328. [CrossRef]
6. Ibrahim, W.A.W.; Nodeh, H.R.; Sanagi, M.M. Graphene-based materials as solid phase extraction sorbent for trace metal ions, organic compounds, and biological sample preparation. *Crit. Rev. Anal. Chem.* **2016**, *46*, 267–283. [CrossRef]
7. Coleman, J.N. Liquid exfoliation of defect-free graphene. *Acc. Chem. Res.* **2013**, *46*, 14–22. [CrossRef] [PubMed]
8. Ciesielski, A.; Samorì, P. Grapheneviasonication assisted liquid-phase exfoliation. *Chem. Soc. Rev.* **2014**, *43*, 381–398. [CrossRef]
9. Xu, Y.; Bai, H.; Lu, G.; Li, C.; Shi, G. Flexible graphene films via the filtration of water-soluble noncovalent functionalized graphene sheets. *J. Am. Chem. Soc.* **2008**, *130*, 5856–5857. [CrossRef] [PubMed]
10. Lotya, M.; Hernandez, Y.; King, P.J.; Smith, R.J.; Nicolosi, V.; Karlsson, L.S.; Blighe, F.M.; De, S.; Wang, Z.; McGovern, I.T.; et al. Liquid phase production of graphene by exfoliation of graphite in surfactant/water solutions. *J. Am. Chem. Soc.* **2009**, *131*, 3611–3620. [CrossRef]
11. Wang, S.; Yi, M.; Shen, Z. The effect of surfactants and their concentration on the liquid exfoliation of graphene. *RSC Adv.* **2016**, *6*, 56705–56710. [CrossRef]
12. Smith, R.J.; Lotya, M.; Coleman, J.N. The importance of repulsive potential barriers for the dispersion of graphene using surfactants. *New J. Phys.* **2010**, *12*, 125008. [CrossRef]
13. Guardia, L.; Fernández-Merino, M.; Paredes, J.; Solís-Fernández, P.; Villar-Rodil, S.; Martínez-Alonso, A.; Tascón, J. High-throughput production of pristine graphene in an aqueous dispersion assisted by non-ionic surfactants. *Carbon* **2011**, *49*, 1653–1662. [CrossRef]
14. Andrés, M.P.S.; Díez-Pascual, A.M.; Palencia, S.; Torcuato, J.S.; Valiente, M.; Vera, S. Fluorescence quenching of α-tocopherol by graphene dispersed in aqueous surfactant solutions. *J. Lumin.* **2017**, *187*, 169–180. [CrossRef]
15. Vera-López, S.; Martínez, P.; Andrés, M.S.; Díez-Pascual, A.; Valiente, M. Study of graphene dispersions in sodium dodecylsulfate by steady-state fluorescence of pyrene. *J. Colloid Interface Sci.* **2018**, *514*, 415–424. [CrossRef]
16. Mateos, R.; Vera, S.; Valiente, M.; Díez-Pascual, A.M.; Andrés, M.P.S. Comparison of anionic, cationic and nonionic surfactants as dispersing agents for graphene based on the fluorescence of riboflavin. *Nanomaterials* **2017**, *7*, 403. [CrossRef] [PubMed]
17. Mateos, R.; García-Zafra, A.; Vera-López, S.; Andrés, M.P.S.; Díez-Pascual, A.M. Effect of graphene flakes modified by dispersion in surfactant solutions on the fluorescence behaviour of pyridoxine. *Materials* **2018**, *11*, 888. [CrossRef]
18. Díez-Pascual, A.M.; García-García, D.; Andrés, M.P.S.; Vera, S. Determination of riboflavin based on fluorescence quenching by graphene dispersions in polyethylene glycol. *RSC Adv.* **2016**, *6*, 19686–19699. [CrossRef]
19. Díez-Pascual, A.M.; Hermosa-Ferreira, C.; San Andrés, M.P.; Valiente, M.; Vera, S. Effect of graphene and graphene oxide dis-persions in poloxamer-407 on the fluorescence of riboflavin: A comparative study. *J. Phys. Chem.* **2017**, *121*, 830–843. [CrossRef]
20. Ahmed, G.H.G.; Laíño, R.B.; Calzón, J.A.G.; García, M.E.D. Fluorescent carbon nanodots for sensitive and selective detection of tannic acid in wines. *Talanta* **2015**, *132*, 252–257. [CrossRef]
21. Nepka, C.; Asprodini, E.; Kouretas, D. Tannins, xenobiotic metabolism and cancer chemoprevention in experimental animals. *Eur. J. Drug Metab. Pharm.* **1999**, *24*, 183–189. [CrossRef] [PubMed]
22. Ghigo, G.; Berto, S.; Minella, M.; Vione, D.; Alladio, E.; Nurchi, V.M.; Lachowicz, J.; Daniele, P.G. New insights into the protogenic and spectroscopic properties of commercial tannic acid: The role of gallic acid impurities. *New J. Chem.* **2018**, *42*, 7703–7712. [CrossRef]
23. Jia, M.-Y.; Xu, L.-S.; Li, Y.; Yao, C.-L.; Jin, X.-J. Synthesis and characterization of graphene/carbonized paper/tannic acid for flexible composite electrodes. *New J. Chem.* **2018**, *42*, 14576–14585. [CrossRef]

24. Zhao, S.; Xie, S.; Zhao, Z.; Zhang, J.; Li, L.; Xin, Z. Green and high-efficiency production of graphene by tannic acid-assisted exfoliation of graphite in water. *ACS Sustain. Chem. Eng.* **2018**, *6*, 7652–7661. [CrossRef]
25. He, X.-P.; Deng, Q.; Cai, L.; Wang, C.-Z.; Zang, Y.; Li, J.; Chen, G.-R.; Tian, H. Fluorogenic resveratrol-confined graphene oxide for economic and rapid detection of Alzheimer's disease. *ACS Appl. Mater. Interfaces* **2014**, *6*, 5379–5382. [CrossRef]
26. Abdullah, A.H.; Ismail, Z.; Abidin, A.S.Z.; Yusoh, K. Green sonochemical synthesis of few-layer graphene in instant coffee. *Mater. Chem. Phys.* **2019**, *222*, 11–19. [CrossRef]
27. Wang, Y.; Shi, Z.; Yin, J. Facile synthesis of soluble graphene via a green reduction of graphene oxide in tea solution and its biocomposites. *ACS Appl. Mater. Interfaces* **2011**, *3*, 1127–1133. [CrossRef]
28. Pinto, J.; Rivlin, R. Riboflavin (Vitamin B2). In *Handbook of Vitamins*, 5th ed.; CRC Press: Boca Raton, FL, USA, 2013; pp. 191–266.
29. Ding, L.; Yang, H.; Ge, S.; Yu, J. Fluorescent carbon dots nanosensor for label-free determination of vitamin B12 based on inner filter effect. *Spectrochim. Acta Part A* **2018**, *193*, 305–309. [CrossRef]
30. Suwannasom, N.; Kao, I.; Pruß, A.; Georgieva, R.; Bäumler, H. Riboflavin: The health benefits of a forgotten natural vitamin. *Int. J. Mol. Sci.* **2020**, *21*, 950. [CrossRef]
31. Drössler, P.; Holzer, W.; Penzkofer, A.; Hegemann, P. Fluoresence quenching of riboflavin in aqueous solution by methionin and cystein. *Chem. Phys.* **2003**, *286*, 409–420. [CrossRef]
32. Lakowicz, J.R. Quenching of fluorescence. In *Principles of Fluorescence Spectroscopy*, 3rd ed.; Springer: New York, NY, USA, 2006; pp. 277–330.
33. Silber, M.L.; Davitt, B.B.; Khairutdinov, R.F.; Hurst, J.K. A mathematical model describing tannin-protein association. *Anal. Biochem.* **1998**, *263*, 46–50. [CrossRef] [PubMed]
34. Kozlovskaya, V.; Kharlampieva, E.; Drachuk, I.; Cheng, D.; Tsukruk, V.V. Responsive microcapsule reactors based on hydrogen-bonded tannic acid layer-by-layer assemblies. *Soft Matter* **2010**, *6*, 3596–3608. [CrossRef]
35. Charlton, A.J.; Baxter, N.J.; Khan, M.L.; Moir, A.J.G.; Haslam, E.; Davies, A.P.; Williamson, M.P. Polyphenol/peptide binding and precipitation. *J. Agric. Food Chem.* **2002**, *50*, 1593–1601. [CrossRef] [PubMed]
36. Abouelmagd, S.; Meng, F.; Kim, B.-K.; Hyun, H.; Yeo, Y. Tannic acid-mediated surface functionalization of polymeric nanoparticles. *ACS Biomater. Sci. Eng.* **2016**, *2*, 2294–2303. [CrossRef] [PubMed]

Article

Graphene Coating Obtained in a Cold-Wall CVD Process on the Co-Cr Alloy (L-605) for Medical Applications

Łukasz Wasyluk [1], Vitalii Boiko [1,2,3], Marta Markowska [1,2], Mariusz Hasiak [4], Maria Luisa Saladino [5], Dariusz Hreniak [1,2], Matteo Amati [6], Luca Gregoratti [6], Patrick Zeller [6,7,8], Dariusz Biały [2,9,†], Jacek Arkowski [2,9] and Magdalena Wawrzyńska [2,9,*]

[1] Division of Optical Spectroscopy, Institute of Low Temperature and Structure Research, Polish Academy of Sciences, Okólna 2, PL-50-422 Wrocław, Poland; lukaszwasyluk76@gmail.com (Ł.W.); v.boiko@intibs.pl (V.B.); m.markowska@intibs.pl (M.M.); d.hreniak@intibs.pl (D.H.)
[2] Carbonmed Spółka z ograniczoną odpowiedzialnością, ul. Okólna 2, PL-50-422 Wrocław, Poland; dariusz.bialy@umed.wroc.pl (D.B.); jacek.arkowski@umed.wroc.pl (J.A.)
[3] Institute of Physics of the National Academy of Science of Ukraine, Prospect Nauky 46, UA-03028 Kyiv, Ukraine
[4] Department of Mechanics, Materials Science and Biomedical Engineering, Wrocław University of Science and Technology, Smoluchowskiego 25, PL-50-370 Wrocław, Poland; mariusz.hasiak@pwr.edu.pl
[5] Department of Biological, Chemical and Pharmaceutical Sciences and Technologies (STEBICEF) and INSTM UdR—Palermo, University of Palermo, Viale delle Scienze, Bld. 17, IT-90128 Palermo, Italy; marialuisa.saladino@unipa.it
[6] Elettra–Sincrotrone Trieste S.C.p.A., SS14—km 163.5 in Area Science Park, IT-34149 Trieste, Italy; matteo.amati@elettra.eu (M.A.); luca.gregoratti@elettra.eu (L.G.); patrick.zeller@posteo.de (P.Z.)
[7] Helmholtz-Zentrum Berlin für Materialien und Energie GmbH, BESSY II, Albert-Einstein-Straße 15, DE-12489 Berlin, Germany
[8] Fritz-Haber-Institut der Max Planck Gesellschaft, Dept. Inorganic Chemistry, Faradayweg 4-6, DE-14195 Berlin, Germany
[9] Department of Preclinical Research, Faculty of Health Sciences, Wrocław Medical University, Ludwika Pasteura 1, PL-50-367 Wrocław, Poland
* Correspondence: magdalena.wawrzynska@umed.wroc.pl
† Deceased 12 July 2020.

Abstract: Graphene coating on the cobalt-chromium alloy was optimized and successfully carried out by a cold-wall chemical vapor deposition (CW-CVD) method. A uniform layer of graphene for a large area of the Co-Cr alloy (discs of 10 mm diameter) was confirmed by Raman mapping coated area and analyzing specific G and 2D bands; in particular, the intensity ratio and the number of layers were calculated. The effect of the CW-CVD process on the microstructure and the morphology of the Co-Cr surface was investigated by scanning X-ray photoelectron microscope (SPEM), atomic force microscopy (AFM), scanning electron microscopy (SEM), and energy dispersive X-ray spectroscopy (EDS). Nanoindentation and scratch tests were performed to determine mechanical properties of Co-Cr disks. The results of microbiological tests indicate that the studied Co-Cr alloys covered with a graphene layer did not show a pro-coagulant effect. The obtained results confirm the possibility of using the developed coating method in medical applications, in particular in the field of cardiovascular diseases.

Keywords: graphene coating; biocompatibility; cobalt-chromium alloy; cold-wall chemical vapor deposition method

1. Introduction

Graphene is one of the most prospective materials in terms of its unique properties for many applications in the field of electronics [1,2] electricity [3], sensors [4], biosensors [5], catalysis, etc. [1–3,6]. At the same time, there are several scientific reports on the use of graphene-based materials in biology and medicine [7–9]. One of the ideas is the utilization of its chemically inert properties for coating medical devices and the use of the external surface as a biologically neutral protective anti-corrosion film [10–13]. Covering implants and medical instruments with graphene-like materials does not lead to the problem of releasing graphene from the human body because the material is attached to the implant. However, for this particular case, the inertness of graphene leads to problems with coating attached to the substrate material; the coating technology must ensure that it is stable and tear-resistant in a biological environment. For effective metal surface coating, it is therefore necessary to prepare the surface of the substrate to be coated and to optimize the chemical vapor deposition (CVD) process, allowing better attachment, e.g., by using the defective interface side of the graphene layer for bonding. Additionally, the high temperatures required for the processes used for deposition of metallic coatings on 3D structures (e.g., scaffolds) are problematic in many cases. This problem becomes critical for alloys, mainly due to the danger of overheating the material (microstructural changes, recrystallization) and possible phase transitions affecting its mechanical properties. In this work, we assumed that these changes can also lead to weakening of the alloy–graphene interactions. To date, a large area of high-quality and uniform graphene grown by the CVD method was obtained for Cu or Ni foils [14,15]. However, the CVD method is also used for the graphene growth on Co [16], Fe [17], Pt [18], and other transition metal substrates [19] or their oxides, including co- or tri-metallic alloys. It has been demonstrated and intensively studied in thousands of papers [19,20]. Besides, a modified cold-walled CVD (CW-CVD) method has been successfully used to obtain high-quality graphene layer for Cu [21,22], cobalt [16], and cobalt-nickel alloy [23]. However, the same coating on the complex, modern medical alloys has not been reported yet, and it still is a technological challenge. In this work, for the first time, a successful graphene-coating of Co-Cr medical alloy, one of the materials commonly used for cardiovascular applications, is presented [24]. In our previously reported research, we showed that graphene coating obtained by poly(methyl methacrylate) (PMMA) method promotes vascular endothelial cells growth on the covered surface and thus shows promising biocompatibility features for cardiovascular applications [25]. For investigated materials, we paid attention to two main aspects affecting the harmfulness of the proposed solution: (1) homogeneity of coating and its mechanical stability on the alloy surface (possibility of de-attaching of graphene flakes) and (2) chemical interaction of tissues with graphene, especially at the interface of graphene and alloy substrate where some chemical bonding between graphene and alloy may be expected—probably via oxygen, forming graphene oxide (GO) regions. In both aspects, tentative answers were provided by the hemocompatibility tests we carried out.

2. Results and Discussion

2.1. Optimization of the Deposition Process

The obtained alloy discs were subject to the graphene coating procedure, optimized by changing the parameters of the CW-CVD process. The initial set of deposition process parameters was based on the results described by Macháč et al. [16] who successfully used the CW-CVD process to prepare graphene on pure cobalt substrates. The first experiments were conducted for both polished and unpolished alloy substrates. However, since Raman spectroscopy used as the first probe testing for graphene confirmed its existence only on unpolished samples, optimization of the procedure and related characterization were performed only for these samples. Optimal results were obtained by subjecting unpolished Co-Cr substrates to the CW-CVD process consisting of four stages: heating (SP1), annealing in argon with hydrogen to remove the oxygen residues at the sample surface (SP2), subsequent heating (SP3), and the stage of graphene formation and growth (SP4) (Table 1).

In order to obtain the optimal parameters for the best quality graphene layer, a series of samples was prepared. At the stage of graphene formation and growth (SP4), the time (1800, 2700 and 3600 s) and the deposition temperature (800, 900 and 1000 °C) changed. At the final stage, the cooling process was supplied in an argon flow atmosphere up to 100 °C.

Table 1. Stages and corresponding process parameters in the cold-wall chemical vapor deposition (CW-CVD) chamber.

	t, s	T, °C	p, Torr	Ar, % (sccm)	H_2, % (sccm)	CH_4, % (sccm)
SP 1	0	90	10	5 (100)	1 (20)	0 (0)
SP 2	300	90	10	5 (100)	1 (20)	0 (0)
SP 3	0	900	10	5 (100)	1 (20)	0 (0)
SP 4	2700	900	10	0 (0)	2 (1.2)	35 (20)

t—time in second; T—heater substrate temperature in °C; sccm—standard cubic centimeters per minute.

2.2. Structure Characterization by SEM and EDS Method

SEM and energy dispersive X-ray spectroscopy (EDS) analyses were carried out for all the alloy samples obtained. The investigations performed allowed us to visualize the influence of surface preparation of the alloy samples on their morphology. Examples of SEM micrographs made for cut and polished surfaces and cut only samples obtained after CW-CVD processing are shown in Figure 1a,b, respectively. As it can be seen, the surfaces of samples differed significantly. The polished sample, apart from the obvious polishing marks, was characterized by the disappearance of all structures on the surface, while the sample without polishing was characterized by an extended surface, indicating a strong influence of spark cutting on the separation of different alloy fractions.

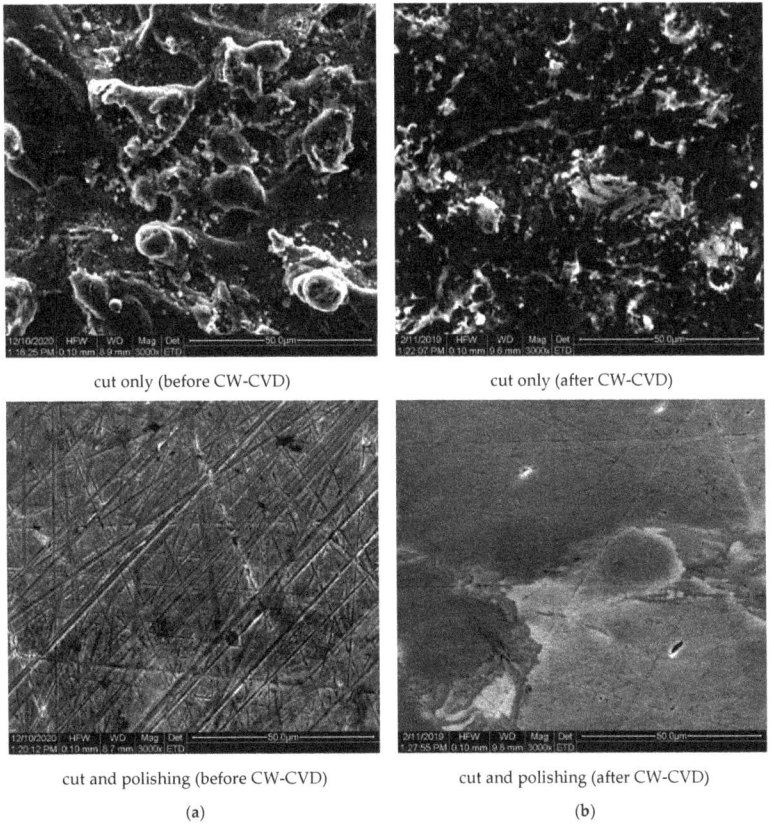

Figure 1. SEM micrographs cut (top) and polished (bottom) sample before (**a**) and after (**b**) CW-CVD.

On comparison of the SEM micrographs before and after the CW-CVD process, a slight homogenization of the surface microstructure was also evident for both unpolished and polished samples, manifesting itself in a slight decrease in the roughness of the unpolished sample and the disappearance of scratches on the polished sample. Additional information is provided by the results of EDS microanalysis performed on the chosen area of these samples (Figure 2). It can be seen, among other things, that polishing had a fundamental influence on the oxygen content of the analyzed surface. This was manifested by the almost ten times higher oxygen content for the unpolished sample compared to the polished one. It can also be seen that subjecting the same sample to an effective graphene coating in the CW-CVD process resulted in only a slight change in the oxygen content, still much higher than for the polished surface. Observed difference between the C content between the sample before and after polishing (Figure 2a) was most probably due to presence of M6C carbides on the surface of the as-cut sample. It is known that carbide precipitates (grain-boundary carbide precipitates [26]) are often present after heat treatment of this type of alloy. The applied polishing of the disc surface may have led to their removal from the surface, which was reflected in the values determined from the EDS. Additionally, since carbon was present in a wide range in the surrounding environment, and the samples were attached to a conductive tape containing carbon during the measurement, it seems that the measurement of the carbon percentage may have been disturbed because of this. Additionally, the identification of carbon content is always subject to a large measurement error due to the low energy on the EDS spectrum compared to the other elements present in the alloy under study. However, the percentage content with respect to the measurement error of the other elements in the alloy was not questionable.

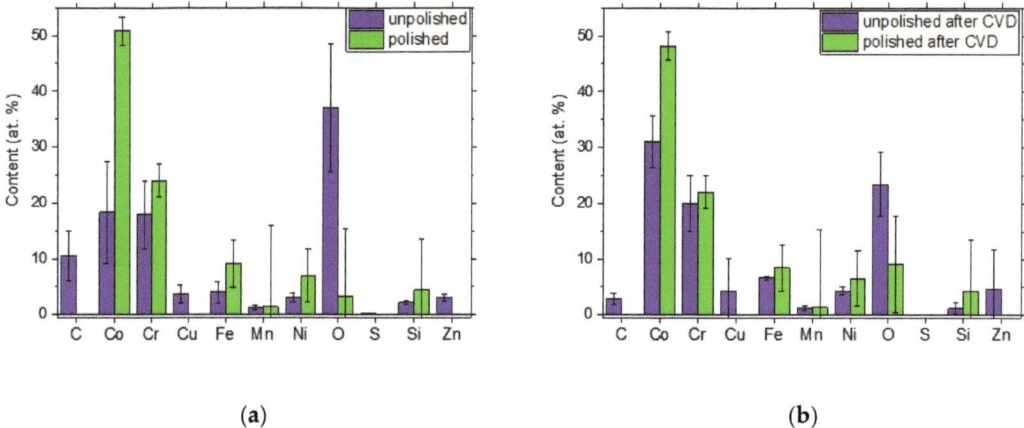

Figure 2. The averaged results of the energy dispersive X-ray spectroscopy (EDS) analysis performed for Co-Cr alloys before (**a**) and after graphene deposition (**b**).

Additional analyses with the SEM micrographs were carried out to establish the influence of modification of the CW-CVD application parameters for non-polished samples with the graphene layer (Figure 3).

As we can see from the micrographs shown in the left panel (Figure 3a), the surface microstructure did not change noticeably with a change in the duration of the process at 900 °C. On the other side, we see some modification of the structure with the temperature increase from 800 °C to 1000 °C (Figure 3b). Most of the few micrometer sized irregular cracks in the surface of the alloy evident in the case of sample after the processing at 800 °C for 45 min disappeared after being treated at 900 °C.

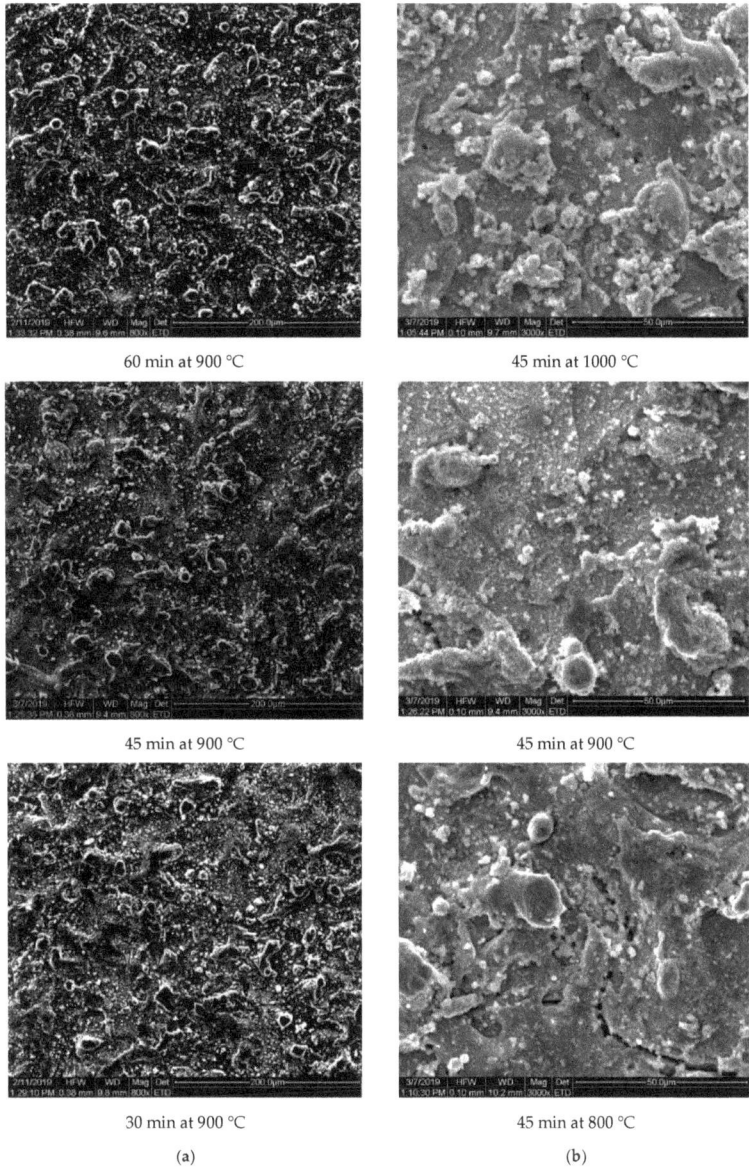

60 min at 900 °C	45 min at 1000 °C
45 min at 900 °C	45 min at 900 °C
30 min at 900 °C	45 min at 800 °C
(a)	(b)

Figure 3. SEM micrographs of Co-Cr alloys after CW-CVD as function time deposition (**a**) and temperature deposition (**b**).

2.3. Investigations of Topography by Atomic Force Microscopy (AFM)

One of the most advanced, versatile, and powerful methods is the measurement of the surface topography on the atomic scale by atomic force microscopy. These investigations deliver 2D/3D topography images as well as various types of surface characterization. The 2D and 3D topography images recorded for samples of Co-Cr alloy polished and cut after CW-CVD process are presented in Figure 4.

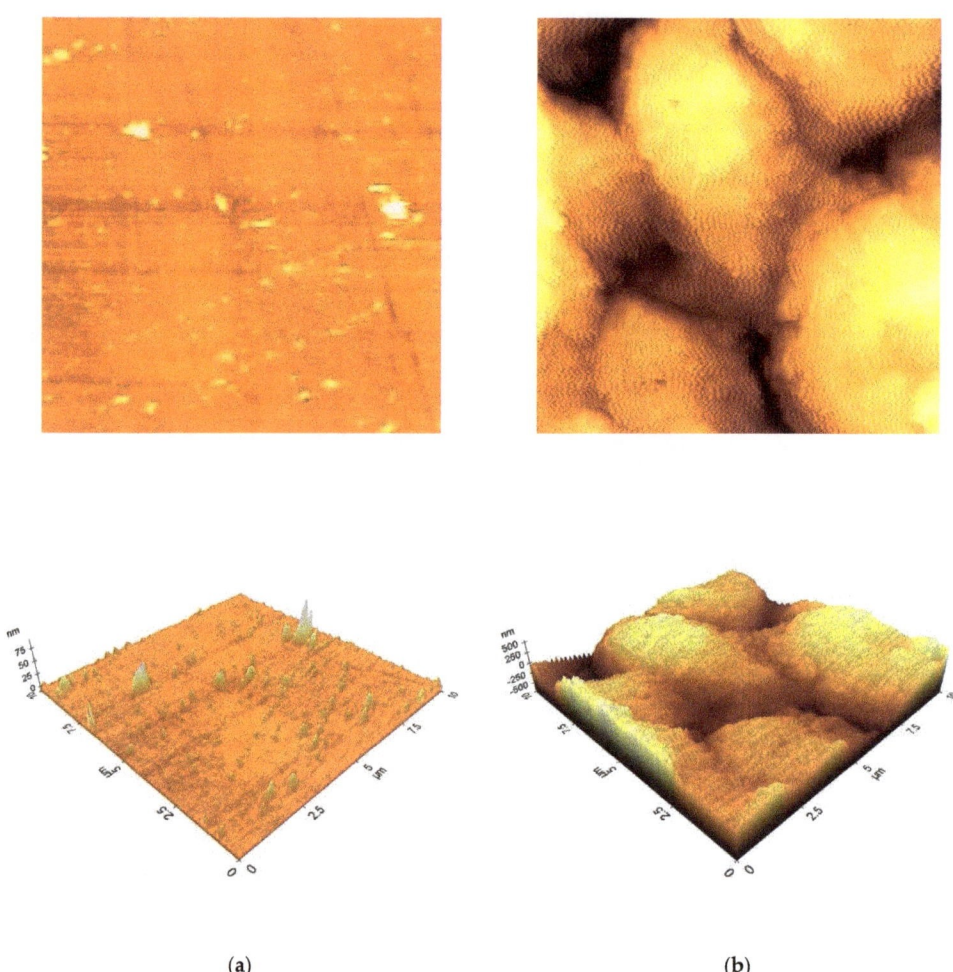

(a)   (b)

Figure 4. 2D/3D atomic force microscopy (AFM) topography images for polished (**a**) and cut after CW-CVD process (**b**) samples of Co-Cr alloy recorded in contact and noncontact mode, respectively.

It is well visible that the topography of the cut sample after CW-CVD process in comparison to the polished sample was not homogeneous, and some features of about 2–3 μm width were present. The difference between the minimum and the maximum values in the recorded surface profile equaled 1273.09 nm. The root mean-squared roughness (Rq), the average roughness (Ra), and the ten point average roughness (Rz) for the image presented in Figure 4a were 215.18 nm, 175.89 nm, and 1241.08 nm, respectively. The complex surface of cut samples allowed us only to record the topography of the surface in noncontact mode. The polished samples of Co-Cr alloy presented quite different surface structure, and only a few objects less than 1 μm wide were registered in the topography scanning process. This measurement was performed in contact mode, which allows for a much more accurate projection of the analyzed surface. The difference between the minimum and the maximum values in the recorded surface profile equaled 101.24 nm, and it was 12.6 times lower than for cut samples. The roughness parameters Rq, Ra, and Rz for the surface presented in Figure 4b were 4.32 nm, 2.27 nm, and 88.54 nm, respectively.

2.4. Raman Spectroscopy

For each deposited sample, the presence of the G band and the quality of the graphene layer were tested by Raman spectroscopy. The presence of such a coating consisting of several overlapping layers of graphene was evidenced by the presence of characteristic peaks for graphene: D, G, and 2D peaks. They were located at around 1350 cm^{-1}, 1580 cm^{-1}, and 2700 cm^{-1}, respectively [27,28]. The G peak was a result of stretching vibrations of the sp^2 pairs in both rings and chains, while the D peak was due to the breathing modes of the sp^2 pairs in rings. The 2D peak was the second order of the D peak [29]. Preliminary measurements of the samples after the CW-CVD process for graphene deposition showed a complete lack of specific bands on each polished substrate (Figure 5). Meanwhile, for the unpolished sample before CW-CVD process, two broad covered bands around 1500 cm^{-1} that can be related to low sp^3 amorphous carbon were present [29].

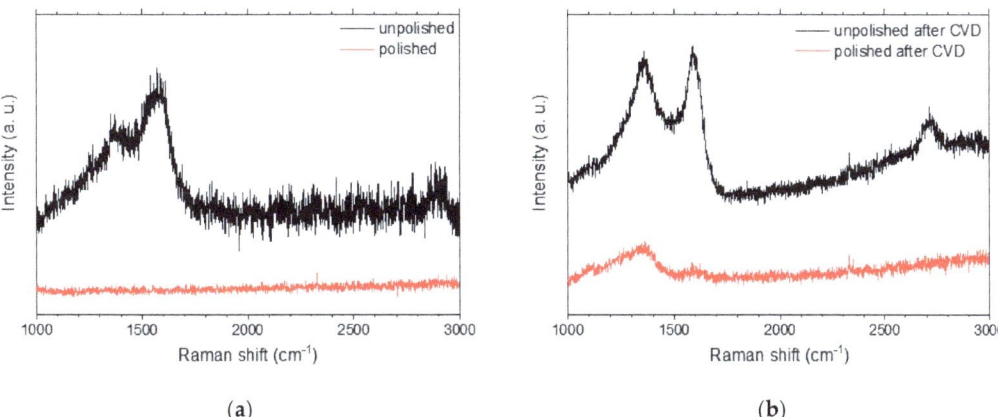

Figure 5. Raman spectra of the unpolished (cut only) and the polished Co-Cr alloy samples (before (**a**) and after CW-CVD (**b**)).

Therefore, in order to understand the factors influencing the quality of the graphene layer and to optimize the deposition process, spectra of samples obtained with different CW-CVD parameters were recorded only for non-polished samples. Representative Raman spectra of the samples obtained in both (temperature and time) series are shown in Figure 6.

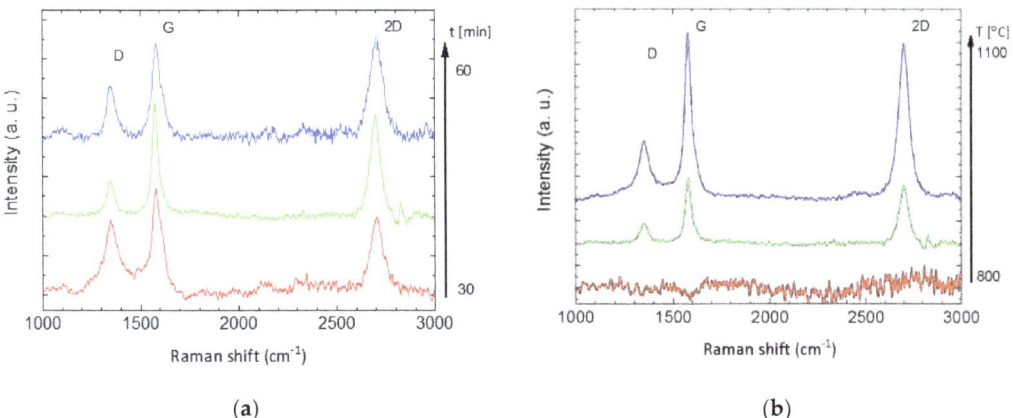

Figure 6. Raman spectra of the Co-Cr alloys as a function of the time deposition (**a**) and the temperature deposition (**b**).

As shown above, the optimal process parameters (t = 2700 s, T = 900 °C) yielded the best results in terms of design and quality of layers. The half-widths of 2D, G, and D bands were 55.5, 45, and 53 cm^{-1}, and the ratios of the intensity of the 2D and the D bands to the G band were 0.94 and 0.35. Both with shorter and longer deposition times of the graphene layer growth stage, more defective layers were obtained, as evidenced by a more intense D peak in relation to the G peak intensity (Figure 6a). At a temperature lower than 900 °C, no graphene layer was formed at all (red line in Figure 6b), while at 1000 °C, a graphene layer was formed, but the sample surface coverage was incomplete, which is clearly visible in Figure 7 below, where the areas are marked with blue color, in which no trace of graphene is present (the ratio of 2D peak intensity to G goes to zero).

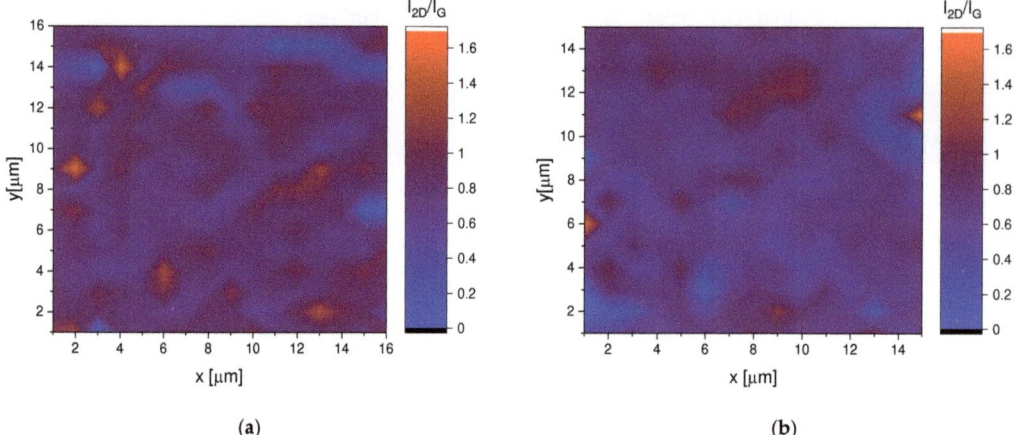

Figure 7. Map made on the basis of Raman spectra collected for a sample deposited at 900 °C (**a**) and 1000 °C (**b**).

The shape and the intensity of the 2D peak were significantly different in graphene compared to bulk graphite. This is because, in bulk graphite, the 2D peak consisted of two components, while graphene had a single, sharp 2D peak with Lorentzian shape [29]. Its intensity, compared to the G band, could vary from three to one-third from graphene to graphite, respectively. When disorder increased, this band broadened, overlapping with other bands, and nearly disappeared. The relative intensity ratio between the 2D and the G bands was also found to be dependent on the number of layers. I_{2D}/I_G was close to three for monolayer graphene and fell to 0.3 for highly oriented pyrolytic graphite [30].

According to the shape of the 2D band and the ratio of its intensity to the intensity of the G band, the number (up to five) of the graphene layers and their quality could be clearly identified [27]. The intensity ratio (I_{2D}/I_G) maps based on the Raman data were constructed (Figure 7). As it can be seen, for alloys with graphene deposited at 900 °C (Figure 7a), the intensity ratio maps had a more homogenous color compared to alloys with graphene deposited at 1000 °C (Figure 7b). The discontinuities in the graphene layer (blue areas), together with a sharper color gradient from point to point, may indicate an increase of disorder in the graphene layers.

2.5. Mechanical Properties

Mechanical properties of Co-Cr discs were investigated for polished and cut (after CW-CVD process) Co-Cr samples. Figure 8 shows the surface of these samples observed in polarized light. In the case of the polished sample, the grain boundaries represented by fine needles were clearly visible. For the cut sample, only a very complex structure was observed. These results are in good agreement with data obtained from AFM investigations, and they determined the mechanical parameters.

Figure 8. Microscope images recorded in polarized light for polished (**a**) and unpolished (**b**) Co-Cr samples.

Instrumented nanoindentation is a measurement method that allows one to investigate the mechanical properties such as hardness, elastic modulus, and deformation energies of the samples where conventional Vickers hardness measurements cannot be performed due to the small sample dimension of the samples or low load. Therefore, the determination of parameters of mechanical properties for Co-Cr discs was performed by measurements of indentation depth as a function of applied load with respect to ISO 14577 standard.

Figure 9 shows the dependence of penetration depth versus a maximum load of 100 mN for polished and cut Co-Cr samples. Moreover, an imprint of the Berkovich diamond indenter for the polished sample is also shown. The results obtained from numerical analysis of F_n-P_d curves with respect to the Oliver–Pharr procedure are presented in Table 2. It can be seen that instrumental hardness, instrumental elastic modulus, as well as plastic and total deformation energies of the polished sample were higher than for the cut samples. This change was related to the formation of new structures on the disk surface during the CVD process.

To determine the quality of Co-Cr disc surface, the scratch tests were performed for polished and cut samples. The pictures of the surface obtained for the investigated materials are presented in Figure 10.

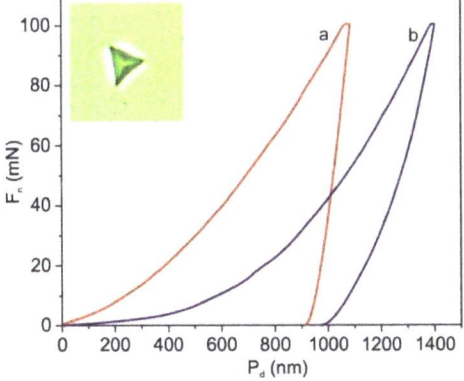

Figure 9. Example of load (F_n) versus indentation penetration depth (P_d) curves for polished (**a**) and cut after CW-CVD process (**b**) Co-Cr discs. The inset shows an imprint obtained for polished sample.

Table 2. Mechanical properties as instrumental hardness (HV_{IT}), instrumental elastic modulus (E_{IT}), elastic deformation energy (W_{elast}), plastic deformation energy (W_{plast}), total deformation energy (W_{total}), and elastic deformation energy to total energy ratio W_{elast}/W_{tot} (n_{IT}) for Co-Cr discs.

	Co-Cr (Polished Disc)	Co-Cr (Cut after CVD Process Disc)
HV_{IT} (Vickers)	406.31	262.82
HV_{IT} (MPa)	4387.30	2837.90
EIT (GPa)	161.49	72.26
W_{elast} (pJ)	6777.72	15,125.11
W_{plast} (pJ)	36,322.14	24,542.69
W_{total} (pJ)	43,099.86	39,667.80
n_{IT} (%)	15.73	38.13

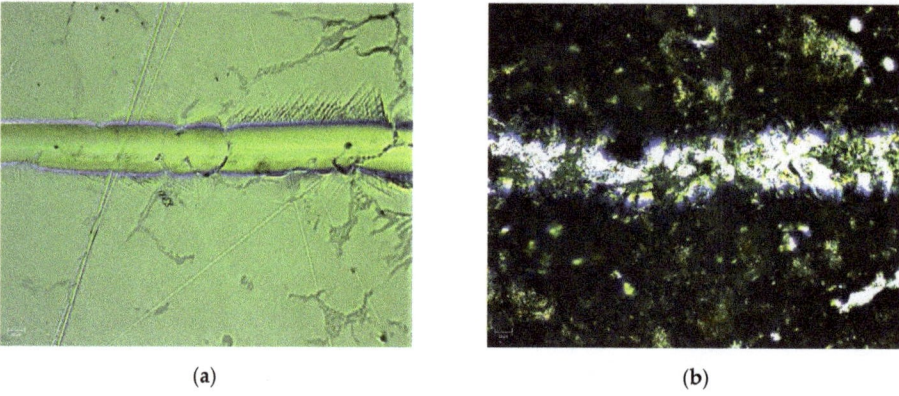

(a)   (b)

Figure 10. Microscopic image of the residual groove after a scratch test on polished (**a**) and unpolished (**b**) Co-Cr disc.

Numerical analyses of the frictional force and the friction coefficient for the polished and the cut Co-Cr discs are presented in Figure 11. All measurements were performed for a length of 3 mm and a maximum load of 10 N.

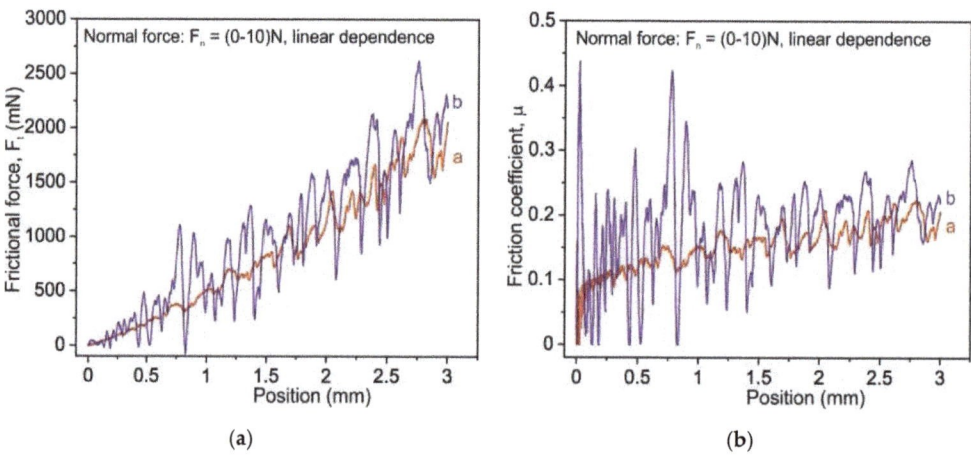

(a)   (b)

Figure 11. Frictional force displacement curves for scratch tests (left side) and friction coefficient displacement curves for polished (curve **a**, red color) and cut (curve **b**, blue color) Co-Cr discs.

It is seen that frictional force and friction coefficient increased with the load applied during measurement for both materials. The large fluctuations of frictional force and friction coefficient (for low load) recorded for cut samples were mostly caused by surface unevenness. Moreover, with increasing normal force, an increase in F_t and μ was observed for both samples.

2.6. Scanning X-Ray Photoelectron Microscope (SPEM)

The image shown in Figure 12 is a SPEM map of the unpolished sample covered by graphene. The contrast present in the image is dominated by surface topography; brighter areas are associated with features oriented towards the hemispherical electron energy analyzer (HEA) while dark regions represent shadowed areas or holes in the surface.

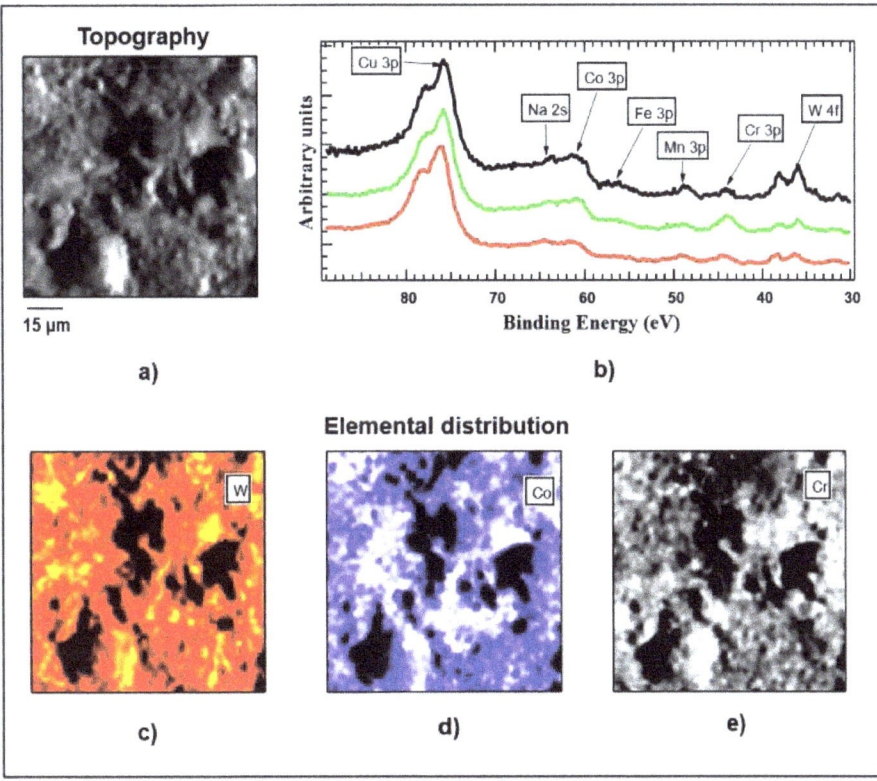

Figure 12. Scanning X-ray photoelectron microscope (SPEM) map of the unpolished sample covered by graphene. The contrast present in the image is dominated by surface topography (**a**) brighter areas are associated with features which are oriented towards the hemispherical electron energy analyzer (HEA), while dark regions represent shadowed areas or holes in the surface. X-ray photoelectron spectroscopy (XPS) spectra (**b**) acquired at different representative positions. (**c–e**)—elemental distribution of W, Co, and Cr, respectively. The black areas visible in the images must be ignored because of the intensity of the photoemission signal detected from these areas, which is too poor to be correctly processed. Changes in the color scale reflect the variations in the elemental concentration.

As a result, the raw data images acquired at different elements appeared similar. A collection of XPS spectra acquired at various regions of the map in Figure 12 are reported in Figure 13; the core levels of distinct elements were present, namely Cu 3p, Na 2s, Co 3p, Fe 3p, Mn 3p, Cr 3p, and W 4f, as indicated in the graph.

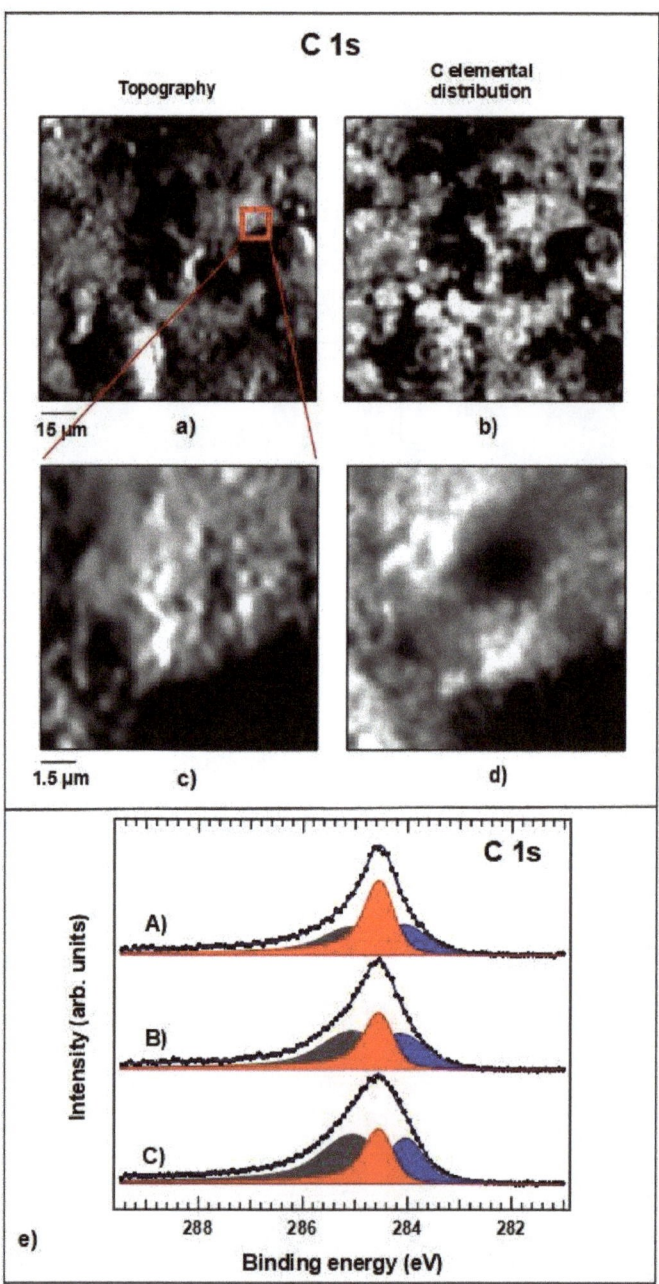

Figure 13. (a,b) raw data SPEM image acquired at the C 1s level and the corresponding C elemental distribution. (c)—same maps of the magnified region indicated by the red square in image and (d) same maps of a magnified region after processed. (e) representative C 1s spectra acquired at different locations. The three color filled lines are the result of the deconvolution procedure, which separates the contribution of the different chemical C moieties. See text for the details.

The extended surveys (not shown here) additionally reported the presence of C and traces of Si. The three spectra shown in the panel were normalized with respect to the Cu 3p intensity for a better comparison; the relative intensities of W, Mn, Cr and Co changed in each spectrum, suggesting a different chemical composition in the probed areas. To investigate deeper the elemental heterogeneity of the surface, we processed the raw data maps acquired at the different core levels and removed the topographic contribution of the contrast, highlighting the elemental one according to well-known algorithms, as described in [31]. The results of this analysis for W, Co, and Cr are shown in the three maps of Figure 12c–e, respectively. The black areas visible in the images must be ignored because of the intensity of the photoemission signal detected from these areas, which is too poor to be correctly processed. Changes in the color scale reflect the variations in the elemental concentration, supporting what is evidenced by the three spectra of Figure 12b. The distribution of the concentration of W showed variations as high as 80% in the different areas, while those associated with Co and Cr did not exceed 30%. The list of detected elements reflected well the EDS data reported in Figure 2. It is important to note that SPEM data reflect the chemical composition of the outermost atomic layers of the surface, since the escape depth of photoelectrons was limited to 1–1.5 nm. Such thickness, including the graphene coating, will be analyzed later. The XPS spectra revealed a high surface concentration of Cu. This was not a surprise, since EDS analysis showed the presence of Cu only in the unpolished samples, suggesting a segregation of Cu in the outermost areas, which were removed with polishing. The BEs of the core levels detected agree with high oxidation states of the same as expected for such alloy.

The distribution of carbon over the surface is shown in Figure 13a,c are the raw data maps recorded at two different levels of magnification. Additionally, in this case, the majority of the contrast in the pictures was generated by the topography of the surface. By applying the same processing algorithms as done for the maps in Figure 6, we isolated the local elemental concentration and show it in maps Figure 13b,d, respectively. In agreement with the Raman micrographs of Figures 5 and 6, the distribution of C appears not uniform at both magnification scales. For this reason, it is not straightforward to evaluate the thickness of the Gr–C layer, but we can state that all spectra that were acquired at different regions allowed the detection of the substrate elements, indicating an overall thickness of the C coating of less than 1–1.5 nm. To investigate more in detail the chemical state of the C coating, high resolution XPS spectra of the C 1 s were acquired at different areas. Three representative spectra are shown in Figure 13e; the three color filled lines are the result of the deconvolution procedure, which separated the contribution of the different chemical C moieties. For this procedure, we used Doniac-Sunjich functions, which describe the photoemission process and are commonly used for XPS peaks deconvolution. The red component centered at 285.5 eV BE shows the typical line shape generated by the graphene two-dimensional layers of carbon atoms with sp^2 hybridization that were connected in a hexagonal lattice; as expected, this component was narrower than the other ones. The blue filled line centered at 284.0 eV BE was typically associated with defects in the C hexagonal lattice of the graphene, while the broad grey filled one at 285.0 eV BE took into account all other C-C, C-O, and organic carbon species, which generally adsorb on a surface. The weights of the components in the three spectra were different, proving that the chemical state of carbon was heterogeneous over our surface; in some areas, the graphene was predominant, while in others, adventitious carbon was more abundant.

2.7. Biocompatibility Studies

2.7.1. Genotoxicity Studies

The tested extracts from Co-Cr discs coated with graphene did not show any mutagenic effect on TA98 and TA100 strains, both used separately and in combination with the S9 fraction.

The tests also showed that the application of the polar (−FBS) and the non-polar (+FBS) solutions did not cause significant differences in the appearance of reversion. The number of reverse wells in all tested strains after incubation with extracts from Co-Cr discs coated with graphene was less than 10. There was also no effect of direct and inversely proportional formation of a greater number of reversions with increasing concentration of the examined extracts. The test was performed in triplicate.

2.7.2. Cytotoxicity Studies

Cytotoxicity studies showed that the extracts of the Co-Cr discs tested showed low toxicity, which was estimated at the level of one according to the cytotoxicity scale. Extracts from Co-Cr discs coated with graphene showed little toxicity only at the highest concentration of 100% used, and at the remaining concentrations tested of 50%, 25%, and 12.5%, no toxicity was observed against mouse fibroblast cells. Moreover, the use of the serum extract (FBS +) or without (FBS −) showed no significant differences in cytotoxicity. Detailed results of cytotoxicity tests are presented in Table 3.

Table 3. Evaluation of the cytotoxicity of extracts from tested Co-Cr disks, extract preparation solutions, and experiment control.

Sample	Concentration [%]	Cell Viability [%]	The Degree of Cytotoxicity
Co-Cr Graphene FBS+	100	77.09	1
	50	93.13	0
	25	98.36	0
	12.5	108.7	0
Co-Cr Graphene FBS−	100	81.12	1
	50	96.75	0
	25	94.32	0
	12.5	98.04	0
Co-Cr Reference FBS+	100	78.92	1
	50	83.59	1
	25	93.69	0
	12.5	91.46	0
Co-Cr Reference FBS−	100	84.12	1
	50	87.63	1
	25	89.92	0
	12.5	96.54	0
Positive control	0.2	15.63	4
	0.15	18.32	4
	0.1	63.17	3
	0.05	82.15	1
Negative control	-	98.02	0

The tests carried out on the discs in direct contact showed a moderate reactivity towards mouse fibroblast cells, estimated according to the reactivity scale at level three. Placement of Co-Cr discs on the BALB/3T3 cell monolayer (Figure 14) resulted in the appearance of shrunken cells around the disc and a free zone up to 1 cm. This effect was observed both in the case of the disc covered with a layer of graphene and the reference disc. In turn, sowing cells directly onto Co-Cr discs also resulted in the appearance of shrunken cells around the disc and a free zone up to 1 cm.

The surface of the tested Co-Cr discs after incubation with cells was also analyzed. It turned out that there were no cells on the surface of Co-Cr discs located on the cell layer of mouse fibroblasts. However, BALB/3T3 cells were seen on the surface of Co-Cr discs plated with cells. Cells were present on the surface of both the Co-Cr disc with a graphene layer and on the surface of the Co-Cr disc of the reference. It was noticed that the degree of cell surface coverage was higher in the Co-Cr disc without the graphene layer (Figure 15).

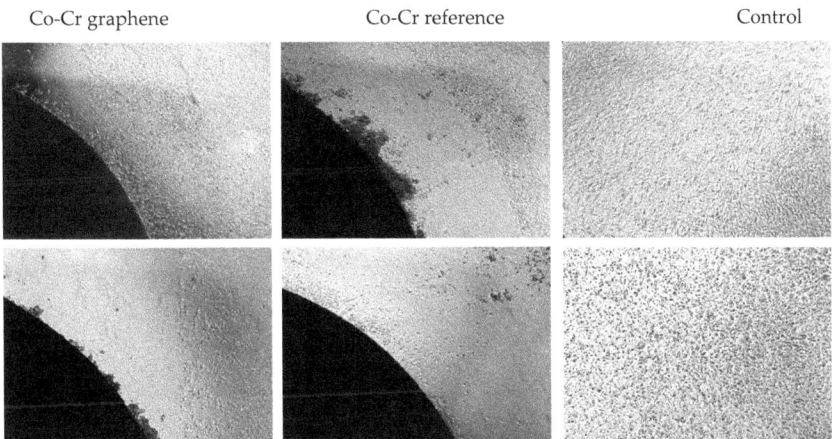

Figure 14. Evaluation of direct contact of BALB/3T3 mouse fibroblast cells with the examined discs. Co-Cr disks are visible as a dark semi-circle in the lower left corner, negative control high-density polyethylene (HDPE) of cells without Co-Cr disks. Magnification 50×.

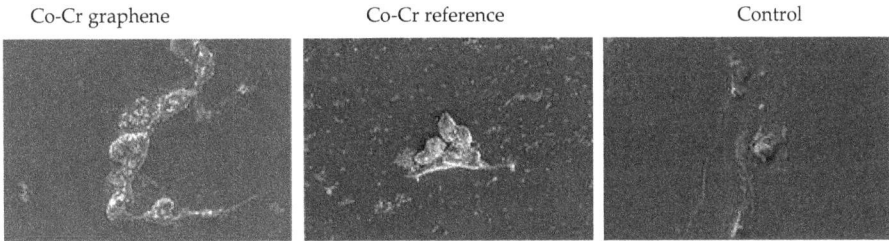

Figure 15. Evaluation of direct contact of BALB/3T3 mouse fibroblast cells with the examined discs. Deposits and cells are marked in orange. Magnification 2000×.

The tested extracts from Co-Cr discs coated with graphene and without a graphene layer showed no toxic effect on the cells of BALB/3T3 mouse fibroblasts. The highest applied concentrations caused a slight cytotoxic effect on a five-point scale, estimated at level one, while for in vitro cytotoxicity, cytotoxicity above grade two classifies an extract as toxic.

In direct contact tests, both Co-Cr discs with and without a graphene layer resulted in the appearance of shrunken cells and a free zone around the disc up to 1 cm, which was assessed on a five-point scale at level three as moderate reactivity. One of the reasons for this reactivity may be the fact that the discs were not permanently attached to the bottom of the plate and could move slightly, causing mechanical damage to the cells. In contrast, the results obtained using the scanning electron microscope showed that, in the wells with Co-Cr discs, there were lumpy structures and deposits that adhered to the cell surface, which could cause a toxic effect. The analysis of the elemental composition of EDS confirmed that these were exfoliating fragments of the Co-Cr discs tested. Interestingly, the analysis of the surface of the examined discs after incubation with the cells showed that the cells of the murine BALB/3T3 fibroblasts attached and grew on the surface of both types of Co-Cr discs coated and uncoated with graphene. This observation suggests that the surface of the discs was not toxic to the cells, while the fragments exfoliating from it caused a cytotoxic effect.

2.7.3. Hemocompatibility Studies

Hemolysis assay: Incubation of tested Co-Cr discs with a concentrate of red blood cells did not cause hemolysis. In the material after incubation with Co-Cr discs with and without graphene, about 0.05 mg/mL of hemoglobin was observed. On the other hand, when red blood cells were incubated with extracts, a slightly lower level of hemoglobin, approximately 0.03 mg/mL, was recorded. The positive control, which was nitrile rubber, caused a significant increase in the level of hemoglobin to the level of approximately 0.3 mg/mL compared to the other tested groups. High-density polyethylene (HDPE) was not observed to significantly increase hemoglobin levels. This level was similar to the level observed after incubation with Co-Cr disks and disk extracts (Figure 16).

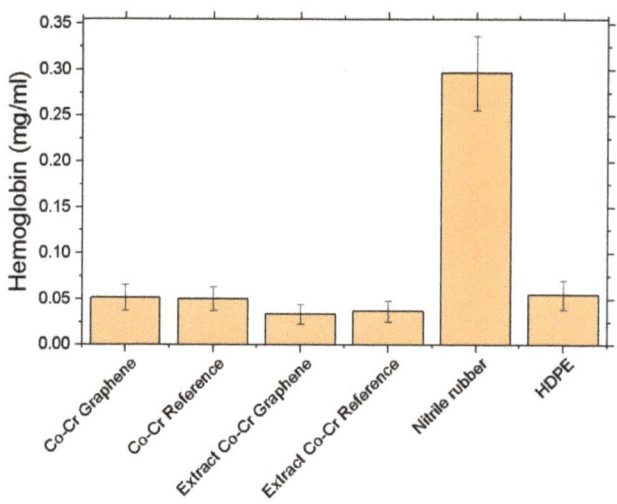

Figure 16. Hemoglobin level after incubation with test materials and extracts from Co-Cr discs.

Clotting time and coagulation factors measurements: Evaluation of the activated partial thromboplastin time (APTT) showed that none of the materials tested increased the APTT. In all studied groups, an increase in fibrinogen levels was observed, but differences in these results were not statistically significant between the groups. No differences were observed for other parameters such as prothrombin time, thrombin time, and recalcination time. Tests of the activity of blood clotting inhibitors such as ATIII showed that, in all the trials, the level of this factor was increased compared to the prescribed norm. However, it was not observed that the differences in ATIII level between the samples were statistically significant. There were no differences between the samples in the activity of plasminogen and protein C. The study of coagulation factors VII, X, and XII showed that neither the tested Co-Cr discs nor the extracts obtained from these discs caused statistically significant changes in the level of the factors.

Blood platelets activation analysis: The analysis of the surface of the tested materials showed that, after incubation with whole blood, all the materials had morphosis elements of blood in the form of red blood cells, white blood cells, and platelets. Red blood cells on all materials showed normal morphology. On the surface of the Co-Cr discs tested, no clot appeared. Numerous activated platelets were observed, but no clot formation was observed. According to the scale for clot formation, both Co-Cr discs with and without graphene layer obtained a result of zero. The situation was definitely different with the use of nitrile rubber fragments as a positive control, where strongly activated plates and clot assessed at level three were visible. For the formation of a clot on high-density polyethylene, a result of zero was obtained. Photos of the surface of the tested materials with blood morphotic elements are shown in Figure 17.

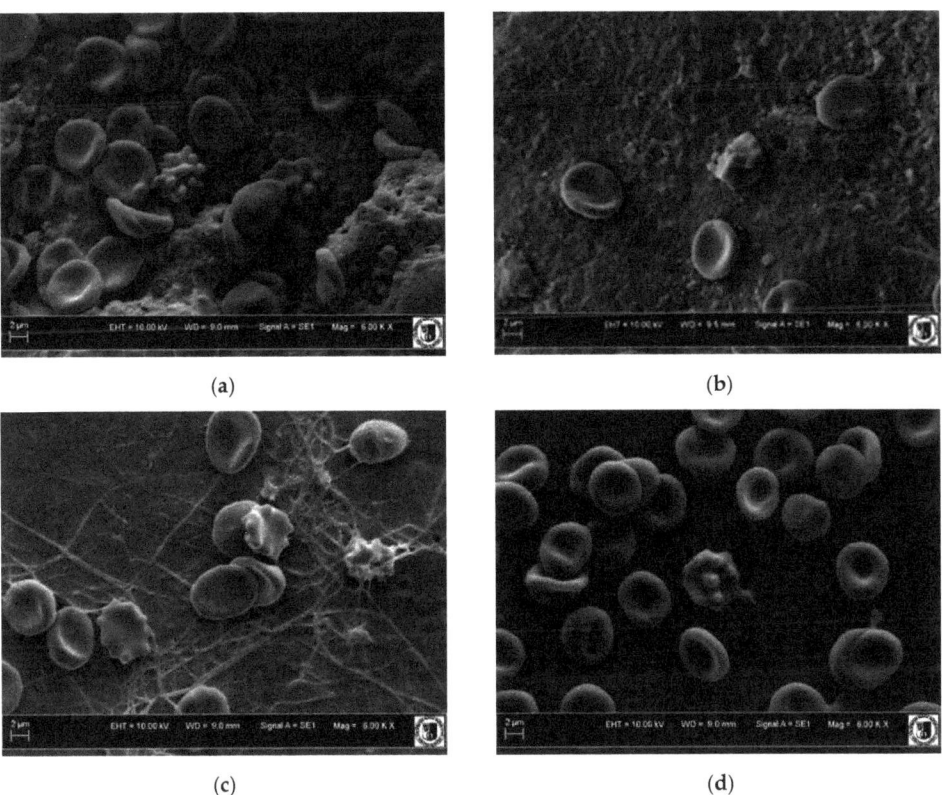

Figure 17. Surface of tested materials after contact with solid blood. Co-Cr disc with graphene (**a**), Co-Cr disc without graphene (**b**), nitrile rubber (**c**), high-density polyethylene (HDPE) (**d**). Magnification 6000×.

The conducted tests show that the tested Co-Cr discs with and without a graphene layer showed good hemocompatibility. No changes in the morphology of blood components were observed, and normal morphology of red blood cells was observed on the surface of the tested materials. Incubation of whole blood with the tested materials showed that they did not cause significant changes in the composition of morphotic elements. No differences were observed between the tested Co-Cr discs with and without graphene and the control materials in the form of nitrile rubber and high-density polyethylene.

3. Materials and Methods

3.1. Materials

The samples of substrates were prepared in the form of disks with a diameter of 7 mm and a thickness of ~1 mm. The discs were cut using an electrical discharge machining (EDM) with electric sparks underwater from an alloy sheet of a cobalt-based L-605 alloy of standardized composition: Co 55.76–65.193%, Cr 27–31%, Fe 2–4.5%, Ni 3%, Mo 0.8–1.5%, Si 0.5–1, 5%, C 0.8–1.4%, Mn 0.6–1.2%, Cu 0.08%, S 0.015–0.03%, P 0.012–0.03%. Before graphene deposition, the obtained samples were divided into two parts. One of them was additionally polished, and both of them were subjected to the following washing process (step by step):

- sonication in acetone (99.5% pure P.A.-Basic) for 10 min in an ultrasonic bath;
- sonication in ethyl alcohol (96% pure P.A.-Basic) for 10 min in an ultrasonic bath;
- sonication in isopropanol (99.7% pure P.A.-Basic) for 10 min in an ultrasonic bath.

The ultrasonic procedure permits efficient removal of all organic contaminations and also any residual of grease or oil coming from the stents production process.

3.2. Methods

3.2.1. Cold-Wall Chemical Vapor Deposition (CW-CVD)

Deposition of graphene on the cobalt-chrome alloys was carried out by the cold-wall CVD method on the commercial nanoCVD-8G system from Moorfield Nanotechnology Ltd. (Manchester, UK). The alloy substrates were placed directly on the heater (4.0 cm × 2.5 cm) embedded thermocouple. Programmable logic controller electronics equipped with a touchscreen interface controlled the hardware, continuously reporting the reactor chamber state: heater and chamber temperatures, pressure, and gas flow. Maximum operating temperature was 1100 °C, and maximum process gas flow was: argon −200 sccm (standard cubic centimeters per minute); hydrogen −20 sccm; methane −20 sccm. Pressure parameters: capacitance manometer 20 Torr (full scale), and valid entries were in the range of 0.0–0.5 Torr.5.

3.2.2. Raman Spectroscopy

Raman spectra were recorded using a Renishaw inVia Raman microscope (100×, magnification) equipped with Argon laser (λ_{ex} − 514 nm, power maximum at sample surface ~9 mW) and a CCD camera under environmental conditions (Renishaw plc, Wotton-under-Edge, Gloucestershire, UK). Detection range was 100–3200 cm^{-1}. Maps of relative intensity values of specific graphene bands were plotted based on a set of Raman spectra recorded in an automatic regime with a step of 5 microns.

3.2.3. Atomic Force Microscopy (AFM)

The topography of the cobalt-chrome alloy (L-605) (measured as a deflection of the cantilever in the vertical direction) was investigated in contact and noncontact modes with scanning frequency of 0.5 Hz by using an atomic force microscope (XE-100, Park Systems, Suwon, Korea). All measurements were performed by scanning an area of 10 µm × 10 µm. The obtained data were analyzed using XEI Software (Version 4.3.4) provided by the microscope's manufacturer. The roughness parameters such as Rq (root-mean-squared roughness), Ra (roughness average), and Rz (ten-point average roughness) were calculated for recorded images.

3.2.4. Scanning Electron Microscopy (SEM)

The surface of coated cobalt-chrome alloy (L-605) samples as well as the surface of reference uncoated L-605 samples were observed using a scanning electron microscope (Quanta 200 W, FEI). The magnification used was from 200× to 3000×, Wehnelt's electrode voltage 25 kV, Hi-Vacuum mode. Chemical composition of surfaces was analyzed with EDS detector (EDAX Element with solid silicon nitride (Si3N4) window and non-nitrogen Silicon Drift Detector) and dedicated software (TEAM by EDAX). Samples were mounted on single plates, with double-sided carbon tape.

3.2.5. Scanning X-ray Photoelectron Microscope (SPEM)

The scanning photoelectron microscope (SPEM) allows spatially resolved XPS measurements in the submicron scale [32,33]. SPEM measurements were performed by using the Escamicroscopy beamline at Elettra synchrotron facility (Trieste, Italy). Samples were annealed in vacuum at about 200 °C for 5 h to remove residual surface contamination.

Spatial resolution in the submicron scale was achieved by de-magnifying the X-ray beam with a zone plate (ZP), a Fresnel type lens, plus an order sorting aperture (OSA) to stop the higher order produced by the ZP. Spot sizes down to 130 nm were obtained. The

X-ray illumination was perpendicular to the sample surface. The emitted photoelectrons were collected by a hemispherical electron energy analyzer (HEA) mounted at a 30° angle with respect to the sample plane to enhance the surface sensitivity of the measurement; a delay line detector grouped the detected electrons in 48 channels. The overall energy resolution was better than 300 meV. The photon energy (hv) used for the measurements was 644.5 eV.

3.2.6. Mechanical Properties

The investigations of mechanical properties were performed as nanoindentation measurements of hardness, elastic modulus, and deformation energies by NHT2 Nanoindentation Tester (CSM Instruments, Needham, MA, USA) equipped with a Berkovich diamond indenter with total included angle of 142.3°. During the measurement, the following parameters were applied: linear loading, maximum load −100 mN, loading/unloading rate −200 mN/min, Poisson's ratio −0.3. The scratch tester (CSM Instruments, Needham, MA, USA) was used to determine the quality of the polished and the cut Co-Cr discs surface. The following parameters were used for measurements: linear scratch, begin load −30 mN, end load −10 N, loading rate −4985. The Rockwell indenter was applied (diamond with the radius of 0.2 mm).

3.2.7. Biocompatibility Study

Sample preparation: The samples of investigated material were prepared in accordance with the guidelines described in the PN-EN ISO 10993-12: 2012 Biological evaluation of medical devices Part 12: Sample preparation and reference materials standard. Co-Cr discs were tested in two adequate ways:
— through direct contact with the tested Co-Cr disc;
— using extracts from the tested Co-Cr discs.

Genotoxicity studies: For this purpose, the Ames test was used (in accordance with OECD guidelines). The bacterial gene mutation test was used (according to PN-EN ISO 10993-3: 2009 Biological evaluation of medical devices Part 3: Testing for Genotoxicity). The study was performed using the Ames MPF TM 98/100 test (Xenometrix, Allschwil, Switzerland) on two strains of *Salmonella typhimurium* TA98 and TA100, both used separately and in combination with the S9 fraction obtained from the livers of Sprague-Dawley rats. The 25-fold more concentrated extracts were used, which were then diluted to obtain concentrations of 100%, 50%, 25%, 12.5%, 6.25%, and 3.125%. The positive controls for TA98 and TA100 strains without S9 fraction were 2-nitrofluorene (2-NF) and 4-nitroquinoline N-oxide (4-NQO). In turn, for strains TA98 and TA100 with fraction S9, 2-aminoanthracene (2-AA) was the control.

Cytotoxicity studies: Cytotoxicity testing was performed in accordance with the PN-EN ISO 10993-5 (2009) Biological evaluation of medical devices Part 5: Tests for in vitro cytotoxicity. BALB/3T3 clone A31 (ATCC® CCL-163 ™) mouse fibroblast cells grown in Dulbecco's Modified Eagle's Medium (DMEM) with 10% fetal bovine serum and 1% antibiotic at a final concentration of 1% (10,000 U/mL of penicillin) were used in the study. The extracts from the Co-Cr discs were tested in four concentrations: 100%, 50%, 25%, and 12.5%, by diluting them in the culture medium. The negative control was high-density polyethylene (HDPE, Greiner, Rastatt, Germany), as a positive control, sodium lauryl sulphate solution (SLS, Sigma Aldrich, St. Louis, MO, USA) was used at concentrations of 0.2, 0.1, 0, 05, or 0.025 mg/mL, dissolved in the culture medium. The culture plates were incubated for 24 h. After this time, the morphology of the cells was assessed using a Leica DMi 1 light microscope (Leica, Wetzlar, Germany). Cytotoxicity was assessed using the MTT assay. For this purpose, the test solutions were removed from the wells of the culture plates, and the MTT reagent was added. Cells were then incubated for 2 h, after which time, the MTT solution was removed, and isopropanol was added to each well. After 30 min, a reading was made at a wavelength of 570 nm using an Epoch 2 spectrophotometric reader (BioTek, Highland Park, MI, USA). Cell viability was assessed using the formula:

$V = (Ab : As) \times 100\%$, where V is cell viability, Ab is mean absorbance value in the test sample, and As is mean absorbance value in the blank sample.

Direct contact test was performed by two methods. The first method was to place sterile Co-Cr disks in separate wells at the bottom of the wells in a 24-well plate (Greiner, Rastatt, Germany) and then seed the cells onto Co-Cr disks. The second method involved seeding the cells, and after sticking to the bottom of the well and reaching 80% confluence, Co-Cr discs were placed on them. In both methods used, the number of cells plated per well was comparable.

After 24 h of incubation, the cell morphology was assessed using a Leica DMi 1 light microscope (Leica, Wetzlar, Germany), and then the cells and the Co-Cr discs were fixed with 2.5% glutaraldehyde (Avantor Performance Materials Poland SA, Gliwice, Poland) for 1 h. After fixation, cells were dehydrated in an increasing series of ethyl alcohol (from 50–100%). It was dried and coated with a 20 nm gold layer using a Leica EM ACE200 vacuum sputter sputtering machine (Leica, Wetzlar, Germany) and imaged on an Evo LS15 scanning electron microscope (Zeiss, Oberkochen, Germany).

All reagents were from Sigma Aldrich, St. Louis, MO, USA. Cells were grown under standard conditions at 37 °C in the presence of 5% CO_2 in a Midi40 incubator (Thermo Fisher Scientific, Waltham, MA, USA).

Hemocompatibility studies: All adequate hemocompatibility testing was performed in accordance with PN-EN ISO 10993-5 (2009): Biological evaluation of medical devices Part 4: Selection of tests for interactions with blood. The performed tests included: hematological tests (assessment of number and morphology of blood cell elements), assessment of hemolytic activity, tests of platelet dysfunction, tests of clotting system activation (determination of activated partial thromboplastin time, prothrombin time, thrombin time, fibrinogen concentration, recalcination time), the activity of coagulation factors (VII, X, XII), activation of the fibrinolysis system (determination of fibrinolysis time, plasminogen concentration), testing of coagulation inhibitors (ATIII activity, protein C). Whole blood collected for EDTA and sodium citrate (BioIVT, West Sussex, Great Britain) and concentrated red blood cells (HaemoScan, Groningen, the Netherlands) were used in the research. Co-Cr discs with a total area of 3 cm^2 were incubated with 1 mL of whole blood and incubated for 60 min at room temperature. Nitrile rubber materials (HaemoScan, Groningen, The Netherlands) with a total area of 0.5 cm^2 were a positive control. The negative control was high-density polyethylene (HDPE, Greiner, Rastatt, Germany). Extracts from the tested Co-Cr discs in the volume of 300 µL were also used for the research.

After the incubation time elapsed, the blood counts were measured, and activated partial thromboplastin time, prothrombin time, thrombin time, fibrinogen concentration, recalcination time, level of clotting inhibitors (ATIII activity, protein C), and plasminogen activity were determined. The analyses were performed using a hematological and biochemical analyzer (Diagnostyka, Wrocław, Poland).

4. Conclusions

Samples with a layer of graphene throughout the large area of the Co-Cr alloy substrate (L-605) by a cold-wall CVD process were obtained and confirmed by the Raman spectroscopy measurements. It was shown by surface microanalysis studies that the surface of unpolished substrate samples with the demonstrated presence of graphene was not homogeneous and had still a relatively high average roughness compared to polished substrates. At the same time, measurements of the samples after the CVD process showed a complete absence of specific graphene bands on the polished Co-Cr substrates. This may indicate a critical influence of the preparation method, the resulting uniformity of the microstructure, and the chemical composition on the effectiveness of the graphene deposition process. The investigation of mechanical properties showed that instrumental hardness and elastic modulus of polished and cut only Co-Cr discs were equal to 406.31 Vickers, 161.49 GPa and 262.82 Vickers, 72.264 GPa, respectively. The results of the hemocompatibility test, understood as the ability to inactivate blood coagulation processes, indicate that Co-Cr alloy samples covered with a graphene layer did not show the pro-coagulant effect. The tested material also showed good biocompatibility. This fact, together with a proof of deposition of good quality layers by use of the relatively cheaper technique (CW-CVD), gives hope for the implementation of the developed technology in the field of coating medical devices.

Author Contributions: Conceptualization D.H., D.B. and M.W.; methodology: M.H., M.A., L.G., M.L.S., D.H., D.B. and M.W.; preparation of samples: Ł.W., V.B. and M.M.; Biocompatibility studies: D.B., J.A. and M.W.; AFM and mechanical tests: M.H.; SEM and EDS: Ł.W.; SPEM: M.A., L.G. and P.Z.; Raman data: V.B. and M.M.; writing—original draft preparation: Ł.W., V.B., M.M., D.H. and M.W.; writing—review and editing: M.L.S., M.A., L.G., D.H. and M.W.; supervision: D.H., D.B., and M.W.; funding acquisition: D.B., D.H. and M.W. All authors have read and agreed to the published version of the manuscript.

Funding: This research received financial support of Operational Program Smart Growth within project no. POIR.01.01.01-00-0319/17-02, Measure 1.1 "Research and Development", Submeasure 1.1.1 "Fast track", co-financed from the European Regional Development Fund. M.L.S., D.B., and M.W. thank the University of Palermo for the CORI2018–Action C2 Project (CORI-2018-C-D15-180463) to give the possibility of visits and cooperation between University of Palermo and Wroclaw Medical University.

Institutional Review Board Statement: This study did not required ethical approval.

Informed Consent Statement: Not applicable.

Data Availability Statement: Data sharing is not applicable to this article.

Acknowledgments: This work is the result of the inspiration and positive motivation by the actions of Dariusz Biały, who sadly passed away before its final completion.

Conflicts of Interest: The authors declare no conflict of interest.

References

1. Wang, C.; Xia, K.; Wang, H.; Liang, X.; Yin, Z.; Zhang, Y. Advanced carbon for flexible and wearable electronics. *Adv. Mater.* **2019**, *31*, 1801072. [CrossRef]
2. Kamyshny, A.; Magdassi, S. Conductive nanomaterials for 2D and 3D printed flexible electronics. *Chem. Soc. Rev.* **2019**, *48*, 1712–1740. [CrossRef] [PubMed]
3. Fang, R.; Chen, K.; Yin, L.; Sun, Z.; Li, F.; Cheng, H.M. The regulating role of carbon nanotubes and graphene in lithiumion and lithium–sulfur batteries. *Adv. Mater.* **2019**, *31*, 1800863. [CrossRef]
4. He, Q.; Wu, S.; Yin, Z.; Zhang, H. Graphene-based electronic sensors. *Chem. Sci.* **2012**, *3*, 1764–1772. [CrossRef]
5. Kovalska, E.; Lesongeur, P.; Hogan, B.T.; Baldycheva, A. Multi-layer graphene as a selective detector for future lung cancer biosensing platforms. *Nanoscale* **2019**, *11*, 2476–2483. [CrossRef] [PubMed]
6. Liu, B.; Zhou, K. Recent progress on graphene-analogous 2D nanomaterials: Properties, modeling and applications. *Prog. Mater. Sci.* **2019**, *100*, 99–169. [CrossRef]
7. Panwar, N.; Soehartono, A.M.; Chan, K.K.; Zeng, S.; Xu, G.; Qu, J.; Coquet, P.; Yong, K.T.; Chen, X. Nanocarbons for biology and medicine: Sensing, imaging, and drug delivery. *Chem. Rev.* **2019**, *119*, 9559–9656. [CrossRef]
8. Orsu, P.; Koyyada, A. Recent progresses and challenges in graphene based nano materials for advanced therapeutical applications: A comprehensive review. *Mater. Today Commun.* **2020**, *22*, 100823. [CrossRef]
9. Saladino, M.L.; Markowska, M.; Carmone, C.; Cancemi, P.; Alduina, R.; Presentato, A.; Scaffaro, R.; Biały, D.; Hasiak, M.; Hreniak, D.; et al. Graphene oxide carboxymethylcellulose nanocomposite for dressing materials. *Materials* **2020**, *13*, 1980. [CrossRef]
10. Chapman, D.A.; Corner, A.; Webster, R.; Markowitz, E.M. Climate visuals: A mixed methods investigation of public perceptions of climate images in three countries. *Glob. Environ. Chang.* **2016**, *41*, 172–182. [CrossRef]
11. Zhang, W.; Lee, S.; Mcnear, K.L.; Chung, T.F.; Lee, S.; Lee, K.; Crist, S.A.; Ratliff, T.L.; Zhong, Z.; Chen, Y.P.; et al. Use of graphene as protection film in biological environments. *Sci. Rep.* **2014**, *4*, 1–8. [CrossRef]
12. Kyhl, L.; Nielsen, S.F.; Čabo, A.G.; Cassidy, A.; Miwa, J.A.; Hornekær, L. Graphene as an anti-corrosion coating layer. *Faraday Discuss.* **2015**, *180*, 495–509. [CrossRef]
13. Weatherup, R.S.; D'Arsié, L.; Cabrero-Vilatela, A.; Caneva, S.; Blume, R.; Robertson, J.; Schloegl, R.; Hofmann, S. Long-term passivation of strongly interacting metals with single-layer graphene. *J. Am. Chem. Soc.* **2015**, *137*, 14358–14366. [CrossRef]
14. Li, X.; Cai, W.; An, J.; Kim, S.; Nah, J.; Yang, D.; Piner, R.; Velamakanni, A.; Jung, I.; Tutuc, E.; et al. Large-area synthesis of high-quality and uniform graphene films on copper foils. *Science* **2009**, *324*, 1312–1314. [CrossRef]

15. Chen, X.; Zhang, L.; Chen, S. Large area CVD growth of graphene. *Synth. Met.* **2015**, *210*, 95–108. [CrossRef]
16. Macháč, P.; Hejna, O.; Slepička, P. Graphene growth by transfer-free chemical vapour deposition on a cobalt layer. *J. Electr. Eng.* **2017**, *68*, 79–82. [CrossRef]
17. An, H.; Lee, W.J.; Jung, J. Graphene synthesis on Fe foil using thermal CVD. *Curr. Appl. Phys.* **2011**, *11*, S81–S85. [CrossRef]
18. Nam, J.; Kim, D.C.; Yun, H.; Shin, D.H.; Nam, S.; Lee, W.K.; Hwang, J.Y.; Lee, S.W.; Weman, H.; Kim, K.S. Chemical vapor deposition of graphene on platinum: Growth and substrate interaction. *Carbon N. Y.* **2017**, *111*, 733–740. [CrossRef]
19. Seah, C.M.; Chai, S.P.; Mohamed, A.R. Mechanisms of graphene growth by chemical vapour deposition on transition metals. *Carbon N. Y.* **2014**, *70*, 1–21. [CrossRef]
20. Thanh, T.D.; Balamurugan, J.; Kim, N.H.; Lee, J.H. Recent advances in metal alloy-graphene hybrids for biosensors. In *Graphene Bioelectronics*; Elsevier Inc.: Amsterdam, The Netherlands, 2018; pp. 57–84. ISBN 9780128133507.
21. Bointon, T.H.; Barnes, M.D.; Russo, S.; Craciun, M.F. High quality monolayer graphene synthesized by resistive heating cold wall chemical vapor deposition. *Adv. Mater.* **2015**, *27*, 4200–4206. [CrossRef] [PubMed]
22. Alnuaimi, A.; Almansouri, I.; Saadat, I.; Nayfeh, A. Toward fast growth of large area high quality graphene using a cold-wall CVD reactor. *RSC Adv.* **2017**, *7*, 51951–51957. [CrossRef]
23. Petr, M.; Ondrej, H. Graphene growth by transfer-free CVD method using cobalt/nickel catalyst layer. *Mater. Sci. Forum* **2018**, *919*, 207–214. [CrossRef]
24. Mani, G.; Feldman, M.D.; Patel, D.; Agrawal, C.M. Coronary stents: A materials perspective. *Biomaterials* **2007**, *28*, 1689–1710. [CrossRef] [PubMed]
25. Wawrzyńska, M.; Bil-Lula, I.; Krzywonos-Zawadzka, A.; Arkowski, J.; Łukaszewicz, M.; Hreniak, D.; Stręk, W.; Sawicki, G.; Woźniak, M.; Drab, M.; et al. Biocompatible carbon-based coating as potential endovascular material for stent surface. *Biomed Res. Int.* **2018**, *2018*. [CrossRef]
26. Tanaka, M.; Kato, R. Control of microstructures by heat treatments and high-temperature properties in high-tungsten colbalt-base superalloys. In *Handbook of Material Science Research*; René, C., Turcotte, E., Eds.; Nova Science Publishers Inc.: Hauppauge, NY, USA, 2010; pp. 445–458. ISBN 978-160741798-9.
27. Ferrari, A.C.; Meyer, J.C.; Scardaci, V.; Casiraghi, C.; Lazzeri, M.; Mauri, F.; Piscanec, S.; Jiang, D.; Novoselov, K.S.; Roth, S.; et al. Raman spectrum of graphene and graphene layers. *Phys. Rev. Lett.* **2006**, *97*, 187401. [CrossRef] [PubMed]
28. Lebedieva, T.; Gubanov, V.; Dovbeshko, G.; Pidhirnyi, D. Quantum-chemical calculation and visualization of the vibrational modes of graphene in different points of the brillouin zone. *Nanoscale Res. Lett.* **2015**, *10*, 287. [CrossRef]
29. Ferrari, A.C. Raman spectroscopy of graphene and graphite: Disorder, electron–phonon coupling, doping and nonadiabatic effects. *Solid State Commun.* **2007**, *143*, 47–57. [CrossRef]
30. Merlen, A.; Buijnsters, J.; Pardanaud, C. A guide to and review of the use of multiwavelength raman spectroscopy for characterizing defective aromatic carbon solids: From graphene to amorphous carbons. *Coatings* **2017**, *7*, 153. [CrossRef]
31. Gregoratti, L.; Barinov, A.; Benfatto, E.; Cautsro, G.; Fava, C.; Lacovig, P.; Lonza, D.; Kiskinova, M.; Tommasini, R.; Mähl, S.; et al. 48-Channel electron detector for photoemission spectroscopy and microscopy. *Rev. Sci. Instrum.* **2004**, *75*, 64–68. [CrossRef]
32. Zeller, P.; Amati, M.; Sezen, H.; Scardamaglia, M.; Struzzi, C.; Bittencourt, C.; Lantz, G.; Hajlaoui, M.; Papalazarou, E.; Marino, M.; et al. Scanning photoelectron spectro-microscopy: A modern tool for the study of materials at the nanoscale. *Phys. Status Solidi* **2018**, *215*, 1800308. [CrossRef]
33. Amati, M.; Barinov, A.; Feyer, V.; Gregoratti, L.; Al-Hada, M.; Locatelli, A.; Mentes, T.O.; Sezen, H.; Schneider, C.M.; Kiskinova, M. Photoelectron microscopy at Elettra: Recent advances and perspectives. *J. Electron. Spectros. Relat. Phenom.* **2018**, *224*, 59–67. [CrossRef]

Article

Benefits in the Macrophage Response Due to Graphene Oxide Reduction by Thermal Treatment

Mónica Cicuéndez [1], Laura Casarrubios [1], Nathalie Barroca [2], Daniela Silva [2], María José Feito [1], Rosalía Diez-Orejas [3], Paula A. A. P. Marques [2,*] and María Teresa Portolés [1,4,*]

1. Departamento de Bioquímica y Biología Molecular, Facultad de Ciencias Químicas, Universidad Complutense de Madrid, Instituto de Investigación Sanitaria del Hospital Clínico San Carlos (IdISSC), 28040 Madrid, Spain; mcicuendez@ucm.es (M.C.); laura.casarrubios.molina@gmail.com (L.C.); mjfeito@ucm.es (M.J.F.)
2. Center for Mechanical Technology & Automation (TEMA), Mechanical Engineering Department, University of Aveiro, 3810-193 Aveiro, Portugal; nbarroca@ua.pt (N.B.); danielas@ua.pt (D.S.)
3. Departamento de Microbiología y Parasitología, Facultad de Farmacia, Universidad Complutense de Madrid, 28040 Madrid, Spain; rdiezore@ucm.es
4. CIBER de Bioingeniería, Biomateriales y Nanomedicina, 28040 Madrid, Spain
* Correspondence: paulam@ua.pt (P.A.A.P.M.); portoles@quim.ucm.es (M.T.P.)

Abstract: Graphene and its derivatives are very promising nanomaterials for biomedical applications and are proving to be very useful for the preparation of scaffolds for tissue repair. The response of immune cells to these graphene-based materials (GBM) appears to be critical in promoting regeneration, thus, the study of this response is essential before they are used to prepare any type of scaffold. Another relevant factor is the variability of the GBM surface chemistry, namely the type and quantity of oxygen functional groups, which may have an important effect on cell behavior. The response of RAW-264.7 macrophages to graphene oxide (GO) and two types of reduced GO, rGO15 and rGO30, obtained after vacuum-assisted thermal treatment of 15 and 30 min, respectively, was evaluated by analyzing the uptake of these nanostructures, the intracellular content of reactive oxygen species, and specific markers of the proinflammatory M1 phenotype, such as CD80 expression and secretion of inflammatory cytokines TNF-α and IL-6. Our results demonstrate that GO reduction resulted in a decrease of both oxidative stress and proinflammatory cytokine secretion, significantly improving its biocompatibility and potential for the preparation of 3D scaffolds able of triggering the appropriate immune response for tissue regeneration.

Keywords: graphene oxide; reduced graphene oxide; macrophage; cytokine; immune response

1. Introduction

Due to their unique physical and chemical properties [1], graphene and its derivatives are very promising nanomaterials for a wide range of biomedical applications [2], such as biosensing [3], bioimaging [4], drug delivery [5], and photothermal therapy [6]. In the last decade, graphene oxide (GO) and reduced graphene oxide (rGO) have proven to be very useful for the preparation of scaffolds for tissue repair, capable of acting as support for growing cells in a suitable microenvironment [7–9]. GO is electrically insulating, owing to its disrupted sp² bonding network due to the presence of oxygen functional groups, such as hydroxyl and epoxy functional groups on the basal plane and carbonyl/carboxylic acids groups on the plane edges [10]. However, rGO derives from the reduction of GO, which restores the π network imparting electrical conductivity, a key feature when designing electroconductive devices [11]. There are different types of reduction processes, such as chemical, thermal, microwave, and UV light reduction, with different advantages and disadvantages related to the reduction degree, improvement of material properties, toxicity, and economic cost, among other aspects [7]. Several factors have been identified as the culprits for GO and rGO cytotoxicity, such as dose, lateral size, and surface charge [12].

Another relevant factor is the variability of GO surface chemistry, namely the type and quantity of oxygen functional groups, which may have an important effect on cell behavior [13]. Although there is a low number of studies comparing cellular responses to GO and rGO, some authors indicate that rGO is less toxic than GO [14,15], while others suggest that rGO may cause more plasma membrane disruption and oxidative stress than GO [12]. Several studies have pointed out that GO induces an inflammatory response and chronic injury by interfering with the functions of important organs such as the respiratory tract, the central nervous system, and blood components [16]. In this context, the response of immune cells to these graphene derivatives appears to be critical due to macrophage functional plasticity between two extremes, designated as proinflammatory (M1) and reparative (M2) phenotypes [17]. Thus, the balance between M1 and reparative M2 macrophages has been related to the role of this cell type in disease processes and tissue remodeling after injury [18,19]. These macrophage phenotypes are characterized by differences in the expression of distinct cell surface markers and particular genes and the secretion of different cytokines, chemokines, and enzymes that allow them to respond to changes in their microenvironment [20,21].

The translation of graphene derivatives to the medical market may rely on their use as single-component devices or their incorporation into natural or synthetic matrices depending on the targeted application. Regardless of the strategy, the fate of soluble GO or rGO from medical devices should be carefully addressed. Considering the family of carbon nanostructures with potential applications in the biomedical field, graphene and its derivatives are considered less cytotoxic than single- and multiwalled carbon nanotubes (CNTs) and fullerenes [22–24]. In the case of carbon nanotubes (CNTs), which have previously received the most attention, it is well-known that length and functionalization can cause very different reactions in cells and that it is critical to distinguish between CNTs in terms of physical and chemical properties [25]. Learning from the CNT literature, it is of huge importance to systematically evaluate the specificities of emerging GBM.

In this work, the response of RAW-264.7 macrophages to different doses of GO and rGO nanostructures with different reduction degrees was evaluated by analyzing the uptake of these nanomaterials, the intracellular content of reactive oxygen species, and specific markers of the proinflammatory M1 phenotype, such as CD80 expression and the secretion of inflammatory cytokines such as TNF-α and IL-6. This comparative study, involving flow cytometry, confocal microscopy, and ELISA methods, highlights the effects of both the degree of GO reduction and the dose delivered on macrophage response.

2. Results and Discussion

2.1. Structural and Morphological Analysis of GO, rGO15, and rGO30 Nanostructures

The GO reduction process employed in this work consisted of vacuum-assisted thermal treatment at 200 °C. We intended to avoid reduction using chemical solvents that are environmentally nonfriendly, so thermal reduction at low temperatures was performed. Furthermore, our group has previously demonstrated the combined favorable outcome of thermally reduced GO microfibers at 220 °C for 2 h with both neural cells and macrophages [26]. While in this previous study, GO-based bulk constructs were thermally reduced for 2 h, in this work, we aimed at studying the effect of particles to be used as additives in tissue engineering scaffolds and opted for much shorter times of reduction. Additionally, as we targeted biological applications wherein a complete reduction of graphene oxide is not desired, shorter time was preferable.

The structural changes of GO after the thermal treatment are shown in the X-ray diffraction (XRD) and X-ray photoelectron spectroscopy (XPS) analyses below (Figures 1 and 2, respectively).

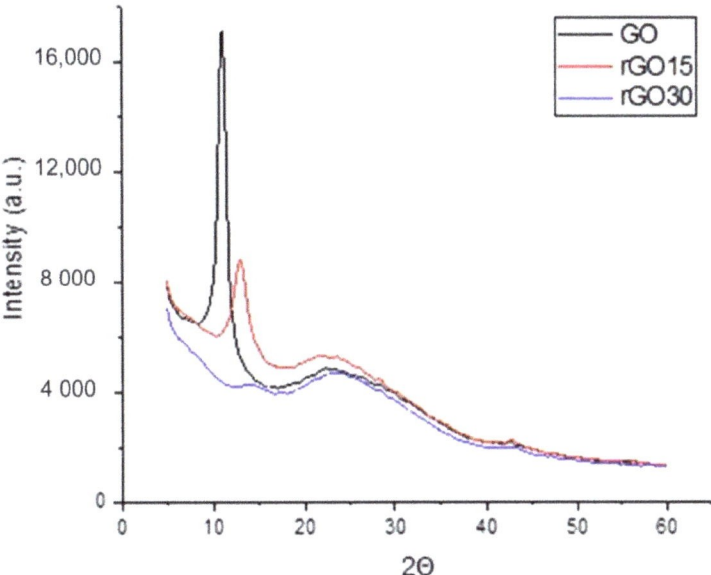

Figure 1. XRD spectra of GO, rGO15, and rGO30.

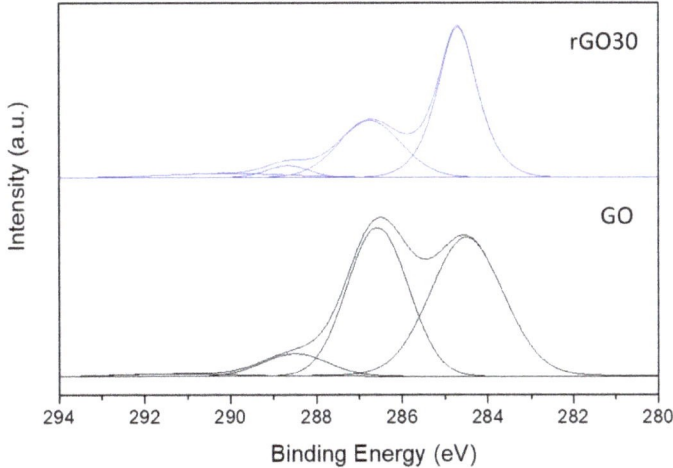

Graphene-based nanomaterials		
Functional groups (%)	GO	rGO30
C-C (284.7 eV)	49.1	58.2
C-O (286.7 eV)	42.4	32.0
C=O (288.5 eV)	7.0	4.7
C=O-O (291.3 eV)	1.5	5.1

Figure 2. C1s XPS spectra of GO and rGO30, and respective percentages of the present functional groups.

XRD structural analysis (Figure 1) revealed that GO presented a sharp and intense crystalline peak at $2\theta = 11.08°$ that corresponded to the (001) diffraction peak. After the vacuum-assisted thermal reduction at 200 °C for 15 min, the peak was less intense and exhibited a shift to the right ($2\theta = 12.98°$). This could be attributed to water deintercalation, removal of oxygen-containing functional groups, and partial restoration of the sp^2 network. Moreover, a second broad peak appeared at $2\theta = 21.18°$, attributed to the (002) plane. A longer thermal reduction time of 30 min induced a further shift of the (001) peak to $2\theta = 13.58°$ and a pronounced broadening due to the partial breakdown of the long-range order of GO [27,28].

These results show that thermal reduction for 15 and 30 min was enough to induce GO reduction. However, the presence of both peaks indicated incomplete reduction. Incomplete reduction of GO is most suited for biomedical applications, as fully reduced GO loses its ability to disperse once most of its oxygen groups are removed, which may make its incorporation into engineered materials for various medical applications difficult. Additionally, residual O-moieties make GO amenable for chemical functionalization, which is valuable for drug delivery [29], cancer therapy [6], and enhancing biocompatibility.

XPS allowed us to evaluate the deoxygenation more comprehensibly. As seen in the C1s XPS spectrum (Figure 2), GO and rGO30 exhibited the four components relative to carbon atoms in different functional groups: nonoxygenated ring C (284.7 eV), C–O bonds (286.7 eV), carbonyl C=O (288.5 eV), and carboxylate C(O)–O (291.3 eV).

The peak intensities of the three oxygenated components in rGO30 were significantly lower than those of GO, demonstrating significant deoxygenation during thermal reduction. The majority of the present oxygen moieties are the C–O bonds of epoxy and hydroxyl groups in the basal plane. More oxidized species such as C=O and C=O (O) are sparser. The C=O species come mainly from single ketones that decorate the edges of GO sheets [30] or are bound to the basal plane as carbonyl groups. Quinones are also located at the edges of GO sheets. Regarding the C=O (O) species, these are mostly found at the edges of GO sheets [31,32]. The spectra show that the thermal reduction of GO induced the removal of unstable in-plane oxygen-containing groups. This observation is consistent with previously reported XPS data on low-temperature thermal reduction of GO, predominantly linked to the reduction of hydroxyl and epoxy groups, further shown by the reduction in the percentages of C–O and C=O present in rGO30 [32,33].

The morphology dependence of the GO sheets on thermal reduction was further assessed by atomic force microscopy (AFM, Figure 3).

The morphology of GO and rGO sheets displayed heterogenous size distribution characterized by the presence of large flakes with smaller sheets piling up on top. Additionally, a notable decrease in sheet size upon reduction could be observed. This is consistent with previous observations on reduced GO, even with different reduction methods, such as a chemical one via hydrazine [34]. Additionally, compared to the more flat-like morphology of GO, both rGO15 and rGO30 sheets exhibited some crumpling. These structures are typically the result of the desorption of H_2O, CO, and CO_2 and the decomposition of oxygen functional groups that leads to graphene-like sheets with disordered stacking observed upon thermal reduction [27,28,35–37].

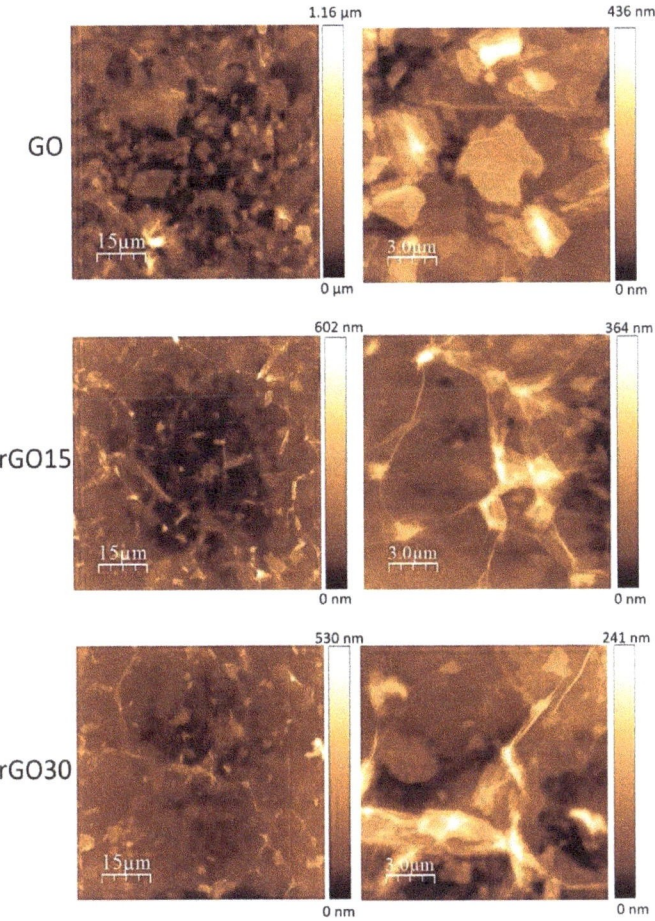

Figure 3. AFM topography images of GO, rGO15, and rGO30.

2.2. Uptake of GO, rGO15, and rGO30 by RAW-264.7 Macrophages

Macrophages are key modulatory and effector cells in the immune response, and their activation influences other components of the immune system in different physiological contexts. These cells perform phagocytic clearance of dead cells during development and adult life and protect the host through innate immunity. Macrophages also play a key role in the removal of nanomaterials or biodegradation products by phagocytosis from scaffolds with potential application in biomedicine, and they are primarily responsible for the uptake and cellular trafficking of nanoparticles in vivo [38]. In this study, cell uptake of different doses of GO, rGO15, and rGO30 nanostructures by RAW-264.7 macrophages was quantified by flow cytometry analyzing 90° light scatter (side scatter, SSC) after 24 h of incubation. This parameter is proportional to the intracellular complexity determined in part by the cellular cytoplasm, mitochondria, and pinocytic vesicles [39]. For this reason, SSC can be used as a measure of the incorporation of these nanostructures inside cells. Figure 4 shows a clear significant dose-dependent increase of the intracellular complexity of macrophages cultured with GO compared to that of the control macrophages. Regarding the results obtained with rGO15, we only observed a significant increase of SSC with macrophages exposed to 10 µg/mL. However, the three assayed doses of rGO30 produced a significant increase in macrophage intracellular complexity compared to that of the control macrophages. Figure 4 also shows the statistical significance among the different nanostructures at the same concentration. Thus, a significant SSC increase (# $p < 0.05$) of macrophages cultured with GO compared to that of rGO15 was observed at the same concentration of 5 μg/mL. Moreover, we also observed a significant increase (# $p < 0.05$) in the intracellular complexity of macrophages exposed to 10 μg/mL of GO compared to that shown by macrophages exposed to the same concentrations of rGO15 and rGO30.

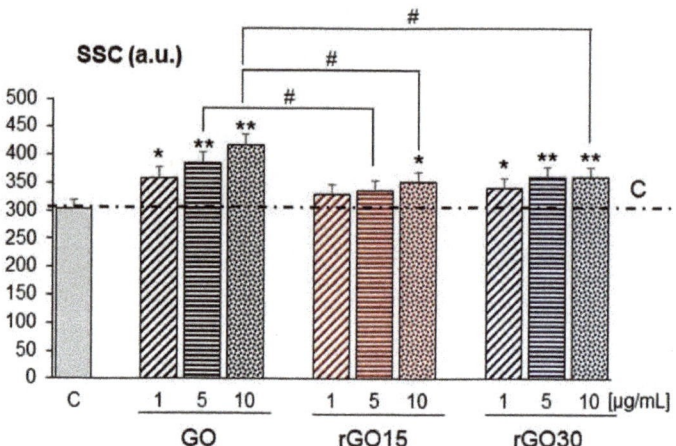

Figure 4. Cell uptake of GO, rGO15, and rGO30 by RAW-264.7 macrophages after 24 h of treatment with 1, 5, and 10 μg/mL of these nanostructures, evaluated by flow cytometry analyzing 90° light scatter (side scatter, SSC). Statistical significance: * $p < 0.05$, ** $p < 0.01$ (compared to control macrophages), # $p < 0.05$ (comparison between nanostructures at the same concentration).

These results show that, in general, GO nanostructures were incorporated by macrophages in greater quantity than those of reduced GO and that the main differences in the cellular incorporation of these nanostructures by RAW-264.7 macrophages were observed at the highest dose (10 μg/mL) evaluated in this study. Cellular uptake of a great variety of nanomaterials is known to be highly dependent on their different physicochemical characteristics such as their lateral dimension, oxidation level, and surface functional groups, as well as on their concentration, purity, and shape, among other factors [40]. Regarding GBM, their internalization into cells is strongly influenced by particle size and surface chemistry [41]. The influence of the thickness of GO on cellular internalization is an open debate. While a few studies reported that GO lateral size is a prime factor at determining cellular uptake, with large lateral size preventing cellular uptake [42], other researches have demonstrated that the saturated uptake amount of GO sheets after 24 h did not vary with the lateral dimension (2 and 350 nm), and identical accumulation occurred in primary macrophages when exposed to doses (2 and 6 μg/mL) similar to those in the current study [43]. Here, although the overall lateral size of the rGO sheets decreased upon reduction, it concomitantly underwent aggregation. Reduction-induced aggregation may have a role in the observed decrease in cellular uptake activity when reducing GO. This has been shown for Fe^{2+}-reduced GO on a murine macrophage cell line [12]. Additionally, it is expected that at a higher dose, aggregation would be more pronounced, consequently reducing internalization, which was observed here. Another factor that may account for the decreased internalization is related to protein adsorption. Vacuum-assisted thermal reduction induces expansion of the GO sheets, along with an increase in surface area [37]. This surface area increase gives more affinity for extracellular proteins, consequently leading to weaker interactions with the cell membrane and lower cellular uptake. Our XPS analysis revealed, besides increased surface area, that the thermal treatment led to an increase in carboxyl groups, representing a much stronger hydrogen bonding moiety than

C-O groups, ensuring stronger hydrogen bonding formation between proteins [12]. The internalization of GBM is also related to the cell type. It has been reported that while GO was internalized by HepG2 cells, by contrast, rGO, which is more hydrophobic than GO, was found to be mostly adsorbed on the cell surface [44]. In addition, it has been demonstrated that different cell types can effectively uptake both GO and rGO nanosheets by different endocytic mechanisms [45,46].

In this work, morphological studies of macrophages after GO, rGO15, and rGO30 uptake were carried out by confocal and phase contrast microscopy. Figure 5 shows the confocal images of RAW-264.7 macrophages with their cytoskeleton intact after 24 h of treatment with 1, 5, and 10 µg/mL of these three nanostructures. GO, rGO15, and rGO30 appear as black deposits inside the cells observed by phase contrast microscopy. Adverse effects of GO on murine peritoneal macrophages have been observed by other authors due to the accumulation of this nanomaterial in macrophage lysosomes, leading to lysosome membrane destabilization, autophagosome accumulation, and reduced autophagic degradation [47]. We have evaluated different parameters (included in the following sections) related to the specific function of macrophages to know if the incorporation of these nanostructures induced oxidative stress and promoted a possible inflammatory response.

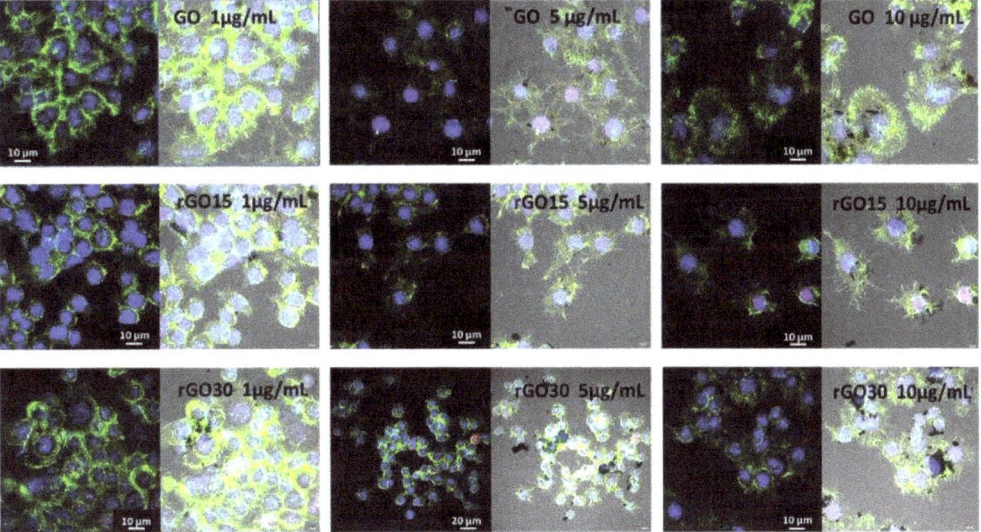

Figure 5. Cell uptake of GO, rGO15, and rGO30 by RAW-264.7 macrophages after 24 h of treatment with 1, 5, and 10 µg/mL, evaluated by confocal and phase contrast microscopy.

2.3. Intracellular Reactive Oxygen Species (ROS) Content of RAW-264.7 Macrophages after GO, rGO15, and rGO30 Uptake

In recent years, numerous GBMs have been developed with the aim of decreasing their toxicity and improving their biocompatibility for use in biomedical applications. Different experimental models have been used in vitro to study the cellular response to graphene and its derivatives, and numerous articles have been published related to the interaction with the components of the immune system [48]. Macrophages represent one of the most useful experimental models, as they are directly involved in the innate immune response and in the uptake of nanoparticles for their elimination from the organism. Concerning the possible adverse effects of these GBMs, several studies propose oxidative stress, mediated by ROS production, as a key mechanism involved in their cytotoxicity [49]. It results from an imbalance between excessive generation of ROS and the limited antioxidant defense capacity of cells, thereby leading to adverse biological effects such as membrane lipid peroxidation, protein denaturation, mitochondrial dysfunction, and DNA damage. Moreover, ROS generation by living cells in response to these kinds of nanomaterials depends greatly on their layer number, lateral size, purity, dose, surface chemistry, dispersibility, and hydrophilicity [45]. Stimulated ROS production was originally described in phagocytic cells such as neutrophils and macrophages. Macrophages are one of the most versatile types of immune cells carrying out a variety of key functions, including phagocytosis of apoptotic cells, bacteria, and viruses, production of reactive nitrogen and oxygen species, antigen processing and presentation, and cytokine and chemokine production. These immune cells also play a central role in directing the host response to implanted biomaterials, including the inflammatory and reparative response related to the M1 and M2 phenotypes, respectively [50,51]. Moreover, it is well-known that M1 proinflammatory macrophages produce and secrete higher ROS levels than M2 reparative cells [52], inducing damage to neighboring cells and promoting the proinflammatory response. Thus, the effects of 1, 5, and 10 µg/mL of GO, rGO15, and rGO30 on the intracellular content of reactive oxygen species (ROS) of RAW-264.7 macrophages were evaluated in the present study by flow cytometry after 24 h of treatment with these nanostructures. Figure 6 shows that GO treatment induced significant increases of intracellular macrophage ROS at all assayed doses, obtaining the most pronounced effect with 5 µg/mL. On the other hand, rGO15 and rGO30 produced a significant elevation of intracellular ROS levels, but less prominent than that observed with GO. These results demonstrate that the GO reduction process employed in this work through a vacuum-assisted thermal treatment of the GO sheets at 200 °C improved its biocompatibility by decreasing its ability to induce oxidative stress. Since the most pronounced effect produced by GO was obtained with 5 µg/mL, and to demonstrate more clearly the benefit of GO reduction, the data of the three nanostructures obtained with the dose of 5 µg/mL are shown in the table included in Figure 6, compared to the value obtained with control macrophages in the absence of material. These results show a progressive decrease of ROS content as the reduction process increased.

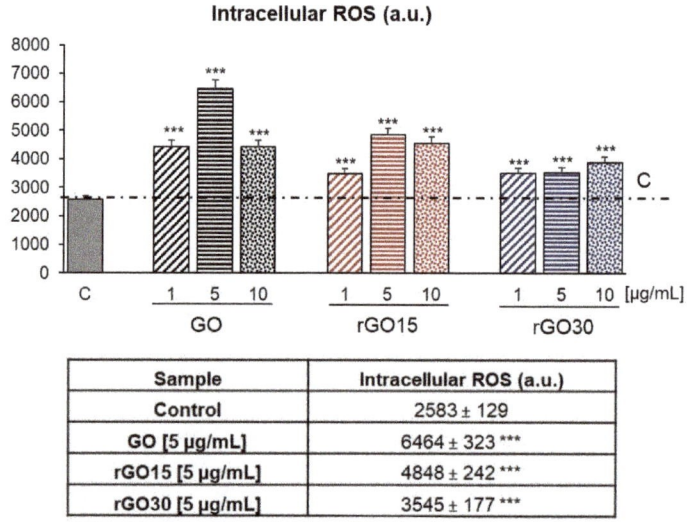

Sample	Intracellular ROS (a.u.)
Control	2583 ± 129
GO [5 µg/mL]	6464 ± 323 ***
rGO15 [5 µg/mL]	4848 ± 242 ***
rGO30 [5 µg/mL]	3545 ± 177 ***

Figure 6. Intracellular content of reactive oxygen species (ROS) of RAW-264.7 macrophages after 24 h of treatment with 1, 5, and 10 µg/mL of GO, rGO15, and rGO30, evaluated by flow cytometry. The table shows the data obtained with the dose of 5 µg/mL of the three nanostructures compared to the value obtained with control macrophages in the absence of material. Statistical significance: *** $p < 0.005$ (compared to control macrophages).

In this context, recent studies indicate the absence of rGO (50 µg/mL) cytotoxicity in HepG2 cells when it is obtained by reduction of GO with hydrazine hydrate. This rGO (25 µg/mL) also showed a protective role against the oxidative stress and toxic effects induced by Cd (2 µg/mL) in this hepatic cell line [53].

2.4. CD80 Expression by Macrophages after GO, rGO15, and rGO30 Uptake

Numerous studies have shown the regulatory role of ROS on the phagocytosis function of macrophages and on their polarization towards M1 or M2 phenotypes, evidencing a dual role in the progression or healing of different diseases. M1 proinflammatory macrophages, also known as classically activated macrophages, are critical for host protection against viruses and intracellular bacteria during acute infections and are involved in helper T cell (Th1) response [54]. This macrophage phenotype is characterized by TLR-2, TLR-4, CD80, CD86, iNOS, and MHC-II surface phenotypes and release various cytokines and chemokines, including tumor necrosis factor (TNF-α) and interleukin IL-1α, IL-1β, IL-6, IL-12, CXCL9, and CXCL10 [55,56]. In addition, M1 macrophages produce microbicidal reagents, such as nitric oxide (NO) and reactive oxygen species (ROS) [57]. In the present study, to find the effects caused by GO, rGO15, and rGO30 nanostructures on the polarization of macrophages towards this M1 phenotype, we have studied different specific markers. In particular, we have evaluated the expression of the CD80 costimulatory molecule on the macrophage surface (CD80$^+$ macrophages) and macrophage secretion of proinflammatory cytokines such as TNF-α and IL-6 after GO, rGO15, and rGO30 uptake. Figure 7 displays the CD80$^+$ macrophages population percentage after 24 h of treatment with 1, 5, and 10 µg/mL of GO, rGO15, and rGO30. The results evidence a significant increase in the CD80+ macrophages population percentage after exposure to the highest dose of GO (10 µg/mL), compared to control macrophages. However, this effect was not observed after exposure to rGO15 and rGO30 at the highest concentration studied. This data shows that, when using particles, thermal reduction as short as 15 min was enough to mitigate the increase in the population of CD80$^+$ when exposed to higher doses of graphene oxide-based particles.

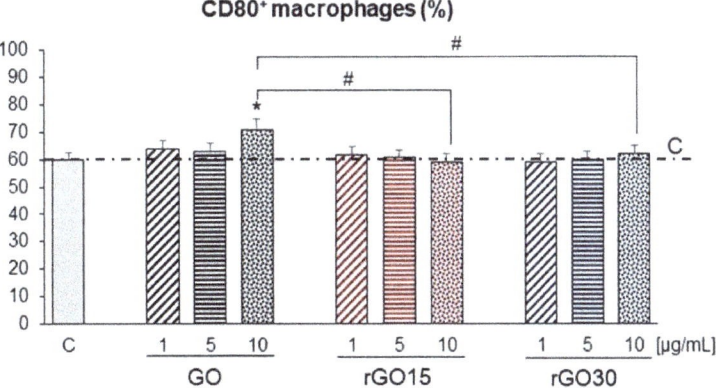

Figure 7. Percentage of the population of CD80$^+$ RAW-264.7 macrophages after 24 h of treatment with 1, 5, and 10 µg/mL of GO, rGO15, and rGO30. Statistical significance: * $p < 0.05$ (compared to control macrophages, horizontal line C), # $p < 0.05$ (comparison between nanostructures at the same concentration).

2.5. Detection of TNF-α and IL-6 as Inflammatory Cytokines

TNF-α and IL-6 are two of the proinflammatory cytokines mainly produced by macrophages polarized towards the M1 phenotype [58]. For this reason, the TNF-α and IL-6 levels released by macrophages after exposure to 1, 5, and 10 µg/mL of GO, rGO15, and rGO30 were evaluated in the present study. Figure 8 clearly shows significant increases in the TNF-α levels secreted by macrophages cultured with 5 and 10 μg/mL of GO, rGO15, and rGO30 compared to those of control macrophages. This effect was more pronounced with GO than with rGO15 or rGO30, showing very high TNF-α secretion induced by GO in a dose-dependent manner. Regarding IL-6, a dose-dependent increase of this cytokine was also detected in the culture medium after treatment with 5 and 10 μg/mL of GO. However, lower levels of IL-6 than that in controls were obtained with the three tested doses of rGO15 (1, 5, and 10 μg/mL), and only 10 μg/mL of rGO30 produced a significant increase of IL-6 compared to that of control macrophages. In this context, when other authors evaluated the effects of 2 μm and 350 nm GO particles on the production of different inflammatory cytokines (IL-6, IL-10, IL-12, TNF-α, MCP-1, IFN-γ) in macrophages, it was shown that the secretion of these mediators was highly dependent on the GO dosage, particularly for the 2 μm GO particles [43]. In thermally reduced rGO microfibers, we previously observed a decrease of TNF-α and IL-6 after 24 h of culturing [26]. Here, the cells were subjected to a distribution of GO sheets rather than contact with bulk fibers and showed higher reactivity to the reduced sheets than to the bulk fibers. This was more evident for a higher concentration of GBM material. Still, thermal reduction as short as 15 min exhibited a significant improvement in terms of decreasing the proinflammatory cues. Our results demonstrate that GO reduction reduced its ability to induce the synthesis and secretion of proinflammatory cytokines, significantly improving its biocompatibility.

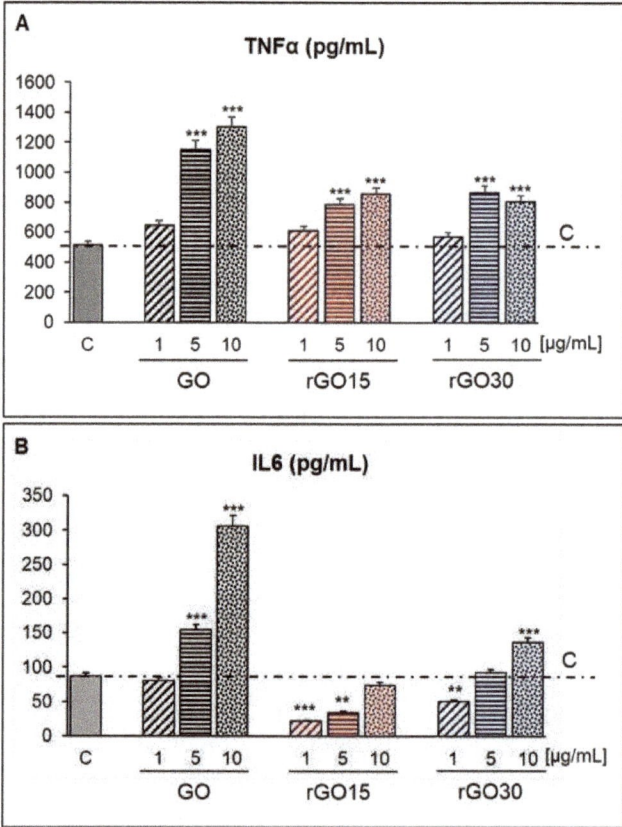

Figure 8. TNF-α (**A**) and IL-6 levels (**B**) (pg/mL) released by RAW-264.7 macrophages after 24 h of treatment with 1, 5, and 10 μg/mL of GO, rGO15, and rGO30 nanostructures, evaluated by ELISA. Statistical significance: ** $p < 0.01$, *** $p < 0.005$ (compared to control macrophages, horizontal line C).

GBM shows great promise for biomedical applications and can be designed with different configurations such as nanosheets, nanoparticles, 2D films, and 3D scaffolds [26,34,59]. The GO and rGO nanostructures evaluated in this study are potentially useful for the preparation of scaffolds for tissue regeneration after assessing the effects that these components could have if they are released during implant degradation once they have been introduced into the body. Thus, when biomedical engineers design graphene-based scaffolds, the aspects related to the targeted biomedical scenario, such as scaffold biodegradation and its kinetics, should be carefully probed and understood to avoid an immune response and allow us to fully exploit the potential of GBM. In this context, numerous in vitro and in vivo studies with rGO-prepared scaffolds have shown promising results for the regeneration of different tissues [7,60–63].

The knowledge of the response of macrophages to GBM is particularly important because these cells are responsible for innate immunity [20] and play a key role in the processing of nanomaterials [38]. In this sense, the possibility of modulating macrophage polarization towards a proinflammatory or reparative phenotype with biomaterials is considered a promising strategy to control inflammatory processes and tissue regeneration at the implant site [18]. In this work, we have evaluated different aspects related to the macrophage response to GO, rGO15, and rGO30, evidencing an active and dose-dependent incorporation of these nanostructures by this cell type without inducing the expression of the proinflammatory marker CD80. On the other hand, the rGO-treated macrophages produced lower amounts of reactive oxygen species and proinflammatory cytokines than GO-treated cells, indicating the benefits of the reduction process of this nanomaterial and further supporting the use of rGO in the preparation of novel scaffolds.

3. Materials and Methods

3.1. Preparation of GO, rGO15, and rGO30 Nanostructures

The GO used in this work was of commercial origin (Graphenea®, San Sebastián, Spain). According to the supplier, in its original source (0.4 wt% aqueous solution) it has monolayer content (measured in 0.05 wt%) higher than 95% and particle lateral size lower than 10 µm. GO is prepared from chemical exfoliation of graphite using a strong oxidant and acidic media. Therefore, pH of the original GO solution is between 2.2 and 2.5, which favors nanosheet dispersion in the media. Other than the acidic residues, other chemical moieties like sulfur and manganese are present due to the exfoliation methodology. The presence of these residues, which can be toxic to cells, can be mitigated to some extent by dialysis treatment. Therefore, the commercial GO dispersion was firstly dialyzed with distilled water that replaced daily for a week. Afterwards, this solution was freeze-dried to obtain chemical-free GO sheets in a Teslar lyoQuest HT-40 freeze-drier (Beijer Electronics Products AB, Sweden). However, it must be considered that this process may result in the agglomeration of GO nanosheets due to rinsing at the pH of the medium and consequent manipulations, such as freeze-drying. Even so, we considered this procedure important to remove chemical impurities that may be related to the often-proclaimed toxic effects of GO on cells [64].

3.2. Morphological and Structural Characterization of GO, rGO15, and rGO30 Nanostructures

Dispersions of GO, rGO15, and rGO30 at 0.5 mg/mL were spin coated on glass coverslips at 800 rpm to analyze the morphology via atomic force microscopy (Bruker Multimode instrument (Bruker Nano Surfaces, Santa Barbara) with a Nanoscope (IV) MMAFM-2 unit) with a conductive Si cantilever (Nanosensors, force constant 15 N/m, Neuchatel, Switzerland).

Structural characterization of GO before and after the 15- and 30-min thermal treatments was performed by X-ray diffraction (XRD) and X-ray photoelectron spectroscopy (XPS). XRD spectra were acquired from 5 to 80° at a scanning speed of 1°/min in a Rigaku SmartLab diffractometer (Rigaku Corporation, Japan) using Cu Kα radiation (λ = 1.5406 Å).

XPS (with a hemispherical electron energy analyzer SPECS Phoibos 150 (Berlin, Germany) and a monochromatic Al Kα (1486.74 eV) X-ray source) was performed in an ultra-high vacuum system (with a base pressure of 2×10^{-8} Pa) at a normal emission take-off angle and 20 eV pass-energy.

3.3. Culture of RAW-264.7 Macrophages for Treatment with GO, rGO15, and rGO30

RAW-264.7 macrophages were seeded with cell density of 1×10^5 cells/mL in Dulbecco's Modified Eagle Medium (DMEM, Gibco BRL, United Kingdom) supplemented with 10% fetal bovine serum (FBS, Gibco BRL, United Kingdom), 1 mM L-glutamine (BioWhittaker Europe, Verviers, Belgium), 800 µg/mL penicillin (BioWhittaker Europe, Verviers, Belgium), and 800 µg/mL streptomycin (BioWhittaker Europe, Verviers, Belgium) in a 5% CO_2 humidified atmosphere at 37 °C for 24 h. Then, the culture medium was replaced by fresh medium containing GO, rGO15, and rGO30 at different concentrations (1, 5, and 10 µg/mL) that was previously sonicated for 5 min to homogenize the mixture. These doses were chosen based on previous studies with macrophage cultures as an in vitro experimental model [19,65] and considering recent toxicity in vivo studies in mice [66]. After culturing the cells for 24 h under these conditions, the macrophages were first washed with phosphate-buffered saline (PBS, Sigma-Aldrich) to remove the nonincorporated nanomaterials and then detached with a scraper before analyzing all cell response-specific studies. Control samples corresponding to macrophages cultured in the absence of nanomaterials were included in all the assays.

3.4. Uptake of GO, rGO15, and rGO30 by RAW-264.7 Macrophages Evaluated by Flow Cytometry and Confocal and Phase Contrast Microscopy

The incorporation of GO, rGO15, and rGO30 by RAW-264.7 macrophages after 24 h of treatment was quantified by flow cytometry analyzing 90° light scatter (side scatter, SSC) that allows to evaluate nanomaterial uptake by mammalian cells [67,68]. The SSC parameter is proportional to the intracellular complexity determined in part by the cellular cytoplasm, mitochondria, and pinocytic vesicles [39]. The conditions for data acquisition and analysis were established using negative and positive controls with the CellQuest Program of Becton Dickinson, and these conditions were maintained in all the experiments. Each experiment was carried out three times, and single representative experiments are displayed. For statistical significance, at least 10,000 cells were analyzed by flow cytometry in each sample. The incorporation of GO, rGO15, and rGO30 by macrophages was observed by confocal and phase contrast microscopy as in previous studies [19]. For these confocal and phase contrast microscopy studies, RAW-264.7 macrophages were cultured on circular glass coverslips under the above-mentioned cell culture conditions. Cells were fixed with 3.7% paraformaldehyde (Sigma-Aldrich Corporation, St. Louis, MO, USA) in PBS for 10 min, washed with PBS, and permeabilized with 0.1% Triton X-100 (Sigma-Aldrich Corporation, St. Louis, MO, USA) for 5 min. The samples were then washed with PBS and preincubated with PBS containing 1% BSA (Sigma-Aldrich Corporation, St. Louis, MO, USA) for 30 min to prevent nonspecific binding. The samples were incubated in 1 mL of staining buffer with PE-conjugated anti-mouse CD80 antibody (2.5 µg/mL, BioLegend, San Diego, CA, USA) for 30 min at 4 °C in the dark. The samples were then washed with PBS, and the cell nuclei were stained with DAPI (4′-6-diamidino-2′-phenylindole, 3 µM in PBS, Molecular Probes, Eugene, OR, USA) for 5 min. The samples were examined in a LEICA SP2 Confocal Laser Scanning Microscope. Fluorescence PE was excited at 488 nm, and the emitted fluorescence was measured at 575–675 nm. DAPI fluorescence was excited at 405 nm and measured at 420–480 nm.

3.5. Measurement of Intracellular Reactive Oxygen Species (ROS) Content of Macrophages by Flow Cytometry after GO, rGO15, and rGO30 Uptake

After exposure to GO, rGO15, and rGO30 for 24 h, RAW-264.7 macrophages were detached, and cell suspensions were incubated with 10 µM of 2′,7′-dichlorodihydro fluorescein diacetate (DCF-H2-DA, Serva, Heidelberg, Germany) for 45 min at 37 °C. The nonfluorescent DCF-H2-DA transforms into 2′,7′-dichlorofluorescein (DCF) after hydrolysis by cellular esterases and oxidation by ROS. When DCF is excited at 488 nm emission wavelengths, it emits green fluorescence that can be detected at 525 nm. DCF fluorescence was measured in a FACScalibur Becton Dickinson flow cytometer with a 530/30 filter, exciting the sample at 488 nm. For statistical significance, at least 10,000 cells were analyzed by flow cytometry in each sample.

3.6. Detection of Macrophage M1 Proinflammatory Phenotype by Flow Cytometry after GO, rGO15, and rGO30 Uptake

The expression of CD80 was used as a specific marker to identify M1 macrophages [69] and quantified by flow cytometry after exposure of RAW-264.7 macrophages to GO, rGO15, and rGO30. Before immunostaining, the cells were detached and incubated in 45 µL of staining buffer (PBS Thermo Fisher Scientific Madrid, Spain, 2.5% FBS Gibco BRL United Kingdom Gibco, and 0.1% sodium azide, Sigma-Aldrich Corporation, St. Louis, MO, USA) with 5 µL of normal mouse serum inactivated for 15 min at 4 °C in order to block the Fc receptors on the macrophage plasma membrane and to prevent nonspecific binding of the primary antibody. Then, the cells were incubated with phycoerythrin (PE)-conjugated anti-mouse CD80 antibody (2.5 µg/mL, BioLegend, San Diego, CA, USA) for 30 min in the dark. Labeled macrophages were then analyzed using a FACSCalibur flow cytometer. The fluorescence was excited at 488 nm and measured at 585/42 nm. The conditions for data acquisition and analysis were established using negative and positive controls with the CellQuest Program of Becton Dickinson, and these conditions were maintained in all the experiments. Each experiment was carried out three times, and single representative experiments are displayed. For statistical significance, at least 10,000 cells were analyzed in each sample.

3.7. Detection of TNF-α and IL-6 as Inflammatory Cytokines

The amount of TNF-α and IL-6 secreted by RAW-264.7 macrophages under the different conditions was quantified in the culture medium by enzyme-linked immunosorbent assay (ELISA, Gen-Probe, Diaclone, Besançon, France) according to the manufacturer's instructions.

3.8. Statistics

Data are expressed as means ± standard deviations of a representative of three experiments carried out in triplicate. Statistical analysis was performed using the Statistical Package for the Social Sciences (SPSS) version 22 software. Statistical comparisons were made by analysis of variance (ANOVA). Scheffé test was used for post hoc evaluations of differences among groups. In all the statistical evaluations, $p < 0.05$ was considered as statistically significant.

4. Conclusions

Our comparative study with RAW-264.7 macrophages that evaluated specific parameters of their in vitro response to different graphene-based nanomaterials has demonstrated the benefits of GO reduction by vacuum-assisted thermal treatment at 200 °C to obtain nanostructures with higher biocompatibility, improving their potential for the preparation of 3D scaffolds that are able to trigger the appropriate immune response for tissue regeneration.

Author Contributions: Conceptualization, P.A.A.P.M. and M.T.P.; data curation, M.C., N.B., P.A.A.P.M. and M.T.P.; formal analysis, M.C., L.C., N.B., P.A.A.P.M. and M.T.P.; funding acquisition, P.A.A.P.M. and M.T.P.; investigation, M.C., L.C., N.B., D.S., M.J.F., R.D.-O., P.A.A.P.M. and M.T.P.; methodology, M.C., L.C., N.B., D.S., M.J.F., R.D.-O., P.A.A.P.M. and M.T.P.; project administration, P.A.A.P.M. and M.T.P.; resources, P.A.A.P.M. and M.T.P.; supervision, P.A.A.P.M. and M.T.P.; validation, P.A.A.P.M. and M.T.P.; visualization, P.A.A.P.M. and M.T.P.; writing—original draft, M.C., N.B., P.A.A.P.M. and M.T.P.; writing—review and editing, M.C., N.B., P.A.A.P.M. and M.T.P. All authors have read and agreed to the published version of the manuscript.

Funding: This work has been supported by the European Union's Horizon 2020 Research and Innovation Programme (H2020-FETOPEN-2018-2020, NeuroStimSpinal Project, Grant Agreement No. 829060).

Institutional Review Board Statement: Not applicable.

Informed Consent Statement: Not applicable.

Data Availability Statement: Data is contained within the article.

Acknowledgments: The authors thank the staff of Centro de Citometría y Microscopía de Fluorescencia (Universidad Complutense de Madrid (Spain)) for the support in the studies of flow cytometry and confocal microscopy. M.C. acknowledges the European Union's Horizon 2020 Research and Innovation Programme for her contract under the NeuroStimSpinal Project. LC is grateful to Universidad Complutense de Madrid for a UCM fellowship. D.S. acknowledges the European Union's Horizon 2020 Research and Innovation Programme for her PhD grant under the NeuroStimSpinal Project.

Conflicts of Interest: The authors declare no conflict of interests. The funders had no role in the design of the study; in the collection, analyses, or interpretation of data; in the writing of the manuscript, or in the decision to publish the results.

References

1. Novoselov, K.S.; Fal'ko, V.I.; Colombo, L.; Gellert, P.R.; Schwab, M.G.; Kim, K.A. A roadmap for graphene. *Nature* **2012**, *490*, 192–200. [CrossRef]
2. Yang, Y.Q.; Asiri, A.M.; Tang, Z.W.; Du, D.; Lin, Y.H. Graphene based materials for biomedical applications. *Mater. Today* **2013**, *16*, 365–373. [CrossRef]
3. Miao, W.; Shim, G.; Kim, G.; Lee, S.; Lee, H.J.; Kim, Y.B.; Byun, Y.; Oh, Y.K. Image-guided synergistic photothermal therapy using photoresponsive imaging agent-loaded graphene-based nanosheets. *J. Control. Release* **2015**, *211*, 28–36. [CrossRef]
4. Bartelmess, J.; Quinn, S.J.; Giordani, S. Carbon nanomaterials: Multi-functional agents for biomedical fluorescence and Raman imaging. *Chem. Soc. Rev.* **2015**, *44*, 4672–4698. [CrossRef] [PubMed]
5. Kim, M.G.; Park, J.Y.; Miao, W.; Lee, J.; Oh, Y.K. Polyaptamer DNA nanothread-anchored, reduced graphene oxide nanosheets for targeted delivery. *Biomaterials* **2015**, *48*, 129–136. [CrossRef]
6. Gonçalves, G.; Vila, M.; Portolés, M.T.; Vallet-Regí, M.; Gracio, J.; Marques, P.A.A.P. Nano-Graphene Oxide: A Potential Multifunctional Platform for Cancer Therapy. *Adv. Healthcare Mater.* **2013**, *2*, 1072–1090. [CrossRef] [PubMed]
7. Raslan, A.; Saenz del Burgo, L.; Ciriza, J.; Pedraz, J.L. Graphene oxide and reduced graphene oxide-based scaffolds in regenerative medicine. *Int. J. Pharm.* **2020**, *580*, 119226. [CrossRef] [PubMed]
8. López-Dolado, E.; González-Mayorga, A.; Portolés, M.T.; Feito, M.J.; Ferrer, M.L.; del Monte, F.; Gutiérrez, M.C.; Serrano, M.C. Subacute tissue response to 3D graphene oxide scaffolds implanted in the injured rat spinal cord. *Adv. Healthc. Mater.* **2015**, *4*, 1861–1868. [CrossRef] [PubMed]
9. Menaa, F.; Abdelghani, A.; Menaa, B. Graphene nanomaterials as biocompatible and conductive scaffolds for stem cells: Impact for tissue engineering and regenerative medicine. *J. Tissue Eng. Regen. Med.* **2015**, *9*, 1321–1338. [CrossRef]
10. Girão, A.F.; Gonçalves, G.; Bhangra, K.S.; Phillips, J.B.; Knowles, J.; Hurietta, G.; Singh, M.K.; Bdkin, I.; Completo, A.; Marques, P.A.A.P. Electrostatic self-assembled graphene oxide-collagen scaffolds towards a three-dimensional microenvironment for biomimetic applications. *RSC Adv.* **2016**, *6*, 49039–49051. [CrossRef]
11. Mohan, V.B.; Brown, R.; Jayaraman, K.; Bhattacharyya, D. Characterization of reduced graphene oxide: Effects of reduction variables on electrical conductivity. *Mater. Sci. Eng. B* **2015**, *193*, 49–60. [CrossRef]
12. Zhang, Q.; Liu, X.; Meng, H.; Liu, S.; Zhang, C. Reduction pathway-dependent cytotoxicity of reduced graphene oxide. *Environ. Sci. Nano* **2018**, *5*, 1361–1371. [CrossRef]
13. Wang, B.; Su, X.; Liang, J.; Yang, L.; Hu, Q.; Shan, X.; Wan, J.; Hu, Z. Synthesis of polymer-functionalized nanoscale graphene oxide with different surface charge and its cellular uptake, biosafety and immune responses in Raw 264.7 macrophages. *Mater. Sci. Eng. C* **2018**, *90*, 514–522. [CrossRef]
14. Tabish, T.A.; Pranjol, M.Z.I.; Hayat, H.; Rahat, A.A.M.; Abdullah, T.M.; Whatmore, J.L.; Zhang, S. In vitro toxic effects of reduced graphene oxide nanosheets on lung cancer cells. *Nanotechnology* **2017**, *28*, 504001. [CrossRef]
15. Palejwala, A.H.; Fridley, J.S.; Mata, J.A.; Samuel, E.L.G.; Luerssen, T.G.; Perlaky, L.; Kent, T.A.; Tour, J.M.; Jea, A. Biocompatibility of reduced graphene oxide nanoscaffolds following acute spinal cord injury in rats. *Surg. Neurol. Int.* **2016**, *7*, 75.
16. Wen, K.; Chen, Y.; Chuang, C.; Chang, H.; Lee, C.; Tai, N. Accumulation and toxicity of intravenously injected functionalized graphene oxide in mice. *J. Appl. Toxicol.* **2015**, *35*, 1211–1218. [CrossRef]
17. Sica, A.; Mantovani, A. Macrophage plasticity and polarization: In vivo veritas. *J. Clin. Investig.* **2012**, *122*, 787–795. [CrossRef] [PubMed]
18. Brown, B.N.; Ratner, B.D.; Goodman, S.B.; Amar, S.; Badylak, S.F. Macrophage polarization: An opportunity for improved outcomes in biomaterials and regenerative medicine. *Biomaterials* **2012**, *33*, 3792. [CrossRef] [PubMed]

19. Feito, M.J.; Díez-Orejas, R.; Cicuéndez, M.; Casarrubios, L.; Rojo, J.M.; Portolés, M.T. Graphene oxide nanosheets modulate peritoneal macrophage polarization towards M1 and M2 phenotypes. *Colloids Surf. B* **2019**, *176*, 96–105. [CrossRef] [PubMed]
20. Mosser, D.M.; Edwards, J.P. Exploring the full spectrum of macrophage activation. *Nat. Rev. Immunol.* **2008**, *8*, 958–969. [CrossRef]
21. Stout, R.D.; Suttles, J. Functional plasticity of macrophages: Reversible adaptation to changing microenvironments. *J. Leukoc. Biol.* **2004**, *76*, 509–513. [CrossRef] [PubMed]
22. Mukherjee, S.P.; Bondarenko, O.; Kohonen, P.; Andón, F.T.; Brzicová, T.; Gessner, I.; Mathur, S.; Bottini, M.; Calligari, P.; Stella, L.; et al. Macrophage sensing of single-walled carbon nanotubes via Toll-like receptors. *Sci. Rep.* **2018**, *8*, 1115. [CrossRef] [PubMed]
23. Bhattacharya, K.; Mukherjee, S.P.; Gallud, A.; Burkert, S.C.; Bistarelli, S.; Bellucci, S.; Bottini, M.; Star, A.; Fadeel, B. Biological interactions of carbon-based nanomaterials: From coronation to degradation. *Nanomedicine* **2016**, *12*, 333–351. [CrossRef]
24. Kinaret, P.A.S.; Scala, G.; Federico, A.; Sund, J.; Greco, D. Carbon Nanomaterials Promote M1/M2 Macrophage Activation. *Small* **2020**, *16*, 1907609. [CrossRef]
25. Harrison, B.S.; Atala, A. Carbon nanotube applications for tissue engineering. *Biomaterials* **2007**, *28*, 344–353. [CrossRef]
26. Serrano, M.C.; Feito, M.J.; González-Mayorga, A.; Diez-Orejas, R.; Matesanz, M.C.; Portolés, M.T. Response of macrophages and neural cells in contact with reduced graphene oxide microfibers. *Biomater. Sci.* **2018**, *6*, 2987–2997. [CrossRef] [PubMed]
27. Dolbin, A.V.; Khlistyuck, M.V.; Esel'son, V.B.; Gavrilko, V.G.; Vinnikov, N.A.; Basnukaeva, R.M.; Maluenda, I.; Maser, W.K.; Benito, A.M. The effect of the thermal reduction temperature on the structure and sorption capacity of reduced graphene oxide materials. *Appl. Surf. Sci.* **2016**, *361*, 213–220. [CrossRef]
28. Huang, H.H.; De Silva, K.K.H.; Kumara, G.R.A.; Yoshimura, M. Structural evolution of hydrothermally derived reduced graphene oxide. *Sci. Rep.* **2018**, *8*, 6849. [CrossRef]
29. Liu, J.; Cui, L.; Losic, D. Graphene and graphene oxide as new nanocarriers for drug delivery applications. *Acta Biomater.* **2013**, *9*, 9243–9257. [CrossRef]
30. Cai, W.; Piner, R.D.; Stadermann, F.J.; Park, S.; Shaibat, M.A.; Ishii, Y.; Yang, D.; Velamakanni, A.; An, S.J.; Stoller, M.; et al. Synthesis and solid-state NMR structural characterization of 13C-labeled graphite oxide. *Science* **2008**, *321*, 1815–1817. [CrossRef] [PubMed]
31. Lerf, A.; He, H.; Forster, M.; Klinowski, J. Structure of graphite oxide revisited. *J. Phys. Chem. B* **1998**, *102*, 4477–4482. [CrossRef]
32. Mattevi, C.; Eda, G.; Agnoli, S.; Miller, S.; Mkhoyan, K.A.; Celik, O.; Mastrogiovanni, D.; Granozzi, G.; Garfunkel, E.; Chhowalla, M. Evolution of electrical, chemical, and structural properties of transparent and conducting chemically derived graphene thin films. *Adv. Funct. Mater.* **2009**, *19*, 2577–2583. [CrossRef]
33. Lipatov, A.; Guinel, M.J.-F.; Muratov, D.S.; Vanyushin, V.O.; Wilson, P.M.; Kolmakov, A.; Sinitskii, A. Low-temperature thermal reduction of graphene oxide: In situ correlative structural, thermal desorption, and electrical transport measurements. *Appl. Phys. Lett.* **2018**, *112*, 053103. [CrossRef]
34. Girão, A.F.; Sousa, J.; Domínguez-Bajo, A.; González-Mayorga, A.; Bdikin, I.; Pujades-Otero, E.; Casañ-Pastor, N.; Hortigüela, M.J.; Otero-Irurueta, G.; Completo, A.; et al. 3D Reduced Graphene Oxide Scaffolds with a Combinatorial Fibrous-Porous Architecture for Neural Tissue Engineering. *ACS Appl. Mater. Interfaces* **2020**, *12*, 38962–38975. [CrossRef]
35. Liu, Y.Z.; Chen, C.M.; Li, Y.F.; Li, X.M.; Kong, Q.Q.; Wang, M.Z. Crumpled reduced graphene oxide by flame-induced reduction of graphite oxide for supercapacitive energy storage. *J. Mater. Chem. A* **2014**, *2*, 5730–5737. [CrossRef]
36. Chang, C.; Song, Z.; Lin, J.; Xu, Z. How graphene crumples are stabilized? *RSC Adv.* **2013**, *3*, 2720–2726. [CrossRef]
37. Zhang, H.B.; Wang, J.W.; Yan, Q.; Zheng, W.G.; Chen, C.; Yu, Z.Z. Vacuum-assisted synthesis of graphene from thermal exfoliation and reduction of graphite oxide. *J. Mater. Chem.* **2011**, *21*, 5392–5397. [CrossRef]
38. Gustafson, H.H.; Holt-Casper, D.; Grainger, D.W.; Ghandehari, H. Nanoparticle Uptake: The Phagocyte Problem. *Nano Today* **2015**, *10*, 487–510. [CrossRef]
39. Udall, J.N.; Moscicki, R.A.; Preffer, F.I.; Ariniello, P.D.; Carter, E.A.; Bhan, A.K.; Bloch, K.J. Flow cytometry: A new approach to the isolation and characterization of Kupffer cells. In *Recent Advances in Mucosal Immunology*; Advances in Experimental Medicine and Biology; Mestecky, J., McGhee, J.R., Bienenstock, J., Ogra, P.L., Eds.; Springer: Boston, MA, USA, 1987; Volume 216A, pp. 821–827.
40. Déciga-Alcaraz, A.; Medina-Reyes, E.I.; Delgado-Buenrostro, N.L.; Rodríguez-Ibarra, C.; Ganem-Rondero, A.; Vázquez-Zapién, G.J.; Mata-Miranda, M.M.; Limón-Pacheco, J.H.; García-Cuéllar, C.M.; Sánchez-Pérez, Y.; et al. Toxicity of engineered nanomaterials with different physicochemical properties and the role of protein corona on cellular uptake and intrinsic ROS production. *Toxicology* **2020**, *442*, 152545. [CrossRef]
41. Gratton, S.E.; Ropp, P.A.; Pohlhaus, P.D.; Luft, J.C.; Madden, V.J.; Napier, M.E.; DeSimone, J.M. The effect of particle design on cellular internalization pathways. *Proc. Natl. Acad. Sci. USA* **2008**, *105*, 11613–11618. [CrossRef]
42. Ma, J.; Liu, R.; Wang, X.; Liu, Q.; Chen, Y.; Valle, R.P.; Zuo, Y.Y.; Xia, T.; Liu, S. Crucial role of lateral size for graphene oxide in activating macrophages and stimulating proinflammatory responses in cells and animals. *ACS Nano* **2015**, *9*, 10498–10515. [CrossRef]
43. Yue, H.; Wei, W.; Yue, Z.; Wang, B.; Luo, N.; Gao, Y.; Ma, D.; Ma, G.; Su, Z. The role of the lateral dimension of graphene oxide in the regulation of cellular responses. *Biomaterials* **2012**, *33*, 4013–4021. [CrossRef]
44. Chatterjee, N.; Eom, H.J.; Choi, J. A systems toxicology approach to the Surface functionality control of graphene–cell interactions. *Biomaterials* **2014**, *35*, 1109–1127. [CrossRef] [PubMed]
45. Zhang, B.; Wei, P.; Zhou, Z.; Wei, T. Interactions of graphene with mammalian cells: Molecular mechanisms and biomedical insights. *Adv. Drug Deliv. Rev.* **2016**, *105*, 145–162. [CrossRef]

46. Linares, J.; Matesanz, M.C.; Vila, M.; Feito, M.J.; Goncalves, G.; Vallet-Regi, M.; Marques, P.A.; Portoles, M.T. Endocytic mechanisms of graphene oxide nanosheets in osteoblasts, hepatocytes and macrophages. *ACS Appl. Mater. Interfaces* **2014**, *6*, 13697–13706. [CrossRef] [PubMed]
47. Wan, B.; Wang, Z.X.; Lv, Q.Y.; Dong, P.X.; Zhao, L.X.; Yang, Y.; Guo, L.H. Single-walled carbon nanotubes and graphene oxides induce autophagosome accumulation and lysosome impairment in primarily cultured murine peritoneal macrophages. *Toxicol. Lett.* **2013**, *221*, 118–127. [CrossRef]
48. Saleem, J.; Wang, L.; Chen, C. Immunological effects of graphene family nanomaterials. *NanoImpact* **2017**, *5*, 109–118. [CrossRef]
49. Liao, C.; Li, Y.; Tjong, S.C. Graphene Nanomaterials: Synthesis, Biocompatibility, and Cytotoxicity. *Int. J. Mol. Sci.* **2018**, *19*, 3564. [CrossRef]
50. Julier, Z.; Park, A.J.; Briquez, P.S.; Martino, M.M. Promoting tissue regeneration by modulating the immune system. *Acta Biomater.* **2017**, *53*, 13–28. [CrossRef]
51. Sridharan, R.; Cameron, A.; Kelly, D.; Kearney, C.; O'Brien, F. Biomaterial based modulation of macrophage polarization: A review and suggested design principles. *Mater. Today* **2015**, *18*, 313–325. [CrossRef]
52. Tan, H.-Y.; Wang, N.; Li, S.; Hong, M.; Wang, X.; Feng, Y. The reactive oxygen species in macrophage polarization: Reflecting its dual role in progression and treatment of human diseases. *Oxid. Med. Cell. Longev.* **2016**, 2795090. [CrossRef] [PubMed]
53. Ahamed, M.; Akhtar, M.J.; Khan, M.A.M.; Alhadlaq, H.A. Reduced graphene oxide mitigates cadmium-induced cytotoxicity and oxidative stress in HepG2 cells. *Food Chem. Toxicol.* **2020**, *143*, 111515. [CrossRef]
54. Italiani, P.; Boraschi, D. From monocytes to M1/M2 macrophages: Phenotypical vs. functional differentiation. *Front. Immunol.* **2014**, *5*, 514. [CrossRef] [PubMed]
55. Lee, K.Y. M1 and M2 polarization of macrophages: A mini-review. *Med. Biol. Sci. Eng.* **2019**, *2*, 1–5. [CrossRef]
56. Marco, O.; Yanal, G.; Bala, P.A.; Klaus, L. Macrophage Polarization: Different Gene Signatures in M1(LPS+) vs. Classically and M2 (LPS–) vs. Alternatively Activated Macrophages. *Front. Immunol.* **2019**, *10*, 1084.
57. Yunna, C.; Mengru, H.; Lei, W.; Weidong, C. Macrophage M1/M2 polarization. *Eur. J. Pharmacol.* **2020**, *877*, 173090. [CrossRef]
58. Jiménez-Uribe, A.P.; Valencia-Martínez, H.; Carballo-Uicab, G.; Vallejo-Castillo, L.; Medina-Rivero, E.; Chacón-Salinas, R.; Pavón, L.; Velasco-Velázquez, M.A.; Mellado-Sánchez, G.; Estrada-Parra, S.; et al. CD80 Expression Correlates with IL-6 Production in THP-1-Like Macrophages Costimulated with LPS and Dialyzable Leukocyte Extract (Transferon®). *J. Immunol. Res.* **2019**, *2019*, 2198508. [CrossRef]
59. Cong, H.P.; Chen, J.F.; Yu, S.H. Graphene-based macroscopic assembles and architectures: An emerging material system. *Chem. Soc. Rev.* **2014**, *43*, 7295–7325. [CrossRef]
60. Thangavel, P.; Kannan, R.; Ramachandran, B.; Moorthy, G.; Suguna, L.; Muthuvijayan, V. Development of reduced graphene oxide (rGO)-isabgol nanocomposite dressings for enhanced vascularization and accelerated wound healing in normal and diabetic rats. *J. Colloid Interface Sci.* **2018**, *517*, 251–264. [CrossRef]
61. López-Dolado, E.; González-Mayorga, A.; Gutiérrez, M.C.; Serrano, M.C. Immunomodulatory and angiogenic responses induced by graphene oxide scaffolds in chronic spinal hemisected rats. *Biomaterials* **2016**, *99*, 72–81. [CrossRef]
62. Gohari, P.H.M.; Nazarpak, M.H.; Solati-Hashjin, M. The effect of adding reduced graphene oxide to electrospun polycaprolactone scaffolds on MG-63 cells activity. *Mater. Today Commun.* **2021**, *27*, 102287. [CrossRef]
63. Magaz, A.; Li, X.; Gough, J.E.; Blaker, J.J. Graphene oxide and electroactive reduced graphene oxide-based composite fibrous scaffolds for engineering excitable nerve tissue. *Mater. Sci. Eng. C* **2021**, *119*, 111632. [CrossRef] [PubMed]
64. Guo, X.; Mei, N. Assessment of the toxic potential of graphene family nanomaterials. *J. Food Drug Anal.* **2014**, *22*, 105–115. [CrossRef] [PubMed]
65. Cicuéndez, M.; Fernandes, M.; Ayán-Varela, M.; Oliveira, H.; Feito, M.J.; Diez-Orejas, R.; Paredes, J.I.; Villar-Rodil, S.; Vila, M.; Portolés, M.T.; et al. Macrophage inflammatory and metabolic responses to graphene-based nanomaterials differing in size and functionalization. *Colloids Surf. B* **2020**, *186*, 110709. [CrossRef] [PubMed]
66. Poulsen, S.S.; Bengtson, S.; Williams, A.; Jacobsen, N.R.; Troelsen, J.T.; Halappanavar, S.; Vogel, U. A transcriptomic overview of lung and liver changes one day after pulmonary exposure to graphene and graphene oxide. *Toxicol. Appl. Pharmacol.* **2021**, *410*, 115343. [CrossRef]
67. Greulich, C.; Diendorf, J.; Simon, T.; Eggeler, G.; Epple, M. Uptake and intracellular distribution of silver nanoparticles in human mesenchymal stem cells. *Acta Biomater.* **2011**, *7*, 347–354. [CrossRef]
68. Suzuki, H.; Toyooka, T.; Ibuki, Y. Simple and easy method to evaluate uptake potential of nanoparticles in mammalian cells using a flow cytometric light scatter analysis. *Environ. Sci. Technol.* **2007**, *41*, 3018–3024. [CrossRef]
69. Burastero, S.E.; Magnani, Z.; Confetti, C.; Abbruzzese, L.; Oddera, S.; Balbo, P.; Rossi, G.A.; Crimi, E. Increased expression of the CD80 accessory molecule by alveolar macrophages in asthmatic subjects and its functional involvement in allergen presentation to autologous TH2 lymphocytes. *J. Allergy Clin. Immunol.* **1999**, *103*, 1136–1142. [CrossRef]

Article

Detoxification of Ciprofloxacin in an Anaerobic Bioprocess Supplemented with Magnetic Carbon Nanotubes: Contribution of Adsorption and Biodegradation Mechanisms

Ana R. Silva [1], Ana J. Cavaleiro [1], O. Salomé G. P. Soares [2], Cátia S.N. Braga [1], Andreia F. Salvador [1], M. Fernando R. Pereira [2], M. Madalena Alves [1] and Luciana Pereira [1,*]

[1] CEB, Centre of Biological Engineering, University of Minho, 4710-057 Braga, Portugal; ana.rita.silva@ceb.uminho.pt (A.R.S.); acavaleiro@deb.uminho.pt (A.J.C.); catia.braga@ceb.uminho.pt (C.S.N.B.); asalvador@ceb.uminho.pt (A.F.S.); madalena.alves@deb.uminho.pt (M.M.A.)

[2] Laboratory of Separation and Reaction Engineering, Laboratory of Catalysis and Materials (LSRE-LCM), Faculty of Engineering, University of Porto, 4200-465 Porto, Portugal; salome.soares@fe.up.pt (O.S.G.P.S.); fpereira@fe.up.pt (M.F.R.P.)

* Correspondence: lucianapereira@deb.uminho.pt

Citation: Silva, A.R.; Cavaleiro, A.J.; Soares, O.S.G.P.; Braga, C.S.N.; Salvador, A.F.; Pereira, M.F.R.; Alves, M.M.; Pereira, L. Detoxification of Ciprofloxacin in an Anaerobic Bioprocess Supplemented with Magnetic Carbon Nanotubes: Contribution of Adsorption and Biodegradation Mechanisms. *Int. J. Mol. Sci.* **2021**, *22*, 2932. https://doi.org/10.3390/ijms22062932

Academic Editor: Ana Maria Diez Pascual

Received: 31 January 2021
Accepted: 10 March 2021
Published: 13 March 2021

Publisher's Note: MDPI stays neutral with regard to jurisdictional claims in published maps and institutional affiliations.

Copyright: © 2021 by the authors. Licensee MDPI, Basel, Switzerland. This article is an open access article distributed under the terms and conditions of the Creative Commons Attribution (CC BY) license (https://creativecommons.org/licenses/by/4.0/).

Abstract: In anaerobic bioreactors, the electrons produced during the oxidation of organic matter can potentially be used for the biological reduction of pharmaceuticals in wastewaters. Common electron transfer limitations benefit from the acceleration of reactions through utilization of redox mediators (RM). This work explores the potential of carbon nanomaterials (CNM) as RM on the anaerobic removal of ciprofloxacin (CIP). Pristine and tailored carbon nanotubes (CNT) were first tested for chemical reduction of CIP, and pristine CNT was found as the best material, so it was further utilized in biological anaerobic assays with anaerobic granular sludge (GS). In addition, magnetic CNT were prepared and also tested in biological assays, as they are easier to be recovered and reused. In biological tests with CNM, approximately 99% CIP removal was achieved, and the reaction rates increased ≈1.5-fold relatively to the control without CNM. In these experiments, CIP adsorption onto GS and CNM was above 90%. Despite, after applying three successive cycles of CIP addition, the catalytic properties of magnetic CNT were maintained while adsorption decreased to $29 \pm 3.2\%$, as the result of CNM overload by CIP. The results suggest the combined occurrence of different mechanisms for CIP removal: adsorption on GS and/or CNM, and biological reduction or oxidation, which can be accelerated by the presence of CNM. After biological treatment with CNM, toxicity towards *Vibrio fischeri* was evaluated, resulting in ≈ 46% detoxification of CIP solution, showing the advantages of combining biological treatment with CNM for CIP removal.

Keywords: anaerobic reduction; adsorption; ciprofloxacin; magnetic carbon nanotubes; redox mediators; toxicity

1. Introduction

Pharmaceuticals are considered emergent micropollutants by the European Commission due to their potential environmental, ecotoxicological, and sociological risk [1]. A wide range of pharmaceuticals, such as antibiotics, anti-inflammatory drugs, anxiolytics and hormones, are not totally metabolized by humans and animals, being excreted to the environment. Hospital and pharmaceutical industries, as well as domestic wastewater, are potential sources of contamination [2]. In wastewater treatment plants (WWTP), these compounds are not completely removed nor mineralized, and end up in natural water bodies or in soils, as well as in drinking waters, and biomagnify in food chains [2–5]. Furthermore, in nature, continuous contact between bacteria and such substances increases the number of multi-drug resistant bacteria [6], with negative consequences for human health.

The most prescribed pharmaceuticals coincide with the ones detected in the WWTP [7,8], as, e.g., ciprofloxacin (CIP), a broad-spectrum fluoroquinolone antibiotic, that is extensively used for the treatment of bacterial infections in humans and animals. Up to 72% of the dosed CIP may exit the target organism in unaltered form, thus reaching the WWTP and the environment [3,4]. Fluorinated antibiotics are generally present in WWTP at low concentrations, ranging from nanograms to micrograms per liter. For CIP, concentrations up to 31 mg L^{-1} (i.e., 0.094 mmol L^{-1}) were detected in the effluent of WWTP treating wastewater from several pharmaceutical industries [9]. The presence of CIP in effluents from WWTP reveals the inefficiency of the implemented processes for treating wastewater containing this antibiotic. Nevertheless, significant amount of CIP is retained in the WWTP by adsorption on the sludge [4,10,11]. Indeed, despite the low K_{ow} value of CIP (i.e., 0.28 [12]), Lindberg et al. [13] showed that more than 70% of the CIP entering a conventional WWTP in Sweden was removed by sorption, and was concentrated in the digested sludge. Digested sewage sludge is commonly added to agricultural soil as fertilizer, what may contribute to disseminating CIP in the environment and facilitate its entrance in the food chains.

Although removal of pharmaceutics in WWTP is attributed mostly to sorption on sludge [4,14], biodegradation also occurs [10]. The aerobic bacteria *Labrys portucalensis* is able to degrade CIP in pure culture, when supplemented with an additional and easily biodegradable substrate [3]. Furthermore, CIP was shown to be used as sole carbon source by a complex microbial community retrieved from a drinking water biofilter [15]. Although biodegradable by microbial communities, CIP was found to be toxic to some microorganisms. For example, in anaerobic sludge, acetoclastic methanogens showed higher sensitivity to the presence of CIP, while hydrogenotrophic methanogens indicated low susceptibility to this compound [16]. In another study, anaerobic microbial communities were found to tolerate CIP concentrations up to 50 mg L^{-1} (0.15 mmol L^{-1}) [17]. The vast physiological diversity of anaerobes is still an open field to explore for the development of novel biotechnological processes. In fact, anaerobic biodegradation of pharmaceuticals seems promising, but it is still poorly explored and little is known about the mechanisms involved. The analgesic acetylsalicylic acid [18], the anti-inflammatories ibuprofen and diclofenac, the beta-blocker metoprolol [19], and the antibiotics benzylpenicillin [20], tetracycline [21], norfloxacin [22], sulfamethoxazole, and trimethoprim [23], were found to be degraded anaerobically, with the last two antibiotics achieving removal efficiencies above 84%. Additionally, in anaerobic bioreactors, the electrons produced during the oxidation of organic matter can potentially be used for the biological reduction of pharmaceuticals, which may represent an alternative way of promoting pharmaceuticals biotransformation. Notwithstanding, the low transformation rates of many recalcitrant compounds in anaerobic bioprocesses represent a drawback to their application [24]. These low rates are mainly due to electron transfer limitations, that may be overcome by the application of redox mediators (RM).

RM are organic molecules that can reversibly be oxidized and reduced, acting as an electron carrier in multiple redox reactions. RM can accelerate the global reaction rates, by lowering the corresponding activation energy [25]. The reduction rates of dyes and aromatic amines were greatly improved, in batch and in continuous anaerobic bioreactors, by adding low amounts of different carbon nanomaterials (CNM) and magnetic nanomaterials (MNM) as RM [26–30]. In some cases, no reduction occurred in the absence of the tested nanomaterials [26–30]. CNM efficiency as RM is mainly due to their high surface area, proper pore size and excellent catalytic properties [25,31,32]. In addition, insoluble materials like CNM can be retained in the bioreactors, avoiding the need of continuous supplementation during the process [27]. In order to facilitate the recovery of these materials from bioreactors, which can then be further reutilized, magnetic composites may be used instead. Magnetic composites—i.e., core(ferrite, FeO)-shell (carbon, C) composites and carbon nanotubes (CNT) impregnated with 2% of Fe (CNT@2%Fe) were proved as very efficient RM in the anaerobic reduction of the recalcitrant azo dye Acid Orange 10 (AO10), where a 76-fold increase of the AO10 reduction rate was obtained with CNT@2%Fe [30]. The magnetic properties of those composites allowed their recovery from the reactors, by using a magnetic field, and enabled their reuse in successive cycles, maintaining the RM characteristic [30,33,34].

The anaerobic removal of pharmaceuticals assisted by nanomaterials thus appear as a promising strategy that deserves to be investigated. In this study, commercial and tailored CNT were evaluated as RM in the anaerobic removal of CIP. Tailored CNT were prepared from the commercial CNT though a set of surface modifications, aiming to obtain materials with different surface chemical groups (acidic and basic) while maintaining their main textural properties. Pristine and tailored CNT were characterized, and first utilized in chemical reduction tests, to evaluate the performance of the CNM as RM in the chemical reduction of CIP. The best CNM was further tested in biological anaerobic assays. Magnetic CNT were also prepared, by incorporation of iron (CNT@2%Fe), characterized and used in the biological experiments, considering that these materials are easier to recover and may be reused, which is important for applied biological treatment processes. The potential contribution of adsorption and biodegradation processes was assessed. Detoxification of CIP solutions was evaluated towards *Vibrio fischeri*, before and after the treatment.

2. Results and Discussion

2.1. Textural and Chemical Characterization of CNM

The results of the textural and elemental analysis of the different CNM are presented in Table 1. CNT are mesoporous nanomaterials, presenting a specific surface area (S_{BET}) of 201 $m^2\ g^{-1}$ and pore volume (Vp) of 0.416 $cm^3\ g^{-1}$. The functionalization procedures applied promoted a slight increase of these two parameters, both in CNT_N and in CNT_HNO$_3$ (Table 1), showing the occurrence of changes on the CNT structure. The oxidative treatment may cause breaks on the walls and open up of the tips of the nanomaterial, leading to a slight increase of the S_{BET} [35]. On the other hand, the CNT_N samples were submitted to a ball milling, which promotes a better dispersion of the CNT in the first stage of the process, and leads to shorter CNT by breaking up the tubes without affecting their diameter [36]. Previous CNM characterization by TEM [37], revealed that pristine CNT structure consisted in aggregates of tubes highly entangled, curved, and twisted with each other, and that the ball-milling (sample CNT_N) reduced significantly this entanglement because the mechanical treatment breaks up the tubes, shortening the CNT, and increasing the surface area (Figure S1; Table 1). Thus, the functionalization procedures applied improved the accessibility of the nanotubes, and the increasing of S_{BET} on the CNT's disentangling could be associated with the increase of the Vp. On the other hand, despite the CNT impregnated with 2% of iron demonstrating a slight decrease in the surface area, this decrease is not considered significant since the iron quantity introduced in the carbon network is low, being the surface area of these CNM maintained similar to the original CNT (Figure 1). Scanning Electron Microscopy with Energy Dispersive Spectroscopy (SEM/EDS) analysis of CNT and CNT@2%Fe samples confirms that these CNM are tubes highly entangled and twisted, and that the impregnation of Fe on CNT structure was successful (Figure 1C).

The surface chemistry of the CNT was also modified by the applied treatments, consequently promoting changes on the surface charge of the nanomaterials [37], as assessed by the pH_{PZC} of the CNM, since the pH_{PZC} is related with the surface groups present in the materials surface. The oxidative treatment caused a decrease in the pH_{PZC} from 6.6 (pristine CNT) to 2.2 (CNT_HNO$_3$) (Table 1), due to the incorporation of a large amount of oxygen containing groups [26,38–40]. The introduction of nitrogen functionalities on the CNT_N by the milling process with melamine followed by thermal treatment only slightly increases the pH_{PZC} when comparing to the original CNT [27]. In addition, it is not expected that the impregnation of Fe in CNT causes changes in the pH_{PZC} of pristine CNT, which was confirmed by the experimental determination of CNT@2%Fe $pH_{PZC,}$ that was 6.5 (Table 1).

From the elemental analysis (Table 1), it is possible to observe that all CNM are mainly composed of carbon. Pristine CNT presented a very low percentage of hydrogen and oxygen, while CNT_HNO$_3$ demonstrated a higher amount of oxygen (1.25%), proving the presence of oxygen-rich groups in this sample. Moreover, the incorporation of N-groups on CNT_N was successful, with 1.69% of N being present in this sample.

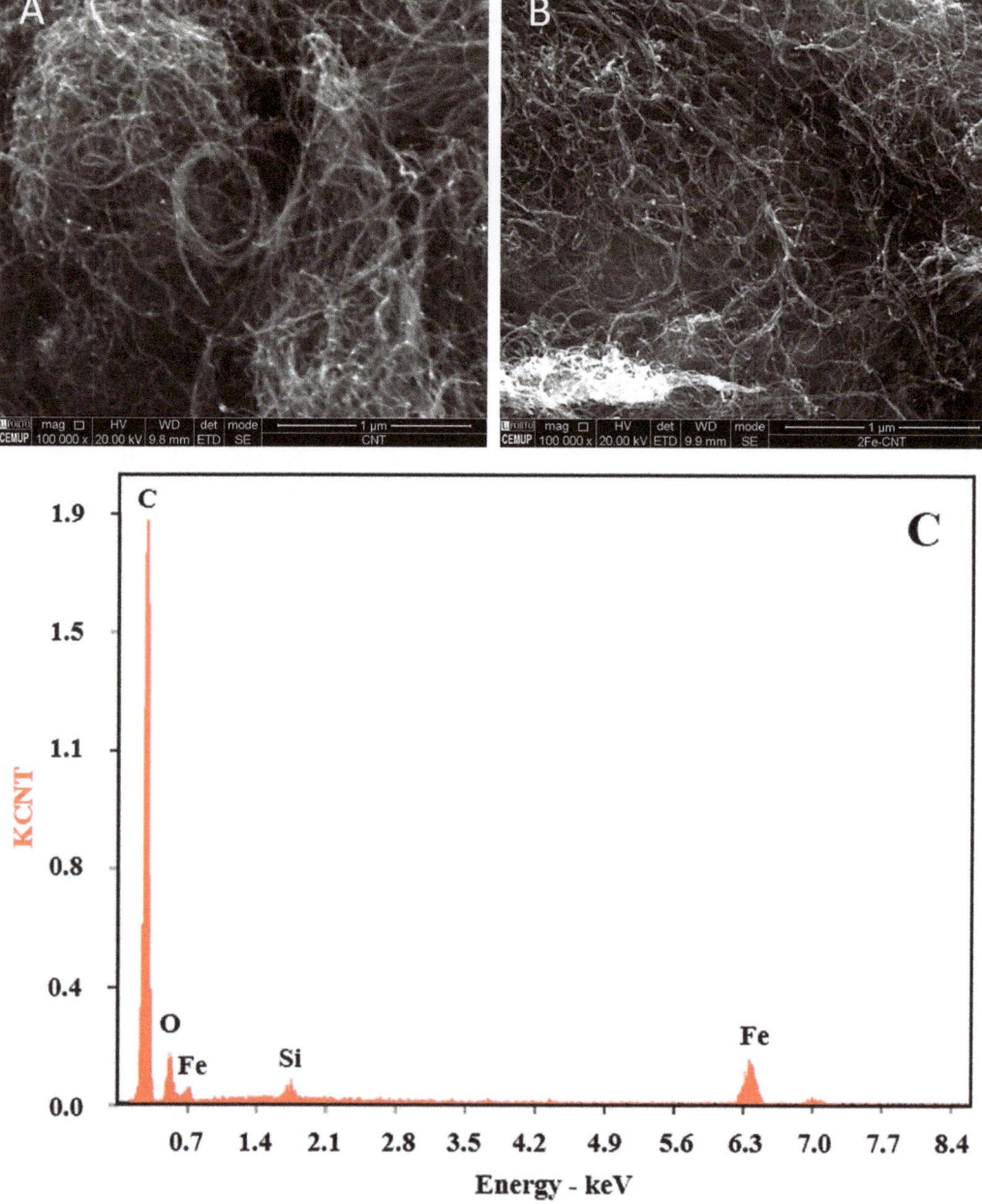

Figure 1. Scanning Electron Microscopy (SEM) images of (**A**) carbon nanotubes (CNT) and (**B**) CNT@2%Fe and (**C**) Energy Dispersive Spectroscopy (EDS) analysis of CNT@2%Fe sample.

Table 1. Surface, textural, and elemental analysis of the different carbon nanotubes (CNT).

Sample	CNT	CNT_N	CNT_HNO$_3$	CNT@2%Fe
S_{BET} (m^2 g^{-1})	201	225	223	196
Vp (cm^3 g^{-1})	0.416	0.503	0.448	0.440
pH$_{PZC}$ (±0.2)	6.6	6.7	2.2	6.5
N (%) *	0.00	1.69	0.00	n.d.
C (%) *	99.8	96.4	98.0	n.d.
H (%) *	0.11	0.18	0.19	n.d.
S (%) *	0.00	0.00	0.15	n.d.
O (%) *	0.06	0.39	1.25	n.d.

* Determined by elemental analysis. n.d.—Not determined. CNT = carbon nanotubes; CNT_N = CNT with N-groups incorporated; CNT_HNO$_3$ = oxidized with HNO$_3$; CNT@2%Fe = CNT impregnated with 2% Fe.

2.2. Effect of CNM on the Chemical Reduction of CIP

The results of the chemical reduction of CIP by sulfide in the absence of oxygen and at pH 7.0, are presented in Table 2. No reduction could be detected in the assays with CNT_HNO$_3$ or in the absence of CNM, revealing the recalcitrant nature of this compound [4,41]. Despite that, in the presence of pristine CNT, 42.6 ± 5.0% of the added CIP was removed, at a reaction rate of 0.082 ± 0.001 mmol L^{-1} d^{-1}. CIP was also removed in the assays with CNT_N, although at a lower extent and rate (i.e., 30.1 ± 8.6% at 0.063 ± 0.001 mmol L^{-1} d^{-1}, respectively), showing the pertinence of screening tailored materials for specific applications, in this case for CIP reduction Indeed, a previous work, on the chemical reduction of azo dyes by Na$_2$S under anaerobic conditions at different pH values (5, 7 or 9), in the presence of pristine and tailored (oxidized or thermal treated) activated carbon (AC) revealed that the pH$_{PZC}$ of the materials, and also the charge of the dyes, played an important role in the reduction efficiency [26]. In that work, thermal treated AC had better efficacy comparatively to pristine and oxidized AC. The same behavior was then proved in the biological experiments for the reduction of the same dyes [26].

Table 2. Chemical reduction of ciprofloxacin (CIP) (1 mmol L^{-1}) by Na$_2$S (1 mmol L^{-1}) in the absence and presence of the different CNM.

Sample	Removal (%)	Rate (mmol L^{-1} d^{-1})
No CNM	0	0
CNT	42.6 ± 5.0	0.082 ± 0.001
CNT_N	30.1 ± 8.6	0.063 ± 0.001
CNT_HNO$_3$	0	0

CNT = carbon nanotubes; CNT_N = CNT with N-groups incorporated; CNT_HNO$_3$ = oxidized with HNO$_3$.

One of the explanation of why the effects of CNM differ according to the pollutant in question and conditions of the process, is related with their amphoteric character, i.e., their surfaces may become positively or negatively charged, depending on their pH$_{PZC}$ and on the pH of the solution. The CNM surface becomes negatively charged at pH > pH$_{PZC}$ and positively charged at pH < pH$_{PZC}$ [30,37].

CIP has amphoteric character as well, due to the bicyclic aromatic ring skeleton with a carboxylic acid group (C-3, pK$_{a1}$ of 5.90 ± 0.15), a keto group, and a basic amino moiety in the piperazine ring (C-7, pK$_{a2}$ of 8.89 ± 0.11). So, depending on the pH conditions, CIP can be in different ionic forms, showing different physicochemical (e.g., on solubility and lipophilicity) and biological behavior [5,42,43]. At pH below 5.90, CIP is in the cationic form (CIP$^+$) due to the protonation of the amine group in the piperazine moiety, and, at pH above 8.89, it is in the anionic form (CIP$^-$), because the carboxylic group lacks a proton. In the range between 5.90 and 8.89, the balance of the two groups stabilizes CIP, which acquires the neutral zwitterionic form (CIP$^\pm$) [44,45]. Among the three ionic species of CIP, CIP$^\pm$ is the most hydrophobic one, owing to the lowest solubility at the neutral pH [42,46,47]. As the assay was conducted at pH 7.3 ± 0.2, CIP$^\pm$ was predominant, and hydrophobic may prevail over electrostatic interactions. Therefore, the main mechanisms proposed are the hydrophobic, hydrogen bond, electrostatic, and/or π-π electron donor–acceptor interactions [48–50].

By knowing the pH_{PZC} of the tailored CNM, it is possible to predict the interaction between the CNM and CIP. At pH 7, CNT (pH_{PZC} 6.6) and CNT_N (pH_{PZC} 6.7) possess pH_{pzc} closer to the neutrality, while CIP is in its neutral zwitterionic form. Thus, the electrostatic interaction between these CNM and $CIP^{\pm}$ may be unfavourable, but hydrophobic interactions are enhanced, which may explain the removal of CIP with these two materials (Table 2). Among them, and contrarily to previous results that indicated the best efficiency of the CNT when doped with N [37], in this work the pristine CNT was shown as the best RM regarding CIP chemical reduction. CNT and CNT_N used in this study have similar pH_{PZC}, but the presence of N group on the surface of tailored CNT seems to hinder the CIP accessibility to the carbon network, and, consequently, its removal from the solution, since the N groups may fill the empty spaces of the carbon structure interfering with the adsorption of large molecules [35,51]. The low adsorption on the material may also decrease the electron transfer and consequently, the reduction of CIP. On the other hand, CNT_HNO_3 possess negative charge at the medium pH 7 and a decreasing tendency to dispersive interactions, revealing some repulsive interactions with CIP, which may justify the lack of CIP reduction under this condition. Similarly, oxidative treatment with HNO_3 worsen the catalytic efficiency of AC as RM in the chemical reduction of the dyes [26].

Adsorption of CIP on nanomaterials was also expected as CNM have been shown as good adsorbents for organic and inorganic compounds, due to their high specific surface area [26,35]. The contribution of the adsorption phenomenon was evaluated in the absence of Na_2S and accounted for circa 3% of CIP removal for all the materials, after reaching the adsorption–desorption equilibrium (Table S1).

Previously, it was stated that higher S_{BET} promotes greater removals of organic and inorganic molecules [26,27,37]. However, S_{BET} is not the only parameter involved in the removal mechanisms, and in this study, despite lower S_{BET} of CNT comparatively to CNT_N, it was more effective on promoting CIP removal, demonstrating the strong influence of the CNM surface chemistry. Based on this observation, CNT were chosen as RM in CIP biological removal experiments.

2.3. Biological Removal of CIP Assisted by CNM under Anaerobic Conditions
2.3.1. CIP Removal under Anaerobic Conditions

The concentration of CIP in the bulk media decreased in the incubations performed with granular sludge (GS), ethanol, CNT, or CNT@2%Fe, but also in the control assays, including abiotic controls, although at a lesser extent (Figure 2). The reactions followed the first-order kinetics and the calculated removal extents and rates are shown in Table 3. In the blank assays (without ethanol) performed in the absence of CNM (GS+CIP), the percentage of CIP removal was $90 \pm 0.1\%$ at a reaction rate of 1.16 ± 0.1 d^{-1}, which suggests a high adsorption of CIP on the anaerobic sludge. However, when ethanol was added as substrate (GS+CIP+E), CIP reduction increased to $95 \pm 1.0\%$, and occurred at the reaction rate of 1.67 ± 0.4 d^{-1}. This improvement pinpoint to the contribution of biological activity in CIP removal. Indeed, the anaerobic sludge consumed ethanol, and the formation of acetate and methane (CH_4) was verified, as it will be further discussed. The rate of biological removal of CIP was upgraded in the presence of CNM: 1.34-fold higher with CNT and 1.53-fold higher with CNT@2%Fe, resulting in removals of $97 \pm 0.7\%$ and $94 \pm 0.5\%$, respectively. This increment suggests stimulation of the biological activity by CNM, so acting as RM on the reductive reactions [30]. In a previous work [27,30], the improvement of the extent and rate of the biological reduction of AO10 obtained with CNT was explained by the CNT's high pore volume and also by the high content of active sites (electron π rich sites on their basal planes), as well as the low concentration of electron-withdrawing groups, which favor the electron transfer and therefore, the reduction of the compounds. Due to the fluorine group present in the molecular structure, CIP is a strong π-acceptor compound [52].

Notwithstanding, the removal of CIP in abiotic controls (without GS) was 98 ± 0.5%, for CNT, and 99 ± 0.4%, for CNT@2%Fe (Table 3), and was likely due to the contribution of CIP adsorption onto the nanomaterials [42,46,47]. Likewise, previous results with an azo dye have also shown that the presence of iron on CNT@2%Fe contributed to enhance the reduction capacity under abiotic conditions, which was attributed to the transfer of electrons first from nanoscale iron to carbon, and finally to the dye [30]. The same process may have occurred in the abiotic assays performed in the presence of this CNM (i.e., CIP + E + CNT@2%Fe), and could have also potentially contributed for the removal of CIP verified both in the biological assays (GS + CIP + E + CNT@2%Fe) and in the blanks (GS + CIP + CNT@2%Fe) containing this magnetic CNT.

The results of the first 24 h suggest the combined contribution of adsorption and biological activity in the removal of CIP. However, regarding the slight differences observed in this first cycle between the blank, abiotic, and biological assays, it was difficult to distinguish between the different phenomena contributing for CIP removal, because in this cycle adsorption may be the main mechanism, once biomass and CNT are not yet saturated (Figure 2).

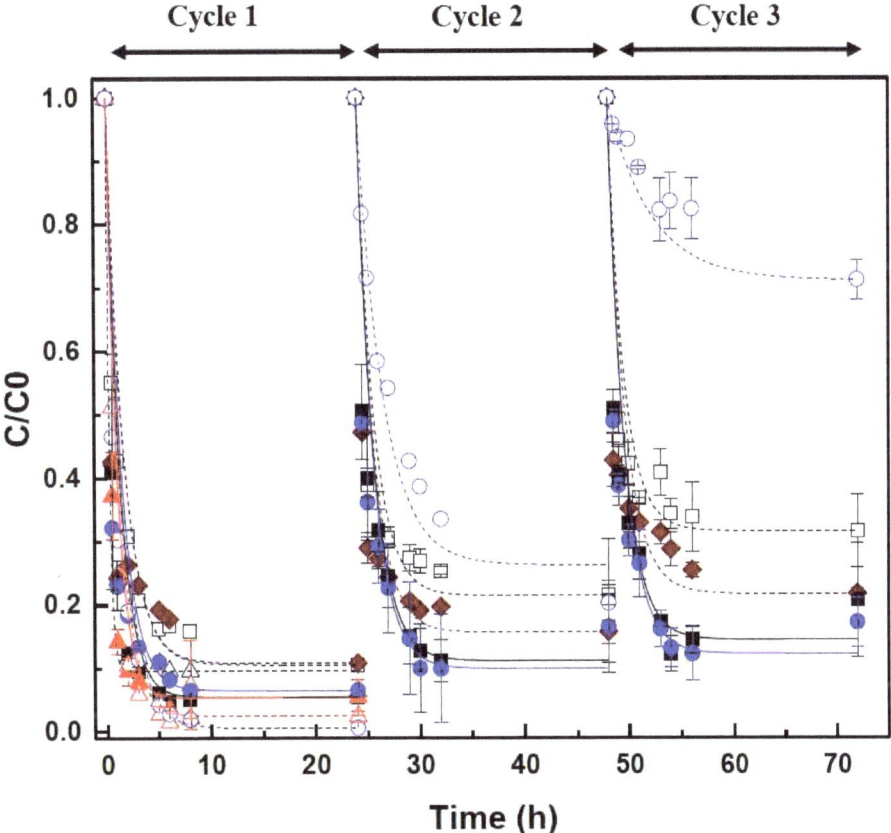

Figure 2. First-order rate curves of CIP removal in the anaerobic assays, performed in the absence of CNM (GS+CIP+E) (■), with CNT (GS+CIP+E+CNT, 1st cycle only) (▲) and in the presence of CNT@2%Fe (GS+CIP+E+CNT@2%Fe) (●). Blank controls (without ethanol) are also shown: without CNM (GS+CIP) (□), with CNT (GS+CIP+CNT, 1st cycle only) (△) and with CNT@2%Fe (GS+CIP+CNT@2%Fe) (◆). Abiotic controls (without granular sludge (GS)) in the presence of CNT (CIP+E+CNT, 1st cycle only) (△) and CNT@2%Fe (CIP+E+CNT@2%Fe) (○) are presented as well.

Table 3. Percentage of CIP removal (%) and rate (d^{-1}) in the anaerobic assays, performed in the absence and presence of CNM. Blank controls without ethanol, as well as abiotic assays without granular sludge are also presented.

	Condition	Cycle 1		Cycle 2		Cycle 3	
		CIP Removal (%)	Rate (d^{-1})	CIP Removal (%)	Rate (d^{-1})	CIP Removal (%)	Rate (d^{-1})
Biotic assays	GS + CIP + E	95 ± 1.0	1.67 ± 0.4	89 ± 3.3	1.39 ± 0.4	86 ± 2.2	1.41 ± 0.2
	GS + CIP + E + CNT	97 ± 0.7	2.24 ± 0.3	n.a.	n.a.	n.a.	n.a.
	GS + CIP + E + CNT@2%Fe	94 ± 0.5	2.55 ± 0.1	90 ± 8.6	1.49 ± 0.2	88 ± 4.1	1.54 ± 0.3
Blank assays	GS + CIP	90 ± 0.1	1.16 ± 0.1	79 ± 2.3	0.92 ± 0.2	68 ± 5.7	1.07 ± 0.1
	GS + CIP + CNT	94 ± 0.1	2.7 ± 0.1	n.a.	n.a.	n.a.	n.a.
	GS + CIP + CNT@2%Fe	89 ± 0.2	2.4 ± 0.1	84 ± 2.6	1.7 ± 0.6	78 ± 0.8	0.99 ± 0.2
Abiotic assays	CIP + E +CNT	98 ± 0.5	1.67 ± 0.4	n.a.	n.a.	n.a.	n.a.
	CIP + E + CNT@2%Fe	99 ± 0.4	1.32 ± 0.6	79 ± 8.3	0.3 ± 0.1	29 ± 3.2	0.13 ± 0.1

n.a.—Not applicable. GS = granular sludge; CIP = ciprofloxacin; E = ethanol; CNT = carbon nanotubes; CNT@2%Fe = carbon nanotubes impregnated with 2% Fe.

In this sense, considering the higher and faster CIP removal achieved in the assays with CNT@2%Fe, and taking into account that these CNM have magnetic properties which favor their recovery and reuse, two additional cycles of 24 h were performed with this material, as well as the blank and abiotic controls (Table 3). The aim was to provide clear evidence on the role of biological degradation in CIP removal, since GS and CNM saturation is expected to occur over the cycles, thus decreasing the contribution of the adsorption phenomenon. At the same time, the reusability and the evolution of the catalytic properties of CNT@2%Fe could be evaluated.

Indeed, in the second and third cycles, lower extents of CIP removal were obtained in all the assays, comparing to the first cycle (Figure 2; Table 3), possibly due to saturation of the adsorbent materials. This decrease was more pronounced in the abiotic assays (CIP + E + CNT@2%Fe), where CIP removal reached only 29 ± 3.2% at the end of the third cycle. In the biological assays with ethanol, a high CIP removal capacity was still verified in the third cycle, both in the presence and absence of CNT@2%Fe (i.e., 88 ± 4.1% and 86 ± 2.2%, respectively), highlighting the importance of the biological activity in this process. In these assays, microorganisms may be oxidizing ethanol and reducing CIP, which acted as final electron acceptor.

The second and third cycles, make clear the contribution of the several removal mechanisms, including adsorption and degradation, occurring simultaneously in the system, but biological reactions might be preponderant in those two last cycles owning the saturation of GS and CNT@2%Fe (Figure 3) [10,53]. In the blank assays without ethanol (GS + CIP and GS + CNT@2%Fe), and after three cycles, 78 ± 0.8% and 68 ± 5.7% of the added CIP was removed in the presence and absence of the CNM, respectively. These values are higher than in the abiotic assay, showing that besides CIP adsorption on CNM and GS, biological removal also occurs in the blank assays, without ethanol as electron donor. This can be justified by the utilization by anaerobic microbial community of other electron donors originated from dead microbial cells, metabolites excreted during cell decay. Alternatively, microbial oxidation of CIP can be hypothesized. As sole carbon source, CIP has only been oxidized in the presence of sulfate or nitrate, and CIP oxidation in the absence of any external electron acceptor other than bicarbonate (i.e., in conditions similar to the ones in this study) was never reported [54].

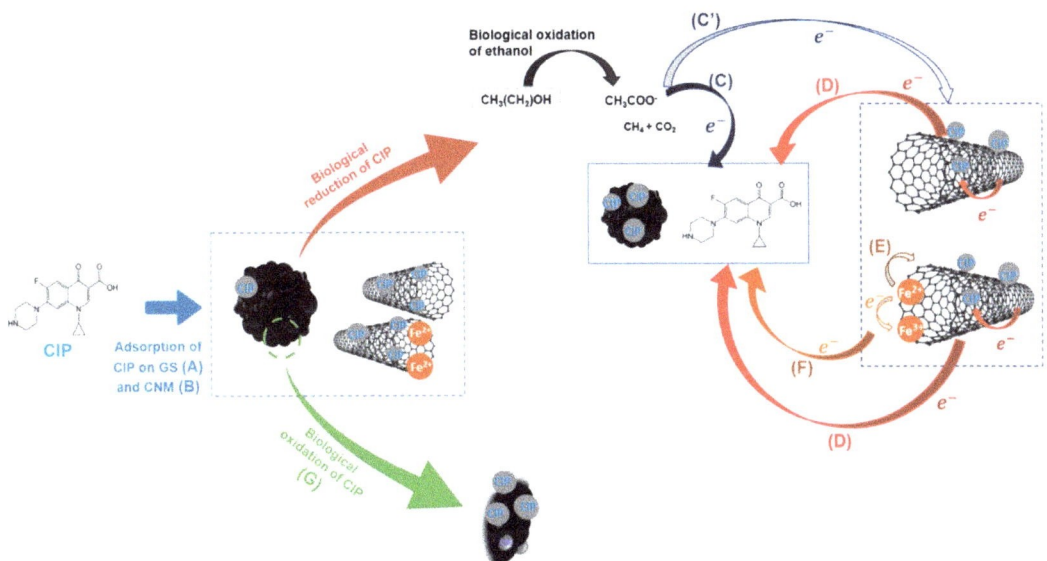

Figure 3. Proposed mechanisms of CIP removal: adsorption of CIP on GS (A) and on CNT or on CNT@2%Fe (B); biological reduction due to electron (e−) flow from the biological oxidation of ethanol to CNM (C′) or CIP (in solution and/or adsorbed on sludge) (C); biological reduction in the presence of CNM (D) due to e− transfer from the oxidation of ethanol to CNM and then to CIP (in solution or adsorbed on sludge or on CNM), and due to e− transfer to CNT from the abiotic oxidation of Fe^{2+} (E) and then to CIP (adsorbed or in solution) (F). Further, oxidation of CIP (in solution and adsorbed on sludge) by the anaerobic microorganisms may occur (G). All these mechanisms may occur independently or combined.

The catalytic properties of CNT@2%Fe were maintained over the cycles, as shown by the higher removal extent and reaction rates verified in the biological assays (GS+CIP+E+CNT@2%Fe), both in the second and third cycles, comparatively to the assay in its absence (GS+CIP+E) (Figure 2; Table 3). Despite the statistically similar reaction rates obtained in the presence and absence of CNT@2%Fe, in the second and third cycles, probably as a result of the adaptation of the microbial community to the substrate and to CIP [55], the presence of CNM could be determinant in the initial stage of the reaction, speeding up the reaction rates and improving the reductive system.

2.3.2. Assessment of the Biological Activity during CIP Removal

The activity of the anaerobic microbial community was assessed in the biological assays by measuring the decrease in ethanol concentrations along the time, coupled to acetate and methane (CH_4) production (Table 4, Figure S2 and Table S2). Ethanol was totally consumed by the anaerobic granular sludge, both in the presence and in the absence of CNM, in all the cycles (Table S2 and Figure S2). The maximum methane concentration produced in all the conditions is in agreement with the value that could be expected from the stoichiometric conversion of ethanol to methane (i.e., 45 mmol L^{-1} CH_4 from 30 mmol L^{-1} ethanol, Equations (1)–(3), Table S2).

$$C_2H_5O^- + H_2O \rightarrow CH_3COO^- + 2H_2 \text{ (acetogenesis)} \tag{1}$$

$$CH_3COO^- + H_2O \rightarrow CH_4 + HCO3^- \text{ (methanogenesis)} \tag{2}$$

$$H_2 + \frac{1}{4}HCO_3^- + \frac{1}{4}H^+ \rightarrow \frac{1}{4}CH_4 + \frac{3}{4}H_2O \text{ (methanogenesis)} \tag{3}$$

Table 4. Rates of ethanol consumption and methane production, over 3 cycles of biological removal of CIP in the presence of CNM.

	Condition	Ethanol Consumption Rate (mmol $L^{-1}h^{-1}$)			Methane Production Rate (mmol $L^{-1}h^{-1}$)		
		Cycle 1	Cycle 2	Cycle 3	Cycle 1	Cycle 2	Cycle 3
Biotic assays	GS + E	3.24 ± 0.62	3.76 ± 1.16	3.26 ± 0.45	2.58 ± 0.05	2.78 ± 0.06	3.03 ± 0.03
	GS + E + CNT	3.66 ± 0.50	n.a.	n.a.	2.62 ± 0.04	n.a.	n.a.
	GS + E + CNT@2%Fe	3.57 ± 0.34	3.72 ± 1.08	3.32 ± 0.51	2.23 ± 0.20	2.84 ± 0.03	3.07 ± 0.03
	GS + CIP + E	3.41 ± 0.46	3.29 ± 0.63	3.21 ± 0.24	2.61 ± 0.03	2.89 ± 0.03	3.00 ± 0.06
	GS + CIP + E + CNT	3.08 ± 0.30	n.a.	n.a.	2.51 ± 0.08	n.a.	n.a.
	GS+ CIP+ E + CNT@2%Fe	3.39 ± 0.47	3.23 ± 0.67	3.27 ± 0.53	2.31 ± 0.20	2.86 ± 0.03	2.92 ± 0.03

n.a.—Not applicable. GS = granular sludge; CIP = ciprofloxacin; E = ethanol; CNT = carbon nanotubes; CNT@2%Fe = carbon nanotubes impregnated with 2% Fe.

Ethanol is converted initially to acetate and H_2 (acetogenesis), and acetate and H_2 are further converted to CH_4 (methanogenesis) [56,57]. The monitoring of acetate concentration over the time in the assay GS+CIP+E+CNT@2%Fe showed a transient accumulation of this compound in the medium, being then almost completely consumed until the end of the cycles (Figure S2). In all the conditions tested, acetate was present at low concentrations (<5 mmol L^{-1}) at the end of each cycle, demonstrating the total conversion of the substrates to CH_4. Indeed, both acetoclastic and hydrogenotrophic methanogens, which perform acetate and hydrogen conversion to methane, respectively, were detected in the inoculum sludge utilized in these experiments. Acetate conversion was probably carried out by *Methanosaeta* species as it was the only acetoclastic genus detected, and in a high relative abundance (~20%) (Table S3). The transient accumulation of acetate may be due to the adaptation of *Methanosaeta* species to the incubation conditions, i.e., the presence of carbon nanomaterials and/or the presence of CIP. On the other hand, different hydrogenotrophic methanogens could be converting hydrogen to methane as several species could be detected in the inoculum sludge, namely, *Methanobacterium* and *Methanolinea* in relative abundances of 9%, and *Methanospirillum* and *Methanobrevibacter* which were less abundant (0.07% and 0.003% relative abundance, respectively). The bacterial community was much more diverse and therefore, it is not possible to confidently infer on the function of specific microorganisms in the assays. Nevertheless, *Geobacter* species were detected in high abundance (over 14%) and once these microorganisms are known as ethanol degraders, they might have had a relevant contribution in the conversion of ethanol to acetate and hydrogen in these experiments.

The presence of CIP did not inhibit the methanogenic activity (Figure S2; Table 4), since there were no significant differences in the CH_4 production rate from ethanol when the sludge was incubated in the presence and absence of CIP (GS+CIP+E and GS+E), in each cycle. These results are in agreement with the ones previously observed by Silva et al. [16].

The consumption rate of ethanol, and the production rate of CH_4, increased at each cycle, probably as a result of the microbial community growth, resulting in higher ethanol conversion rates (Table 4).

2.3.3. Mechanisms of CIP Removal

The obtained results, taken all together, point to the occurrence of different mechanisms of CIP removal, namely adsorption on sludge and/or on CNM, and biological removal by oxidation and/or reduction, which are accelerated by the presence of CNM (Figure 3). The results suggest that adsorption phenomena likely occurred in the beginning until saturation of GS and CNM, and biological reactions prevailed after reaching the adsorption/desorption balance. In fact, in the biological assay with ethanol and without CNM, adsorption of CIP on GS and its reduction due to electrons generated by the oxidation of ethanol, or biological oxidation of CIP, possibly occurred (Figure 3). The event of biological reduction of CIP by the electrons generated from the oxidation of ethanol (Figure 3C,C') probably explain the higher percentage of removal as compared with the assay without substrate where only adsorption (A) and biological oxidation may occur (Figure 3G). When CNM are present, besides absorption of CIP on GS and on the nanomaterials (Figure 3A,B), respectively), improvement of the reaction rates by the CNM, which act as electron shuttles, may justify the high extent of removal (Figure 3C'–D).

In addition, the high rates obtained with the CNT@2%Fe in the biotic and abiotic condition, may result from the fact that besides CIP adsorption and reduction by electrons generated from ethanol oxidation, electrons may flow from Fe^{2+} to CNT and then to CIP (adsorbed on CNT@2%Fe and on GS, and free in solution), as represented in Figure 3E,F [30]. It is important to note that a dynamic adsorption/desorption process to GS and nanomaterials is probably occurring during the incubation period. Adsorption phenomena are required for the success of biological degradation of micropollutants, since the flow of electrons is favored by the proximity between the microorganisms, the catalyst and the pollutant [30,58]. On the other hand, Salvador et al. [58] observed a good binding of CNM on anaerobic microorganisms, which resulted in the improvement of microbial activity.

Adsorption on GS was expected based on previous studies reporting that the removal of CIP is mainly due to adsorption on activated sludge and CNM, rather than biodegradation [13,59–61]. The adsorption of CIP onto the sludge is a spontaneous, exothermic and a linear process that includes both physisorption and chemisorption [10,11]. As mentioned above, at neutral pH, CIP mainly presents zwitterionic form ($CIP^{\pm}$) with $-NH^{2+}$ and $-COO^-$ groups [49,62,63]. The functional groups present on anaerobic sludge, such as C–O, C–O–C, N–H, O–H and COOH provide binding sites for $CIP^{\pm}$ adsorption [10]. These functional groups act as strong electron acceptors and conjugated with the π electron-donating groups of CIP (N–H and O–H) form a π-π electron donor–acceptor system [10]. On the other hand, the O–H groups present on the sludge can be conjugated with the COOH and C=O groups of CIP and the COOH and N–H groups on sludge surface may also form hydrogen bonds with O–H group in CIP molecule [10]. Additionally, the negative surface charge of sludge at neutral pH could also stimulate the $CIP^{\pm}$ adsorption onto the sludge via electrostatic attraction and cation exchange [11]. Thus, the high CIP adsorption onto the sludge could be attributed to the multiple adsorption mechanisms, including hydrophobic interaction, electrostatic attraction, cation exchange and bridging, π-π interaction, and hydrogen bond effect [64].

Furthermore, the adsorption of CIP on CNT is spontaneous when Gibbs free energy (ΔG^0) is negative [42,65,66], where the binding mechanisms mainly associated to this phenomenon is physisorption [42,49,67].

On the other hand, the sorption energy decreased with the increasing of CIP loading, hence, CIP molecules first occupied the high-energy sorption sites at low concentration and then spread to low-energy sorption sites [42,68]. Furthermore, recent studies have reported that the removal of pharmaceuticals by anaerobic sludge occurs initially by sorption, but after the equilibrium being reached, the mass-transfer driving force no longer affects the pharmaceutical uptake due to the absence of a concentration gradient. Being the biodegradation mechanism the major removal route in the system [10,53,68–72].

2.4. Toxicity Assessment with Vibrio fischeri

Evaluation of the toxicity of the samples was performed after the biological anaerobic treatment proposed, to assess the detoxification extent (Table 5). The initial CIP solution (0.015 mmol L^{-1}) led to an inhibition of 56 ± 10% of the luminescence of *V. fischeri*. This inhibition decreased after three cycles (72 h) of biological treatment, to 30 ± 4% and 26 ± 7% in the assays without CNM and with CNT@2%Fe, respectively (Table 5). These values reflect a 46% detoxification of the solutions by the anaerobic process. The luminescence inhibition still measured after the anaerobic treatment may be related to the presence of CIP still existing, in the treated solutions, even though at lower amount, as the removal extent in these assays was 86–88% (Table 4). Alternatively, it may be linked to the possible by-products formed during the degradation cycles, which seems most probable considering the high CIP removal verified in the biological assays [73].

Table 5. Luminescence inhibition (INH) of *V. fischeri* in all the tested samples, after 30 min of exposure.

Samples		INH (%)
CIP solution (0.015 mmol L^{-1})		56 ± 10
GS + CIP + E (Treatment of 72 h)		30 ± 4
GS + CIP + E + CNT@2%Fe (Treatment of 72 h)		26 ± 7
Positive control (ZnSO$_4$.7H$_2$O)		83 ± 8
	Anaerobic medium	4.9 ± 0.9
Medium after incubation with 0.1 g L^{-1} of CNM	CNT	4.7 ± 0.7
	CNT@2%Fe	18.1 ± 1.7

The possible contribution of CNM to the toxicity of the treated solution may not be neglected as the treated solutions may contain traces of small amorphous materials from CNM or even impurities that remained in the solutions after removing the CNM [16,74]. In this sense, the toxic extent of the anaerobic medium, previously incubated with CNM under anaerobic conditions, was also assessed. The INH (%) obtained for the medium incubated with CNT and CNT@2%Fe, was 4.7 ± 0.7% and 18.1 ± 1.7, respectively (Table 5). According to Mendonça et al. [75], the toxicological effects of CNM used in this study are considered negligible, since the luminescence variations could be associated to the adaption of the microorganisms to the presence of the pollutant [75,76]. Moreover, the luminescence inhibition caused by the medium itself, without CNM, was 4.9 ± 0.9%, a value similar to that obtained with the medium incubated with CNT, which confirms that CNT do not contributed for the toxicity obtained with treated solutions. It is important to state that the amount of CNM needed to act as RM is very low, only 0.1 g L^{-1}, so the amount of possible amorphous CNM released to the medium will also be very low, another advantage of using these materials.

Some authors have reported that the iron in CNM have toxic effects, and that the toxic mechanisms are related to the fact that iron can be leachate from the CNT during the incubation time, and due to the high affinity of iron oxides to the cells membrane, generating reactive oxygen species, which could lead to cells death [77,78]. However, because the efficiency of the material was maintained during the cycles, the material probably maintained the initial structure. Nevertheless, due to its magnetic properties, CNT@2%Fe can be easily removed from the treated water by applying a magnetic field. Therefore, it is not expected that the solutions treated in the presence of this CNM will constitute a toxicity problem when discharged.

3. Materials and Methods

3.1. Chemicals

CIP was obtained from Sigma-Aldrich, at the purity of 98%. A stock solution of CIP was prepared in deionized water at a concentration of 0.15 mmol L^{-1}. Due to the low solubility of CIP, a few drops of hydrochloric acid (2 mol L^{-1}) were added, under constant magnetic stirring. CIP has a water solubility of 30 g L^{-1} (0.091 mol L^{-1}) at 20 °C and its solubility is enhanced when it is in the ionic form as explained in sub-Section 3.2. Sodium sulfide (Na$_2$S.9H$_2$O) was purchased from Fluka. Fe(NO$_3$)$_3$, used for the CNT impregnation with 2% Fe (wt.%), was purchased from Sigma-Aldrich. Formic acid and acetonitrile (ACN) for High Performance Liquid Chromatography (HPLC) analysis were purchased from Merk and Fluka, respectively, at the highest analytic grade purity commercially available (98%). All the reagents used for the preparation of the anaerobic basal medium [79]. were purchased from Sigma-Aldrich. ZnSO$_4$.7H$_2$O, obtained from ACS, Panreac, was used in the toxicity assessment.

3.2. Carbon Nanomaterials

Commercial multiwalled CNT (NC3100TM, Nanocyl SA., Sambreville, Belgium), with 1.5 μm average length, 9.5 nm average diameter and more than 95% carbon purity (according to the supplier's technical data sheet) were used in the experiments. In order to obtain CNT with N-groups incorporated (sample CNT_N), commercial CNT were mixed with 0.26 g of N using melamine as nitrogen precursor, and the mixture was ball milled in a closed flask without any gas flow in a Retsch MM200 equipment, during 4 h at a constant vibration frequency of 15 vibrations s^{-1}. Following, the CNT_N were subjected to a thermal treatment under N_2 flow (100 cm^3 min^{-1}), until 600 °C and kept at this temperature during 1 h, as previously reported by Soares et al. [51]. A CNT sample with high amount of oxygen-containing surface groups, and consequently strong acid character (sample CNT_HNO$_3$) was also prepared through oxidative treatment of the commercial CNT with 7 mmol L^{-1} of HNO_3, in liquid phase, at boiling temperature, during 3 h as described by Gonçalves et al. [35]. Subsequently, CNT were washed with distilled water to neutral pH, and dried in an oven at 110 °C for 24 h.

Commercial CNT were also impregnated with a metal phase (2%Fe), thus originating a magnetic carbon-based nanocomposite (CNT@2%Fe). CNT were supplemented with 2% Fe by incipient wetness impregnation from aqueous solution of the corresponding metal salt (Fe(NO$_3$)$_3$). Then, samples were dried at 100 °C for 24 h and placed under nitrogen flow at 400 °C for 1 h, and reduced at 400 °C in hydrogen flow for 3 h [30,80].

Textural properties of CNM, such as the specific surface area (S_{BET}) and total pore volume (Vp), were analyzed, as well as the pH at point of zero charge (pH$_{PZC}$). Elemental analysis and oxygen analysis were also carried out.

Scanning Electron Microscopy with Energy Dispersive Spectroscopy (SEM/EDS) analyses were obtained by using a Schottky scanning electron microscope of high resolution with microanalysis with X-rays and analysis of patterns of diffuse scattering electrons: Quanta 400FEG ESEM/EDAX Genesis X4M.

3.3. Effect of CNM on the Chemical Reduction of CIP

In this assay, pristine or tailored CNT (CNT_N and CNT_HNO$_3$) were tested to verify which was the best catalyst to be used posteriorly as RM on the biological assays. Chemical reduction of CIP was performed in 70 mL serum bottles with 25 mL basal medium, buffered at a pH of 7.3 ± 0.2 with NaHCO$_3$ (2.5 g L^{-1}), as described by Angelidaki et al. [79]. This pH was selected based on the fact that it is the required for the biological assays. CNM were added to the vials at a concentration of 0.1 g L^{-1}. The bottles were sealed with butyl rubber stoppers and aluminum caps and flushed with N_2:CO_2 (80:20% v/v). Na$_2$S was added as reducing agent from a partially neutralized stock solution (0.1 mol L^{-1} Na$_2$S), to obtain an initial concentration of 1 mmol L^{-1}. The flasks were incubated overnight (14 h) at 37 °C in a rotary shaker (120 rpm), after which CIP was added at a concentration of 1 mmol L^{-1}. This relatively high CIP concentration was chosen to facilitate ascertaining whether it is susceptible to being reduced under anaerobic conditions. Further, controls without Na$_2$S were prepared. CIP concentration was analyzed by HPLC over 96 h of incubation.

3.4. Anaerobic Removal of CIP Assisted by CNM and Characterization of the Inoculum Sludge

Anaerobic assays were performed in 200 mL serum bottles containing 100 mL basal medium, supplemented with micro and macro nutrients, salts, and vitamins, as described by Angelidaki et al. [79]. Anaerobic medium was buffered at a pH of 7.3 ± 0.2 with NaHCO$_3$ (2.5 g L^{-1}). The anaerobic granular sludge (GS) used as inoculum, was originated from a brewery plant, collected and transported in a closed container of 25 L and preserved at 4 °C, under anaerobic conditions (by flushing the headspace with nitrogen). GS was used at a final volatile solids (VS) concentration of 3 g L^{-1}. The bottles were supplemented with the CNM (CNT or CNT@2%Fe) at a concentration of 0.1 g L^{-1}. CNT were selected to be tested in biological assays based on the results obtained from the screening of CIP chemical reduction. Because conferring a magnetic character to CNT is beneficial to facilitate their removal after the process, CNT impregnated with 2% were prepared and also used in the biological assays. Bottles were sealed with butyl rubber stoppers and aluminum caps, flushed with N_2/CO_2 (80:20% v/v) and incubated overnight (14 h) at 37 °C and 120 rpm, for the consumption of any residual substrate. After that pre-incubation period, CIP was added at a concentration of 0.015 mmol L^{-1}, as well as ethanol (as primary electron donor) at the concentration of 30 mmol L^{-1}, from a stock solution of 3 mol L^{-1}. Control assays without CNM were also prepared, as well as blank assays without ethanol. Abiotic controls, set up with CNM and ethanol but without sludge, were also included. All the assays were made in triplicate and were incubated at 37 °C, 120 rpm. To verify the reusability and the evolution of the catalytic properties of the materials, CNT@2%Fe, two additional cycles of 24 h were performed in the bottles containing this CNM (Figure S3). For that, after each 24 h, the bottles were flushed with N_2/CO_2 (80:20% v/v) to remove the methane produced, and ethanol (30 mmol L^{-1}) and CIP (0.015 mmol L^{-1}) were added again to each condition. Furthermore, biological controls without CIP, in the presence and absence of CNM were prepared to better understand the effect of CIP on the acetogenic bacteria which consume ethanol and on methanogenic archaea, producing methane. CIP, ethanol, acetate and methane concentrations were monitored by HPLC and Gas Chromatography (GC), over the time in the experiments.

In order to assess the microbial composition of the anaerobic sludge, aliquots of the inoculum sludge were taken in duplicate and preserved with RNA later (Sigma-Aldrich) at −20 °C. RNA extraction, 16S rRNA sequencing and bioinformatics analysis were performed as described by Salvador et al. [81], with minor changes namely the utilization of primer Uni1492r [82] in the cDNA synthesis step, and the universal primer set 515F/806R [83] targeting the prokaryotic 16S rRNA gene, in sequencing amplification by Illumina MiSeq. FASTQ files containing the 16S rRNA sequences, were deposited in the European Nucleotide Archive (ENA), under the study accession number PRJEB43083.

3.5. Analytical Methods

Textural properties such as total specific surface area (S_{BET}) and total pore volume (Vp) at P/P0 = 0.95 were analyzed by N_2 adsorption isotherms at −196 °C using a Quantachrome NOVA 4200e multi-station equipment, where the samples were previously degassed in vacuum for 3 h at 150 °C. S_{BET} was calculated from the nitrogen adsorption data in the relative pressure range of 0.05–0.3 [84]. Thermogravimetric analysis was performed in a NetzschSTA 409 PC Luxx®. The analyses were carried out under a helium flow, at a heating rate of 10 °C min^{-1} from 50 to 900 °C, using two isothermal steps at 900 °C: 7 min under helium flow and 13 min under air flow.

The pH at point of zero charge (pH_{PZC}) was also determined for each CNM. For that propose, 50 cm^3 of 0.01 M NaCl solution was placed in closed Erlenmeyer flask and the pH was adjusted to a value between 2 and 10 with the solutions of 0.1 M HCl or 0.1 M NaOH. Then, 0.15 g of each CNM was added and the final pH measured after 24 h under agitation at room temperature. The pH_{PZC} was obtained by the intersection of the curve pH_{final} vs. $pH_{initial}$ with the line $pH_{initial} = pH_{final}$ [85].

Each element (CHNS) was determined on a vario MICRO cube analyzer from Elemental GmbH in CHNS mode, by combustion of the sample at 1050 °C and calculated by the mean of three independent measurements, using a per-day calibration with a standard compound. Oxygen composition was determined a rapid OXY cube analyzer from Elemental GmbH, by pyrolysis of the sample at 1450 °C and calculated by the mean of three independent measurements, using a per-day calibration with a standard compound [37].

The vs. were determined gravimetrically as described in Standard Methods [86].

Removal of CIP was assessed by HPLC analysis, based on the disappearance of its corresponding peak at retention time of 12.5 min (Figure S4). The analyses were performed as previously reported by Silva et al. [73]. An Ultra HPLC (Nexera XZ, Shimadzu, Japan) equipped with a Diode Array Detector (SPD-M20A), an autosampler (SIL-30AC), degassing unit (DGU-20A5R), LC-30AD solvent delivery unit, a Labsolutions software and a RP-18 endcapped Purospher Star column (250 × 4 mm, 5 µM particle size, from MERK, Germany) were used. The mobile phase was composed by 0.1% formic acid solution (solution A) and ACN (solution B). Prior to analysis, samples were centrifuged (10 min at 10,000 rpm) and filtered (Whatman SPARTAN syringe filters, regenerated cellulose, 0.2 µm pore size). The compounds were eluted at a flow rate of 0.8 mL min^{-1} at 40 °C, with the following gradient: increase of ACN from 5 to 15%, over 6 min, followed by an isocratic step during 12 min, then from 15% to 40% of ACN during 12 min and 40% was then maintained for 10 min. A calibration curve at increasing CIP concentrations from 0.0002 to 0.03 mmol L^{-1} was made.

The percentage of CIP removal (*PR*) was calculated according to Equation (4):

$$PR\ (\%) = \frac{(C_0 - C_t)}{C_0} \times 100 \tag{4}$$

where C_0 is the initial CIP concentration and C_t the CIP concentration at time t.

First-order reduction rate constants were calculated in OriginPro 6.1. software, applying Equation (5):

$$C_t = C_i + Ae^{-t/k} \tag{5}$$

where C_t is defined in equation 1, C_i is the offset, a value closed to the asymptotic of the Y variable (C) for larger time (*t*) values and k is the first-order rate constant (d^{-1}).

Ethanol and acetate were also monitored by HPLC (Jasco, Tokyo, Japan), with a RI and UV detector (at 210 nm), respectively, using a Rezex ROA Organic Acid H$^+$ (8%), (300 mm × 7.8 mm) column. The elution was made at 60 °C using sulfuric acid (5 mmol L^{-1}) as mobile phase, at a flow rate of 0.6 mL min^{-1}.

The concentration of CH_4 produced in each bottle, over each degradation cycle, was assessed by gas chromatography (GC), using a Shimadzu GC-2014 gas chromatograph fitted with Porapak Q 80/100 mesh, packed stainless-steel column (2 m × 1/8 inch, 2 mm) and a flame ionization detector (FID). Nitrogen was the carrier gas at a flow rate of 30 mL min^{-1} and the column, injection port and detector temperatures were respectively 35, 110, and 220 °C. Headspace gas was sampled by a 500 µL pressure-lock syringe (Hamilton). The values of CH_4 production were corrected for the standard temperature and pressure conditions (STP). A standard sample composed of 40% of CH_4 was injected firstly, followed by samples injection.

3.6. Statistical analysis

Statistical significance of the differences in the biological CIP reduction rates and methane production rates obtained after each degradation cycle, was evaluated using single factor analysis of variances (ANOVA). Statistical significance was established at the $p < 0.05$ level.

3.7. Toxicity Assessment with Vibrio fischeri

Toxicity assays were performed with *V. fischeri* strain NRRL-B-1117, purchased as freeze-dried reagent, BioFix® *Lumi*, from Macherey-Nagel (Düren, Germany) and grown under aerobic conditions, as described in the international standard ISO 11348-1 "Water quality–Determination of the inhibition effect of water samples on the light emission of *Vibrio fischeri* (Luminescent bacteria test)" method, using freshly prepared bacteria [87].

Toxicity was assessed for samples collected at the end of the third cycle (72 h of incubation) in the bottles containing CNT@2%Fe, since this is the condition that better represents the contribution of all the mechanisms for CIP removal. Samples were centrifuged (10 min at 10,000 rpm) and filtered (Whatman SPARTAN syringe filters, regenerated cellulose, 0.2 µm pore size) prior to the toxicity assay. CIP solution (0.015 mmol L^{-1}) and anaerobic medium were also tested. Moreover, solutions containing 0.1 g L^{-1} of CNT and CNT@2%Fe were prepared in anaerobic medium and placed at 37 °C and 120 rpm, during 72 h. After that period, samples were collected and centrifuged, and the toxicity of the supernatant was evaluated. Evaluation of pristine CNT, besides of CNT@2%Fe, allows assessing whether iron impregnation makes the material more toxic or not. Negative controls were prepared with the bacterial suspension and a solution of 2% NaCl. Zinc sulfate heptahydrate at a concentration of 19.34 mg L^{-1} was used as positive control [88]. The salinity of all the samples and solutions was adjusted to 2% NaCl. The samples pH was adjusted to values between 6 and 9 with hydrochloric acid or sodium hydroxide. Oxygen concentration was higher than 3 mg L^{-1}, and turbidity was avoided by samples centrifugation and filtration.

Toxicity evaluation was performed according to the standard ISO 11348-1 and 11348-3, using a microplate reader (Biotek® Cytation3, Fisher Scientific, Korea) in kinetic mode to evaluate the bacteria luminescence changes when exposed to potentially toxic substances [87,89]. For that propose, a 96 well optical Btm Plt polymer base Blk plate, from Nalge Nunc™ International, was used, where each sample (100 µL) was mixed with the bacteria test suspension (100 µL), according to the ISO 11348-3.

Luminescence inhibition (*INH* %) was calculated after 30 min [87–89], according to Equation (6):

$$INH\ (\%) = 100 - \frac{IT_t}{KF \times IT_0} \times 100 \tag{6}$$

with

$$KF = \frac{IC_t}{IC_0} \tag{7}$$

where IT_t is luminescence intensity of the sample after the contact time (30 min), IT_0 is the luminescence intensity at the beginning of the assay (time 0), *KF* is the correction factor and characterizes the natural loss of luminescence of the negative control, IC_t is the luminescence intensity of the control after the contact time and IC_0 is the initial luminescence intensity of the negative control. The luminescence signal was recorded in relative light units (RLU s^{-1}).

4. Conclusions

In this work, high extent of CIP removal was obtained by applying an anaerobic treatment supplemented with CNM. CIP removal was attained either by adsorption on GS and CNM, or by a combined effect of sorption and biological removal in anaerobic conditions. The presence of CNM increased the rates of CIP removal ≈ 1.5-fold, highlighting the potential of these nanomaterials to improve the efficiency of the processes. The anaerobic treatment applied, both in the absence and in the presence of the CNM, caused a significant decrease in the toxicity (around 50%), with all the treated solutions being considered slightly toxic, while the initial CIP solution was toxic. Therefore, the use of CNM may be advantageous to increase the removal efficiency of this pharmaceutical compound, and still water detoxification. The application of magnetic nanomaterials, like CNT@2%Fe, facilitates their separation and removal from the process after treatment, by applying a magnetic field, which is an advantage relatively to soluble and other insoluble materials. Furthermore, this magnetic CNM, maintained its good catalytic properties over three treatment cycles, demonstrating its recycling and reusability in anaerobic systems.

Supplementary Materials: The following are available online at https://www.mdpi.com/1422-0067/22/6/2932/s1.

Author Contributions: Conceptualization, L.P.; methodology, A.R.S., C.S.N.B. and O.S.G.P.S.; software, A.R.S., C.S.N.B. and O.S.G.P.S.; validation, L.P., M.M.A., A.J.C. and M.F.R.P.; formal analysis, A.R.S., C.S.N.B. and O.S.G.P.S.; investigation, A.R.S.; resources, L.P., M.M.A. and M.F.R.P.; data curation, A.R.S., C.S.N.B. and A.F.S.; writing—original draft preparation, A.R.S. and L.P.; writing—review and editing,. Cavaleiro, O.S.G.P.S., A.F.S., M.F.R.P., M.M.A. and L.P.; visualization, M.M.A. and L.P.; supervision, M.M.A. and L.P.; project administration, M.M.A.; L.P.; funding acquisition, M.M.A. and M.F.R.P. All authors have read and agreed to the published version of the manuscript.

Funding: This study was supported by the Portuguese Foundation for Science and Technology (FCT) under the scope of the strategic funding of UIDB/04469/2020 unit and BioTecNorte operation

(NORTE-01-0145-FEDER-000004) funded by the European Regional Development Fund under the scope of Norte2020-Programa Operacional Regional do Norte. Ana Rita Silva holds an FCT grant SFRH/BD/131905/2017. Cátia S.N. Braga holds an FCT grant SFRH/BD/132003/2017. This work was also financially supported by: Base Funding-UIDB/50020/2020 of the Associate Laboratory LSRE-LCM-funded by national funds through FCT/MCTES (PIDDAC). OSGPS acknowledges FCT funding under the Scientific Employment Stimulus-Institutional Call CEECINST/00049/2018.References.

Institutional Review Board Statement: Not applicable.

Informed Consent Statement: Not applicable.

Data Availability Statement: The data that support the findings of this study are available from the corresponding author upon reasonable request.

Conflicts of Interest: The authors declare no conflict of interest.

References

1. Carvalho, R.N.; Ceriani, L.; Ippolito, A. *Development of the first watch list under the environmental quality standards directive water policy*; Publications Office of the European Union: Luxembourg, 2015; ISBN 9789279462009.
2. Barbosa, M.O.; Moreira, N.F.F.; Ribeiro, A.R.; Pereira, M.F.R.; Silva, A.M.T. Occurrence and removal of organic micropollutants: An overview of the watch list of EU Decision 2015/495. *Water Res.* **2016**, *94*, 257–279. [CrossRef]
3. Amorim, C.L.; Moreira, I.S.; Maia, A.S.; Tiritan, M.E.; Castro, P.M.L. Biodegradation of ofloxacin, norfloxacin, and ciprofloxacin as single and mixed substrates by Labrys portucalensis F11. *Appl. Microbiol. Biotechnol.* **2014**, *98*, 3181–3190. [CrossRef] [PubMed]
4. Pereira, A.M.P.T.; Silva, L.J.G.; Meisel, L.M.; Lino, C.M.; Pena, A. Environmental impact of pharmaceuticals from Portuguese wastewaters: Geographical and seasonal occurrence, removal and risk assessment. *Environ. Res.* **2015**, *136*, 108–119. [CrossRef] [PubMed]
5. Rusu, A.; Hancu, G.; Uivaroşi, V. Fluoroquinolone pollution of food, water and soil, and bacterial resistance. *Environ. Chem. Lett.* **2015**, *13*, 21–36. [CrossRef]
6. Manaia, C.M.; Novo, A.; Coelho, B.; Nunes, O.C. Ciprofloxacin resistance in domestic wastewater treatment plants. *Water Air Soil Pollut.* **2010**, *208*, 335–343. [CrossRef]
7. Jameel, Y.; Valle, D.; Kay, P. Spatial variation in the detection rates of frequently studied pharmaceuticals in Asian, European and North American rivers. *Sci. Total Environ.* **2020**, *724*, 137947. [CrossRef]
8. Ramírez-Morales, D.; Masís-Mora, M.; Montiel-Mora, J.R.; Cambronero-Heinrichs, J.C.; Briceño-Guevara, S.; Rojas-Sánchez, C.E.; Méndez-Rivera, M.; Arias-Mora, V.; Tormo-Budowski, R.; Brenes-Alfaro, L.; et al. Occurrence of pharmaceuticals, hazard assessment and ecotoxicological evaluation of wastewater treatment plants in Costa Rica. *Sci. Total Environ.* **2020**, *746*, 141200. [CrossRef]
9. Larsson, D.G.J.; De Pedro, C.; Paxeus, N. Effluent from drug manufactures contains extremely high levels of pharmaceuticals. *J. Hazard Mater.* **2007**, *148*, 751–755. [CrossRef]
10. Zhang, H.; Khanal, S.K.; Jia, Y.; Song, S.; Lu, H. Fundamental insights into ciprofloxacin adsorption by sulfate-reducing bacteria sludge: Mechanisms and thermodynamics. *Chem. Eng. J.* **2019**, *378*, 122103. [CrossRef]
11. Yan, B.; Niu, C.H. Modeling and site energy distribution analysis of levofloxacin sorption by biosorbents. *Chem. Eng. J.* **2017**, *307*, 631–642. [CrossRef]
12. Vieno, N.M.; Härkki, H.; Tuhkanen, T.; Kronberg, L. Occurrence of pharmaceuticals in river water and their elimination in a pilot-scale drinking water treatment plant. *Environ. Sci. Technol.* **2007**, *41*, 5077–5084. [CrossRef]
13. Lindberg, R.H.; Olofsson, U.; Rendahl, P.; Johansson, M.I.; Tysklind, M.; Andersson, B.A.V. Behavior of fluoroquinolones and trimethoprim during mechanical, chemical, and active sludge treatment of sewage water and digestion of sludge. *Environ. Sci. Technol.* **2006**, *40*, 1042–1048. [CrossRef]
14. Kim, S.; Eichhorn, P.; Jensen, J.N.; Weber, A.S.; Aga, D.S. Removal of antibiotics in wastewater: Effect of hydraulic and solid retention times on the fate of tetracycline in the activated sludge process. *Environ. Sci. Technol.* **2005**, *39*, 5816–5823. [CrossRef]
15. Liao, X.; Li, B.; Zou, R.; Dai, Y.; Xie, S.; Yuan, B. Biodegradation of antibiotic ciprofloxacin: Pathways, influential factors, and bacterial community structure. *Environ. Sci. Pollut. Res.* **2016**, *23*, 7911–7918. [CrossRef] [PubMed]
16. Silva, A.R.; Gomes, J.C.; Salvador, A.F.; Martins, G.; Alves, M.M.; Pereira, L. Ciprofloxacin, diclofenac, ibuprofen and 17α-ethinylestradiol differentially affect the activity of acetogens and methanogens in anaerobic communities. *Ecotoxicology* **2020**, *29*, 866–875. [CrossRef]
17. Liu, Z.; Sun, P.; Pavlostathis, S.G.; Zhou, X.; Zhang, Y. Inhibitory effects and biotransformation potential of ciprofloxacin under anoxic/anaerobic conditions. *Bioresour. Technol.* **2013**, *150*, 28–35. [CrossRef] [PubMed]
18. Musson, S.E.; Campo, P.; Tolaymat, T.; Suidan, M.; Townsend, T.G. Assessment of the anaerobic degradation of six active pharmaceutical ingredients. *Sci. Total Environ.* **2010**, *408*, 2068–2074. [CrossRef] [PubMed]
19. Butkovskyi, A.; Hernandez Leal, L.; Rijnaarts, H.H.M.; Zeeman, G. Fate of pharmaceuticals in full-scale source separated sanitation system. *Water Res.* **2015**, *85*, 384–392. [CrossRef] [PubMed]

20. Gartiser, S.; Urich, E.; Alexy, R.; Kümmerer, K. Anaerobic inhibition and biodegradation of antibiotics in ISO test schemes. *Chemosphere* **2007**, *66*, 1839–1848. [CrossRef] [PubMed]
21. Lu, M.; Niu, X.; Liu, W.; Zhang, J.; Wang, J.; Yang, J.; Wang, W.; Yang, Z. Biogas generation in anaerobic wastewater treatment under tetracycline antibiotic pressure. *Sci. Rep.* **2016**, *6*, 28336. [CrossRef] [PubMed]
22. de Souza Santos, L.V.; Teixeira, D.C.; Jacob, R.S.; Amaral, M.C.S.; do Lange, L.C. Evaluation of the aerobic and anaerobic biodegradability of the antibiotic norfloxacin. *Water Sci. Technol.* **2014**, *70*, 265–271. [CrossRef]
23. Alvarino, T.; Suárez, S.; Garrido, M.; Lema, J.M.; Omil, F. A UASB reactor coupled to a hybrid aerobic MBR as innovative plant configuration to enhance the removal of organic micropollutants. *Chemosphere* **2016**, *144*, 452–458. [CrossRef]
24. van der Zee, F.P.; Bouwman, R.H.M.; Strik, D.P.B.T.B.; Lettinga, G.; Field, J.A. Application of redox mediators to accelerate the transformation of reactive azo dyes in anaerobic bioreactors. *Biotechnol. Bioeng.* **2001**, *75*, 691–701. [CrossRef]
25. van der Zee, F.P.; Cervantes, F.J. Impact and application of electron shuttles on the redox (bio)transformation of contaminants: A review. *Biotechnol. Adv.* **2009**, *27*, 256–277. [CrossRef] [PubMed]
26. Pereira, L.; Pereira, R.; Pereira, M.F.R.; van der Zee, F.P.; Cervantes, F.J.; Alves, M.M. Thermal modification of activated carbon surface chemistry improves its capacity as redox mediator for azo dye reduction. *J. Hazard. Mater.* **2010**, *183*, 931–939. [CrossRef]
27. Pereira, R.A.; Pereira, M.F.R.; Alves, M.M.; Pereira, L. Carbon based materials as novel redox mediators for dye wastewater biodegradation. *Appl. Catal. B Environ.* **2014**, *144*, 713–720. [CrossRef]
28. Pereira, R.A.; Salvador, A.F.; Dias, P.; Pereira, M.F.R.; Alves, M.M.; Pereira, L. Perspectives on carbon materials as powerful catalysts in continuous anaerobic bioreactors. *Water Res.* **2016**, *101*, 441–447. [CrossRef] [PubMed]
29. Pereira, L.; Pereira, R.; Pereira, M.F.R.; Alves, M.M. Effect of different carbon materials as electron shuttles in the anaerobic biotransformation of nitroanilines. *Biotechnol. Bioeng.* **2016**, *113*, 1194–1202. [CrossRef]
30. Pereira, L.; Dias, P.; Soares, O.S.G.P.; Ramalho, P.S.F.; Pereira, M.F.R.; Alves, M.M. Synthesis, characterization and application of magnetic carbon materials as electron shuttles for the biological and chemical reduction of the azo dye Acid Orange 10. *Appl. Catal. B Environ.* **2017**, *212*, 175–184. [CrossRef]
31. van der Zee, F.P.; Bisschops, I.A.E.; Lettinga, G.; Field, J.A. Activated carbon as an electron acceptor and redox mediator during the anaerobic biotransformation of azo dyes. *Environ. Sci. Technol.* **2003**, *37*, 402–408. [CrossRef] [PubMed]
32. Wu, Y.; Wang, S.; Liang, D.; Li, N. Conductive materials in anaerobic digestion: From mechanism to application. *Bioresour. Technol.* **2020**, *298*, 122403. [CrossRef]
33. Ji, Z.; Shen, X.; Yue, X.; Zhou, H.; Yang, J.; Wang, Y.; Ma, L.; Chen, K. Facile synthesis of magnetically separable reduced graphene oxide/magnetite/silver nanocomposites with enhanced catalytic activity. *J. Colloid Interface Sci.* **2015**, *459*, 79–85. [CrossRef] [PubMed]
34. Toral-Sánchez, E.; Rangel-Mendez, J.R.; Hurt, R.H.; Ascacio Valdés, J.A.; Aguilar, C.N.; Cervantes, F.J. Novel application of magnetic nano-carbon composite as redox mediator in the reductive biodegradation of iopromide in anaerobic continuous systems. *Appl. Microbiol. Biotechnol.* **2018**, *102*, 8951–8961. [CrossRef] [PubMed]
35. Gonçalves, A.G.; Figueiredo, J.L.; Órfão, J.J.M.; Pereira, M.F.R. Influence of the surface chemistry of multi-walled carbon nanotubes on their activity as ozonation catalysts. *Carbon N. Y.* **2010**, *48*, 4369–4381. [CrossRef]
36. Rocha, R.P.; Soares, O.S.G.P.; Gonçalves, A.G.; Órfão, J.J.M.; Pereira, M.F.R.; Figueiredo, J.L. Different methodologies for synthesis of nitrogen doped carbon nanotubes and their use in catalytic wet air oxidation. *Appl. Catal. A Gen.* **2017**, *548*, 62–70. [CrossRef]
37. Silva, A.R.; Soares, O.S.G.P.; Pereira, M.F.R.; Alves, M.M.; Pereira, L. Tailoring carbon nanotubes to enhance their efficiency as electron shuttle on the biological removal of acid orange 10 under anaerobic conditions. *Nanomaterials* **2020**, *10*, 2496. [CrossRef]
38. Rivera-Utrilla, J.; Sánchez-Polo, M.; Gómez-Serrano, V.; Álvarez, P.M.; Alvim-Ferraz, M.C.M.; Dias, J.M. Activated carbon modifications to enhance its water treatment applications. An overview. *J. Hazard. Mater.* **2011**, *187*, 1–23. [CrossRef]
39. Amezquita-Garcia, H.J.; Razo-Flores, E.; Cervantes, F.J.; Rangel-Mendez, J.R. Activated carbon fibers as redox mediators for the increased reduction of nitroaromatics. *Carbon N. Y.* **2013**, *55*, 276–284. [CrossRef]
40. Amezquita-Garcia, H.J.; Rangel-Mendez, J.R.; Cervantes, F.J.; Razo-Flores, E. Activated carbon fibers with redox-active functionalities improves the continuous anaerobic biotransformation of 4-nitrophenol. *Chem. Eng. J.* **2016**, *286*, 208–215. [CrossRef]
41. Gonzalez-Gil, L.; Papa, M.; Feretti, D.; Ceretti, E.; Mazzoleni, G.; Steimberg, N.; Pedrazzani, R.; Bertanza, G.; Lema, J.M.; Carballa, M. Is anaerobic digestion effective for the removal of organic micropollutants and biological activities from sewage sludge? *Water Res.* **2016**, *102*, 211–220. [CrossRef]
42. Li, H.; Zhang, D.; Han, X.; Xing, B. Adsorption of antibiotic ciprofloxacin on carbon nanotubes: pH dependence and thermodynamics. *Chemosphere* **2014**, *95*, 150–155. [CrossRef]
43. Bizi, M.; El Bachra, F.E. Evaluation of the ciprofloxacin adsorption capacity of common industrial minerals and application to tap water treatment. *Powder Technol.* **2020**, *362*, 323–333. [CrossRef]
44. Genç, N.; Dogan, E.C.; Yurtsever, M. Bentonite for ciprofloxacin removal from aqueous solution. *Water Sci. Technol.* **2013**, *68*, 848–856. [CrossRef] [PubMed]
45. Li, X.; Chen, S.; Fan, X.; Quan, X.; Tan, F.; Zhang, Y.; Gao, J. Adsorption of ciprofloxacin, bisphenol and 2-chlorophenol on electrospun carbon nanofibers: In comparison with powder activated carbon. *J. Colloid Interface Sci.* **2015**, *447*, 120–127. [CrossRef]
46. Chen, Y.; Lan, T.; Duan, L.; Wang, F.; Zhao, B. Adsorptive Removal and Adsorption Kinetics of Fluoroquinolone by Nano-Hydroxyapatite. *PLoS ONE* **2015**, *10*, 1–13. [CrossRef] [PubMed]

47. Carabineiro, S.A.C.; Thavorn-amornsri, T.; Pereira, M.F.R.; Serp, P.; Figueiredo, J.L. Comparison between activated carbon, carbon xerogel and carbon nanotubes for the adsorption of the antibiotic ciprofloxacin. *Catal. Today* **2012**, *186*, 29–34. [CrossRef]
48. Pan, B.; Xing, B. Adsorption mechanisms of organic chemicals on carbon nanotubes. *Environ. Sci. Technol.* **2008**, *42*, 9005–9013. [CrossRef] [PubMed]
49. Veclani, D.; Melchior, A. Adsorption of ciprofloxacin on carbon nanotubes: Insights from molecular dynamics simulations. *J. Mol. Liq.* **2020**, *298*, 111977. [CrossRef]
50. Ma, Y.; Yang, L.; Wu, L.; Li, P.; Qi, X.; He, L.; Cui, S.; Ding, Y.; Zhang, Z. Carbon nanotube supported sludge biochar as an efficient adsorbent for low concentrations of sulfamethoxazole removal. *Sci. Total Environ.* **2020**, *718*, 137299. [CrossRef]
51. Soares, O.S.G.P.; Rocha, R.P.; Gonçalves, A.G.; Figueiredo, J.L.; Órfão, J.J.M.; Pereira, M.F.R. Easy method to prepare N-doped carbon nanotubes by ball milling. *Carbon N. Y.* **2015**, *91*, 114–121. [CrossRef]
52. Peng, H.; Pan, B.; Wu, M.; Liu, Y.; Zhang, D.; Xing, B. Adsorption of ofloxacin and norfloxacin on carbon nanotubes: Hydrophobicity- and structure-controlled process. *J. Hazard. Mater.* **2012**, *233–234*, 89–96. [CrossRef] [PubMed]
53. Jia, Y.; Khanal, S.K.; Zhang, H.; Chen, G.-H.; Lu, H. Sulfamethoxazole degradation in anaerobic sulfate-reducing bacteria sludge system. *Water Res.* **2017**, *119*, 12–20. [CrossRef] [PubMed]
54. Martins, M.; Sanches, S.; Pereira, I.A.C. Anaerobic biodegradation of pharmaceutical compounds: New insights into the pharmaceutical-degrading bacteria. *J. Hazard. Mater.* **2018**, *357*, 289–297. [CrossRef] [PubMed]
55. Yang, Z.; Xu, X.; Dai, M.; Wang, L.; Shi, X.; Guo, R. Accelerated ciprofloxacin biodegradation in the presence of magnetite nanoparticles. *Chemosphere* **2017**, *188*, 168–173. [CrossRef] [PubMed]
56. Aquino, S.F.; Chernicharo, C.A.L. Build up of volatile fatty acids (VFA) in anaerobic reactors under stress conditions: Causes and control strategies. *Eng. Sanit. Ambient* **2005**, *10*, 152–161. [CrossRef]
57. Deublein, D.; Steinhauser, A. *Biogas from Waste and Renewable Resources: An Introduction*; Wiley-VCH Verlag GmbH & Co. KGaA: Weinheim, Germany, 2008; Volume 91, ISBN 9783527318414.
58. Salvador, A.F.; Martins, G.; Melle-Franco, M.; Serpa, R.; Stams, A.J.M.; Cavaleiro, A.J.; Pereira, M.A.; Alves, M.M. Carbon nanotubes accelerate methane production in pure cultures of methanogens and in a syntrophic coculture. *Environ. Microbiol.* **2017**, *19*, 2727–2739. [CrossRef]
59. Wu, C.; Spongberg, A.L.; Witter, J.D. Sorption and biodegradation of selected antibiotics in biosolids. *J. Environ. Sci. Health Part A* **2009**, *44*, 454–461. [CrossRef]
60. Zazouli, M.A.; Susanto, H.; Nasseri, S.; Ulbricht, M. Influences of solution chemistry and polymeric natural organic matter on the removal of aquatic pharmaceutical residuals by nanofiltration. *Water Res.* **2009**, *43*, 3270–3280. [CrossRef]
61. Martín, J.; Camacho-Muñoz, D.; Santos, J.L.; Aparicio, I.; Alonso, E. Occurrence of pharmaceutical compounds in wastewater and sludge from wastewater treatment plants: Removal and ecotoxicological impact of wastewater discharges and sludge disposal. *J. Hazard. Mater.* **2012**, *239–240*, 40–47. [CrossRef]
62. Ndoun, M.C.; Elliott, H.A.; Preisendanz, H.E.; Williams, C.F.; Knopf, A.; Watson, J.E. Adsorption of pharmaceuticals from aqueous solutions using biochar derived from cotton gin waste and guayule bagasse. *Biochar* **2020**. [CrossRef]
63. Carabineiro, S.A.C.; Thavorn-Amornsri, T.; Pereira, M.F.R.; Figueiredo, J.L. Adsorption of ciprofloxacin on surface-modified carbon materials. *Water Res.* **2011**, *45*, 4583–4591. [CrossRef] [PubMed]
64. Ferreira, V.R.A.; Amorim, C.L.; Cravo, S.M.; Tiritan, M.E.; Castro, P.M.L.; Afonso, C.M.M. Fluoroquinolones biosorption onto microbial biomass: Activated sludge and aerobic granular sludge. *Int. Biodeterior. Biodegrad.* **2016**, *110*, 53–60. [CrossRef]
65. Sahmoune, M.N. Evaluation of thermodynamic parameters for adsorption of heavy metals by green adsorbents. *Environ. Chem. Lett.* **2019**, *17*, 697–704. [CrossRef]
66. Yin, J.; Cui, C.; Chen, H.; Duoni; Yu, X.; Qian, W. The Application of carbon nanotube/graphene-based nanomaterials in wastewater treatment. *Small* **2020**, *16*, 1902301. [CrossRef] [PubMed]
67. Balarak, D.; McKay, G. Utilization of MWCNTs/Al$_2$O$_3$ as adsorbent for ciprofloxacin removal: Equilibrium, kinetics and thermodynamic studies. *J. Environ. Sci. Health Part A* **2021**, *56*, 324–333. [CrossRef]
68. Liu, X.; Lu, S.; Liu, Y.; Meng, W.; Zheng, B. Adsorption of sulfamethoxazole (SMZ) and ciprofloxacin (CIP) by humic acid (HA): Characteristics and mechanism. *RSC Adv.* **2017**, *7*, 50449–50458. [CrossRef]
69. Sharifpour, N.; Moghaddam, F.M.; Mardani, G.; Malakootian, M. Evaluation of the activated carbon coated with multiwalled carbon nanotubes in removal of ciprofloxacin from aqueous solutions. *Appl. Water Sci.* **2020**, *10*, 140. [CrossRef]
70. Unuabonah, E.I.; Omorogie, M.O.; Oladoja, N.A. 5—Modeling in adsorption: Fundamentals and applications. In *Composite Nanoadsorbents*; Kyzas, G.Z., Mitropoulos, A.C., Eds.; Micro and Nano Technologies; Elsevier Inc.: Oxford, UK, 2019; pp. 85–118. ISBN 978-0-12-814132-8.
71. Li, B.; Zhang, T. Biodegradation and adsorption of antibiotics in the activated sludge process. *Environ. Sci. Technol.* **2010**, *44*, 3468–3473. [CrossRef]
72. Qiu, L.-Q.; Zhang, L.; Tang, K.; Chen, G.; Kumar Khanal, S.; Lu, H. Removal of sulfamethoxazole (SMX) in sulfate-reducing flocculent and granular sludge systems. *Bioresour. Technol.* **2019**, *288*, 121592. [CrossRef] [PubMed]
73. Silva, A.R.; Martins, P.M.; Teixeira, S.; Carabineiro, S.A.C.; Kuehn, K.; Cuniberti, G.; Alves, M.M.; Lanceros-Mendez, S.; Pereira, L. Ciprofloxacin wastewater treated by UVA photocatalysis: Contribution of irradiated TiO$_2$ and ZnO nanoparticles on the final toxicity as assessed by *Vibrio fischeri*. *RSC Adv.* **2016**, *6*. [CrossRef]

74. Arias, L.R.; Yang, L. Inactivation of Bacterial Pathogens by Carbon Nanotubes in Suspensions. *Langmuir* **2009**, *25*, 3003–3012. [CrossRef] [PubMed]
75. Mendonça, E.; Picado, A.; Paixão, S.M.; Silva, L.; Cunha, M.A.; Leitão, S.; Moura, I.; Cortez, C.; Brito, F. Ecotoxicity tests in the environmental analysis of wastewater treatment plants: Case study in Portugal. *J. Hazard. Mater.* **2009**, *163*, 665–670. [CrossRef]
76. Fernández-Alba, A.R.; Hernando, M.D.; Piedra, L.; Chisti, Y. Toxicity evaluation of single and mixed antifouling biocides measured with acute toxicity bioassays. *Anal. Chim. Acta* **2002**, *456*, 303–312. [CrossRef]
77. Auffan, M.; Achouak, W.; Rose, J.; Roncato, M.-A.; Chanéac, C.; Waite, D.T.; Masion, A.; Woicik, J.C.; Wiesner, M.R.; Bottero, J.-Y. Relation between the redox state of iron-based nanoparticles and their cytotoxicity toward *Escherichia coli*. *Environ. Sci. Technol.* **2008**, *42*, 6730–6735. [CrossRef] [PubMed]
78. Lee, C.; Kim, J.Y.; Lee, W.I.; Nelson, K.L.; Yoon, J.; Sedlak, D.L. Bactericidal effect of zero-valent iron nanoparticles on *Escherichia coli*. *Environ. Sci. Technol.* **2008**, *42*, 4927–4933. [CrossRef] [PubMed]
79. Angelidaki, I.; Alves, M.; Bolzonella, D.; Borzacconi, L.; Campos, J.; Guwy, A.; Kalyuzhnyi, S.; Jenicek, P. Defining the biomethane potential (BMP) of solid organic wastes and energy crops: A proposed protocol for batch assays. *Water Sci. Technol.* **2009**, *59*, 927–934. [CrossRef] [PubMed]
80. Tessonnier, J.-P.; Rosenthal, D.; Hansen, T.W.; Hess, C.; Schuster, M.E.; Blume, R.; Girgsdies, F.; Pfänder, N.; Timpe, O.; Su, D.S.; et al. Analysis of the structure and chemical properties of some commercial carbon nanostructures. *Carbon N. Y.* **2009**, *47*, 1779–1798. [CrossRef]
81. Salvador, A.F.; Cavaleiro, A.J.; Paulo, A.M.S.; Silva, S.A.; Guedes, A.P.; Pereira, M.A.; Stams, A.J.M.; Sousa, D.Z.; Alves, M.M. Inhibition studies with 2-bromoethanesulfonate reveal a novel syntrophic relationship in anaerobic oleate degradation. *Appl. Environ. Microbiol.* **2019**, *85*. [CrossRef]
82. Lane, D.J. 16S/23S rRNA sequencing. In *Nucleic Acid Techniques in Bacterial Systematics*; Stackebrandt, E., Goodfellow, M., Eds.; John Wiley and Sons: New York, NY, USA, 1991.
83. Caporaso, J.G.; Lauber, C.L.; Walters, W.A.; Berg-Lyons, D.; Lozupone, C.A.; Turnbaugh, P.J.; Fierer, N.; Knight, R. Global patterns of 16S rRNA diversity at a depth of millions of sequences per sample. *Proc. Natl. Acad. Sci. USA* **2011**, *108*, 4516–4522. [CrossRef]
84. Rodriguez-Reinoso, F.; Martin-Martinez, J.M.; Prado-Burguete, C.; McEnaney, B. A standard adsorption isotherm for the characterization of activated carbons. *J. Phys. Chem.* **1987**, *91*, 515–516. [CrossRef]
85. Figueiredo, J.L.; Pereira, M.F.R. The role of surface chemistry in catalysis with carbons. *Catal. Today* **2010**, *150*, 2–7. [CrossRef]
86. APHA. *Standard Methods for the Examination of Water and Wastewater*; American Public Health Association: Washington, DC, USA, 1999.
87. ISO 11348-1 International Standard. *Water quality—Determination of the Inhibitory Effect of Water Samples on the Light Emission of Vibrio Fischeri (Luminescent Bacteria Test)—Part1: Method Using Freshly Prepared Bacteria*; ISO-International Organization for Standardization: Geneva, Switzerland, 1998.
88. Heinlaan, M.; Ivask, A.; Blinova, I.; Dubourguier, H.-C.; Kahru, A. Toxicity of nanosized and bulk ZnO, CuO and TiO_2 to bacteria *Vibrio fischeri* and crustaceans *Daphnia magna* and *Thamnocephalus platyurus*. *Chemosphere* **2008**, *71*, 1308–1316. [CrossRef] [PubMed]
89. ISO 11348-3 International Standard. *Water Quality—Determination of the Inhibitory Effect of Water Samples on the Light Emission of Vibrio Fischeri (Luminescent Bacteria Test)—Part3: Method Using Freeze-Dried Bacteria*; ISO-International Organization for Standardization: Geneva, Switzerland, 2007.

Article

Reduced Graphene Oxides Modulate the Expression of Cell Receptors and Voltage-Dependent Ion Channel Genes of Glioblastoma Multiforme

Jaroslaw Szczepaniak [1], Joanna Jagiello [2], Mateusz Wierzbicki [1], Dorota Nowak [3], Anna Sobczyk-Guzenda [4], Malwina Sosnowska [1], Slawomir Jaworski [1], Karolina Daniluk [1], Maciej Szmidt [5], Olga Witkowska-Pilaszewicz [6], Barbara Strojny-Cieslak [1,*] and Marta Grodzik [1,*]

[1] Department of Nanobiotechnology, Institute of Biology, Warsaw University of Life Sciences, 02-787 Warsaw, Poland; jaroslaw_szczepaniak@sggw.edu.pl (J.S.); mateusz_wierzbicki@sggw.edu.pl (M.W.); malwina_sosnowska@sggw.edu.pl (M.S.); slawomir_jaworski@sggw.edu.pl (S.J.); karolina_daniluk@sggw.edu.pl (K.D.)
[2] Department of Chemical Synthesis and Flake Graphene, Łukasiewicz Research Network–Institute of Microelectronics and Photonics, 01-919 Warsaw, Poland; joanna.jagiello@itme.edu.pl
[3] Tricomed SA, 5/9 Swientojanska St., 93-493 Lodz, Poland; nowak.dora@gmail.com
[4] Institute of Materials Science and Engineering, Lodz University of Technology, 90-924 Lodz, Poland; anna.sobczyk-guzenda@p.lodz.pl
[5] Department of Morphologic Sciences, Institute of Veterinary Medicine, Warsaw University of Life Sciences, 02-787 Warsaw, Poland; maciej_szmidt@sggw.edu.pl
[6] Department of Pathology and Veterinary Diagnostics, Institute of Veterinary Medicine, Warsaw University of Life Sciences, 02-787 Warsaw, Poland; olga_witkowska_pilaszewicz@sggw.edu.pl
* Correspondence: barbara_strojny@sggw.edu.pl (B.S.-C.); marta_grodzik@sggw.edu.pl (M.G.)

Citation: Szczepaniak, J.; Jagiello, J.; Wierzbicki, M.; Nowak, D.; Sobczyk-Guzenda, A.; Sosnowska, M.; Jaworski, S.; Daniluk, K.; Szmidt, M.; Witkowska-Pilaszewicz, O.; et al. Reduced Graphene Oxides Modulate the Expression of Cell Receptors and Voltage-Dependent Ion Channel Genes of Glioblastoma Multiforme. *Int. J. Mol. Sci.* **2021**, *22*, 515. https://doi.org/10.3390/ijms22020515

Received: 8 December 2020
Accepted: 4 January 2021
Published: 6 January 2021

Publisher's Note: MDPI stays neutral with regard to jurisdictional claims in published maps and institutional affiliations.

Copyright: © 2021 by the authors. Licensee MDPI, Basel, Switzerland. This article is an open access article distributed under the terms and conditions of the Creative Commons Attribution (CC BY) license (https://creativecommons.org/licenses/by/4.0/).

Abstract: The development of nanotechnology based on graphene and its derivatives has aroused great scientific interest because of their unusual properties. Graphene (GN) and its derivatives, such as reduced graphene oxide (rGO), exhibit antitumor effects on glioblastoma multiforme (GBM) cells in vitro. The antitumor activity of rGO with different contents of oxygen-containing functional groups and GN was compared. Using FTIR (fourier transform infrared) analysis, the content of individual functional groups (GN/exfoliation (ExF), rGO/thermal (Term), rGO/ammonium thiosulphate (ATS), and rGO/ thiourea dioxide (TUD)) was determined. Cell membrane damage, as well as changes in the cell membrane potential, was analyzed. Additionally, the gene expression of voltage-dependent ion channels (*clcn3*, *clcn6*, *cacna1b*, *cacna1d*, *nalcn*, *kcne4*, *kcnj10*, and *kcnb1*) and extracellular receptors was determined. A reduction in the potential of the U87 glioma cell membrane was observed after treatment with rGO/ATS and rGO/TUD flakes. Moreover, it was also demonstrated that major changes in the expression of voltage-dependent ion channel genes were observed in *clcn3*, *nalcn*, and *kcne4* after treatment with rGO/ATS and rGO/TUD flakes. Furthermore, the GN/ExF, rGO/ATS, and rGO/TUD flakes significantly reduced the expression of extracellular receptors (uPar, CD105) in U87 glioblastoma cells. In conclusion, the cytotoxic mechanism of rGO flakes may depend on the presence and types of oxygen-containing functional groups, which are more abundant in rGO compared to GN.

Keywords: graphene; reduced graphene oxide; glioblastoma multiforme; voltage-gated ion channel; cell membrane receptor; membrane potential

1. Introduction

Glioblastoma (GBM) is the most common primary malignant brain tumor. GBM is associated with poor prognosis and a life expectancy of approximately 15 months despite optimal therapy, which includes surgery, chemotherapy, and radiotherapy [1]. The lack of therapeutic success has been attributed to a variety of factors, including rapid infiltration of brain tumor cells, inter- and intratumor heterogeneity, limited diffusion of therapeutic drugs across the blood–brain barrier and brain parenchyma/tumor, and the presence of GBM stem cells (GSC) in the tumor, which are resistant to radiotherapy and chemotherapy and are capable of tumor formation and indefinite self-renewal [2].

Graphene is made up of a layer of carbon atoms arranged in a hexagonal pattern and consists purely of sp2 hybridized bonds. It has gained enormous interest in various fields owing to its unique electrochemical properties, which include high thermal conductivity, high current, density, chemical volume, optical transmittance, and very high hydrophobicity [3,4]. It is the simplest form of carbon and the thinnest material ever produced [5]. The graphene family includes sheets and flakes of graphene as well as graphene oxide (GO) and reduced graphene oxide (rGO) [6,7]. GO is highly hydrophilic because of the presence of a large number of oxygen groups on the surface (hydroxyl, carboxyl, epoxy). rGO has more oxygen functional groups than graphene (GN), but less than GO. Therefore, rGO is less hydrophilic than GO and, therefore, has higher electrical conductivity [8]. In addition to the numerous applications of graphene materials in electronics, their derivatives are also assessed for their use in medicine, for example, in anticancer therapy [9,10]. Our previous studies have shown that graphene and its derivatives can be cytotoxic to glioblastoma cells in vitro and in vivo. GN, rGO, and GO induce apoptosis and lead to the reduction of viability and proliferation in U87 and U118 glioblastoma cell lines [11]. Jaworski et al. also showed that graphene flakes were too large (0.45–1.5 μm) to enter the glioblastoma cells [11,12]. However, both GN and rGO can activate the mitochondrial-dependent apoptotic pathway by reducing the mitochondrial membrane potential in U87 glioma cells. On the contrary, GO can regulate the expression of mitochondrial oxidative phosphorylation (OXPHOS) genes in GBM, thus leading to a decrease in the invasion potential of cancer cells [13]. The tested materials (GN/exfoliation (ExF), rGO/thermal (Term), rGO/ammonium thiosulphate (ATS), and rGO/ thiourea dioxide (TUD)) showed no effect on the cell cycle [14]. Despite the fact that even minimal oxygen content on the surface of the flakes may reduce the proapoptotic abilities of graphene and its derivatives, oxygen-containing derivatives have a better affinity for GBM cells than pure graphene, which allows for better targeting of the intended effect [15].

An important finding seems to be the reduction of oxygen content in rGO in relation to the starting material, GO, which results in an increase of delocalized electrons on the surface of the graphene flakes. This can lead to a disruption of the signaling pathways in the plasma membrane or to direct interaction with cell structures, which are sensitive to electrochemical potential. Therefore, graphene and its derivatives are characterized by strong bioelectric properties; because of the presence of delocalized electrons and oxygen-containing functional groups on the surface of the flakes, this material can interact with structures that receive electrical signals, for example, receptors located on the cell surface, as well as proteins building voltage-dependent ion channels, ultimately causing changes in the potential of the cell membrane. Cell surface charge is a key biophysical parameter that depends on the composition of the cytoplasmic membrane and the physiological state of cells. In addition to the presence of ion channels and transporters, negative values of cell membrane potential at physiological pH values are also caused by the presence of nonionic groups in phospholipids (phosphatidylcholine; −62 mV), proteins, and their polysaccharide conjugates [16]. Direct comparisons of the in vitro and in vivo Vm levels of normal and cancer cells showed that the cancer cells were more depolarized (negative Vm) than their normal counterparts [17]. Therefore, the studied materials of graphene origin (GN and rGO) may prefer adhesion to neoplastic cells with depolarized cell membranes. Moreover, Fiorillo et al. proved that GO, characterized by the presence of many functional groups containing oxygen, preferred adhesion to U87 tumor cells rather than stem cells and normal fibroblasts [18]. Therefore, it has been hypothesized that graphene-derived flakes exert modulating and transducing effects on proteins contained in the membrane of U87 glioma cancer cells. These effects result from the unique structure and type of functional groups on the surface of graphene. GN flakes are characterized by several functional groups on the surface, and they possess a larger electron cloud than rGO.

Ion channels play an important role in the regulation of electrical excitability in normal and cancer cells [19]. Ion channels and transporters are also associated with GBM tumor growth and malignancy [20]. Genomic analysis revealed that the genes involved in Na, K, Ca transmission or transport belong to the most frequently mutated functional groups affecting GBM in 90% of the tested samples [21,22]. Therefore, the article discusses the subject of studying the expression of individual subunits that build ion channels (Na^+, K^+, Ca^{2+}, Cl^-). The functioning of ion channels and pumps influences the migration and proliferation of GBM cells. For example, deregulated K^+ and Cl^- channels regulate the osmotic drive, allowing for cell shape and volume changes that promote glioblastoma cell migration [23], and Ca-activated K^+ (BK) channels, which control glioblastoma cell growth [24]. In this study, we hypothesize that flakes of graphene (GN) and reduced graphene oxides (rGOs) may affect the expression of voltage-dependent ion channel genes Cl^- (*clcn3* and *clcn6*), Ca^{2+} (*cacna1b* and *cacna1d*), Na^+ (*nalcn*), K^+ (*kcne4*, *kcnj10*, and *kcnb1*), resulting in an alteration of the potential of the glioblastoma cell membrane. It is also suspected that GN and rGO flakes may reduce the expression of extracellular receptors and further reduce the invasiveness of glioblastoma in vitro.

The aim of this study is to determine changes in the cell membrane caused by direct or indirect contact with bioelectric flakes of rGO with different degrees of reduction compared to GN in U87 glioblastoma multiforme cells in vitro.

2. Results

2.1. Physicochemical Characterization of GN and rGO Flakes

The physicochemical properties (inter alia, transmission electron microscopy, Raman spectroscopy, atomic force microscopy, zeta potential, oxygen content, and electrical resistance) of graphene (GN) and reduced graphene oxides (rGOs) have recently been published [14]. In this study, the physicochemical properties were complemented with an additional analysis of the tested materials. FTIR analysis allowed us to obtain detailed information about the functional groups.

SEM images were used to confirm the morphology of the graphene flakes. Microphotographs of thin layers of GN and rGO graphene sheets are shown in Figure 1. In GN/ExF (a1), irregular edges of graphene flakes were observed. Photos showing rGO/Term (b1), rGO/ATS (c1), and rGO/TUD (d1) samples showed wrinkled and complex textures with rough surfaces, resulting from collapse. SEM images, taken in the AEE (active emission element) mode (Figure 1(a2,b2,c2,d2)), allowed the scanning of graphene with the table current. Owing to this, the quality of the images was good for conductive materials.

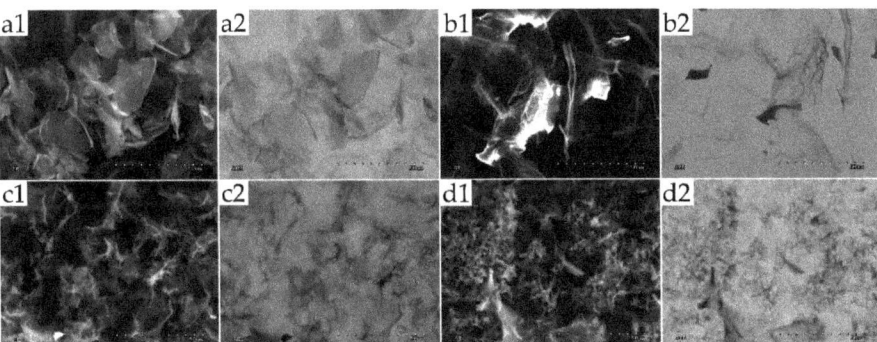

Figure 1. Morphology of GN and rGOs was examined using a scanning electron microscope (SEM). (**a1**) GN/ExF, (**b1**) rGO/Term, (**c1**) rGO/ATS, (**d1**) rGO/TUD—standard SEM images show 2D nanosheets morphologies. Scale bare: 50 μm. (**a2**) GN/ExF, (**b2**) rGO/Term, (**c2**) rGO/ATS, (**d2**) rGO/TUD—AEE mode showing conductive materials. Scale bare: 50 μm. **Abbreviations**: GN, graphene; rGO, reduced graphene oxide; SEM, scanning electron microscope; AEE, active emission element; ExF, exfoliation; Term, thermal; ATS, ammonium thiosulphate; TUD, thiourea dioxide.

The presence of functional groups was confirmed by FTIR analysis (Figure 2). Spectrum analysis showed the presence of aliphatic groups containing C-H bonds in their chemical structure. From the presented spectra, it can be concluded that the purest was graphene GN/ExF. The spectrum of this sample lacked absorption bands in the range of wave numbers between 3000–2800 cm^{-1} and 1470–1440 cm^{-1}. Graphene rGO/Term and rGO/ATS had similar amounts of these impurities. The most C-H bonds occurring in both the form of methyl and methylene groups were in graphene rGO/TUD. In the obtained spectra, bands originating mainly from O-H, C-H, C=O, and C-O bond vibrations were visible. The location and assignment of individual bands to specific bonds are shown and summarized in Table 1.

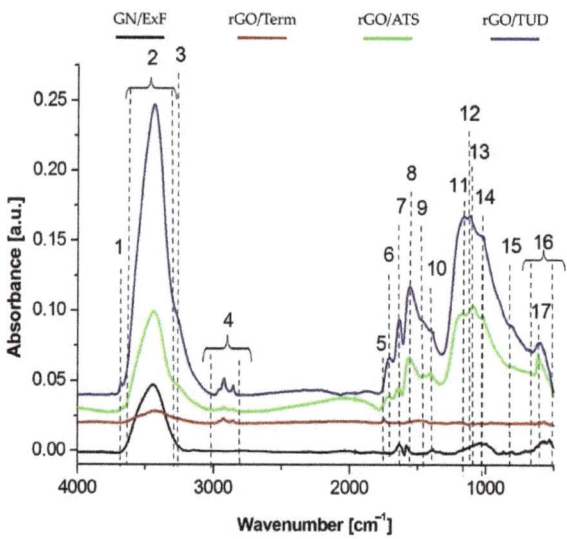

Figure 2. Fourier transform infrared (FTIR) spectra of graphene (GN/ExF; black line) and reduced graphene oxides (rGO/Term, red line; rGO/ATS, green line; rGO/TUD, blue line). The individual digital markings on the graph show the bands associated with the occurrence of individual functional groups; 1 (OH-), 2 (OH-), 3 (N-H), 4 (C-H), 5 (C=O), 6 (C=O), 7 (C=C), 8 (COO-), 9 (C-H), 10 (-OH), 11 (C-O-C), 12 (C-NH_2/C-N), 13 (C=S), 14 (carbon ring), 15 (CH_2=CH), 16 (carbon ring), 17 (C-S/S-O). **Abbreviations**: GN, graphene; rGO, reduced graphene oxide; FTIR, Fourier transform infrared; a.u., absorbance unit; ExF, exfoliation; Term, thermal; ATS, ammonium thiosulphate; TUD, thiourea dioxide.

Considering the quantity of oxidized carbon in its chemical structure as the criterion for assessing the quality of the tested material, the largest number of such connections was found in graphene rGO/TUD. Slightly less connection was observed in graphene rGO/ATS. Graphene GN/ExF had two characteristic low-intensity peaks belonging to C=O and COO- bonds, which were also present in the spectra of samples rGO/ATS and rGO/TUD. The least number of C=O bonds occurred in the graphene rGO/Term. Only one peak with a maximum at 1747 cm^{-1} appeared in the spectrum of this material. Therefore, the rGO/Term material turned out to be the most hydrophobic, which was confirmed by macroscopic observation of the behavior of graphene flakes in this solution. The sharp maximum in graphene rGO/TUD was observed at 3680 cm^{-1}. Additionally, bands from the vibration of N-H and C-NH_2 bonds appear in the spectra of graphene rGO/ATS and rGO/TUD, which were found at 3280 and 1124 cm^{-1}, respectively. The sulfur component was also incorporated into the graphene structure, giving peaks at wave numbers 1100 and 617 cm^{-1}.

Table 1. Location of absorption bands exhibited by the functional groups in FTIR spectra obtained for individual GN and rGO samples.

No.	Wavenumber [cm^{-1}]	Type of Bond	Sample			
			GN/ExF	rGO/Term	rGO/ATS	rGO/TUD
1	3680	O-H (stretch)	-	-	-	+ (w)
2	3600–3000	O-H (stretch)	+ (m)	+ (w)	+ (m)	+ (s)
3	3280	N-H (stretch)	-	-	+ (m)	+ (m)
4	3000–2800	C-H (stretch)	-	+ (m)	+ (m)	+ (s)
5	1747	C=O (stretch)	-	+ (w)	-	-
6	1711	C=O (stretch)	-	-	+ (m)	+ (s)
7	1640	C=C (stretch)	+ (m)	-	+ (m)	+ (s)
8	1563	COO- (stretch)	+ (w)	-	+ (m)	+ (s)
9	1465	C-H (deformation)	-	+ (w)	+ (w)	+ (m)
10	1400	-OH (in-plane bendig)	+ (w)	+ (w)	+ (w)	+ (s)
11	1174	C-O-C (stretch)	+ (w)	-	+ (m)	+ (s)
12	1124	C-NH$_2$/C-N (stretch)	-	-	+ (m)	+ (w)
13	1100	C=S (stretch)	-	-	+ (w)	+ (w)
14	1018	Carbon ring	-	+ (w)	+(m)	+(m)
15	800	CH$_2$=CH (deformation)	-	-	+ (w)	+ (m)
16	659–511	carbon ring	+ (w)	+ (m)	+ (m)	+ (m)
17	617	C-S (stretch)/S-O (scissors)	-	-	+ (s)	+ (m)

Notes: designations used in the table: w—weak; m—medium; s—strong; "+"—current; "-"—absent. Abbreviations: GN, graphene; rGO, reduced graphene oxide; ExF, exfoliation; Term, thermal; ATS, ammonium thiosulphate; TUD, thiourea dioxide.

2.2. Membrane Potential

In the next stage, changes in cell membrane potential were analyzed. The results are presented in the form of line graphs depicting changes in cell membrane potential (Figure 3). In glioblastoma U87 cells, a statistically significant decrease in cell membrane potential was observed only after treatment with rGO/ATS and rGO/TUD flakes. The first potential peak (decrease; 516.7-563 RFU (relative fluorescence units)) after rGO/TUD treatment was observed after 2 and 4 h ($p < 0.05$). Maximum potential reduction after rGO/TUD treatment (714.7 RFU) occurred after 24 h of treatment ($p < 0.05$). In U87 cells after rGO/ATS treatment (567 RFU), the maximum reduction in cell membrane potential was observed after 22 h. The strongest reduction was observed after 18 h for both rGO/ATS and rGO/TUD ($p < 0.05$). In Hs5 cells, a significant reduction in cell membrane potential was observed only for rGO/ATS flakes (maximum 297.7 RFU after 4 h of treatment; p-value <0.05). The strongest reduction was observed after 8 h of rGO/ATS flake treatment ($p < 0.05$). Greater changes in the potential were observed in cancer cells (421 RFU, 561.7 RFU) than in healthy cells (258.4 RFU, 384.3 RFU) after rGO/ATS and rGO/TUD treatment.

2.3. Gene Expression of Voltage-Gated Ion Channels

Analysis of the expression of genes involved in the transport of chlorine, calcium, sodium, and potassium voltage-dependent ions was performed. The results are shown in Figure 4 and Table S1. Gene expression that was studied using real-time PCR was focused on the genes coding the individual subunits of proteins that cocreate ion channels, responsible for the transport of the aforementioned ions.

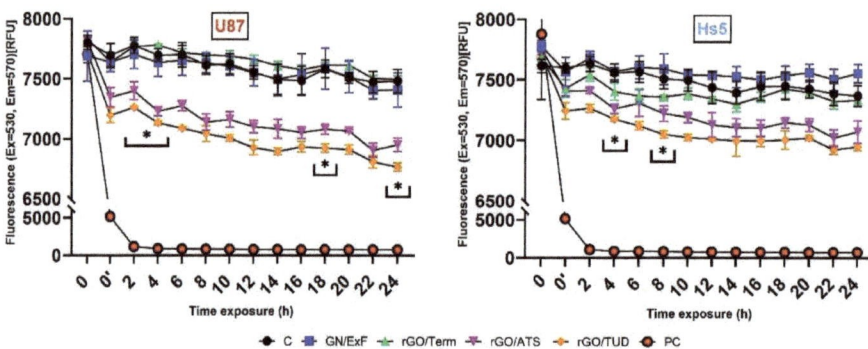

Figure 3. Effect of graphene flakes (GN/ExF) and three types of reduced graphene oxide flakes (rGO/Term, rGO/ATS, and rGO/TUD) on the membrane potential of U87 and Hs5 cell lines. The immediate reading after the addition is marked as 0 on the x-axis of the membrane potential sensor. PC represents the group after H_2O_2 (hydrogen peroxide) treatment as positive control. The reading after 30 min incubation with a membrane potential sensor is marked as 0′ on the x-axis. The readings were taken for 24 h at 2 h intervals. Values are expressed as mean ± standard deviation. Statistical significance between the control and treated cells is indicated by an asterisk and was assessed using Bonferroni's multiple comparisons test. Differences with p-value < 0.05 were considered significant. One asterisk (*) means p-value < 0.05. **Abbreviations**: rGO, reduced graphene oxide; GN, graphene; C, control group (untreated group); PC, positive control; h, hours; ExF, exfoliation; Term, thermal; ATS, ammonium thiosulphate; TUD, thiourea dioxide.

2.3.1. Chloride Channels

The expression of two subunits of chloride channels, *clcn3* and *clcn6*, was measured. In the U87 line, there was an increase in *clcn3* and *clcn6* expression 24 h after rGO/ATS (log2RQ = 4.67 +/− 0.7158) and rGO/TUD (log2RQ = 6.89 +/− 0.7158) treatment. There was an increase in *clcn3* expression after 6 h of treatment in all groups. The highest increase was observed in the rGO/ATS group (log2RQ = 0.42 +/− 0.1063). In Hs5 cells, there was no increase in *clcn3* and *clcn6* expression after 24 h treatment; however, there was higher expression after 6 h of treatment with rGO/ATS flakes (log2RQ = 0.42 +/− 0 1063).

2.3.2. Calcium Channels

In the noncancerous Hs5 cell line, there was an increase in *cacna1d* expression after 6 h of graphene rGO/Term treatment (log2RQ = 0.7957 +/− 0.2863), rGO/ATS (log2RQ = 1.160 +/− 0.2863), and rGO/TUD (log2RQ = 1.387 +/− 0.2863). After 24 h, there was a decrease in *cacna1d* expression in the rGO/ATS (log2RQ = −0.7671 +/− 0.2863) and rGO/TUD (log2RQ = −1.029 +/− 0.2863) groups. There was no expression of the *cacna1d* gene in the U87 glioblastoma line even in the control sample.

Comparing the expression of the *cacna1b* gene in both cell lines, a much higher expression was observed in the noncancerous Hs5 cells (log2RQ = 7.757 +/− 1.492). Additionally, in this cell line, under the influence of graphene flakes, a significant reduction in *cacna1b* expression was observed in all groups after 6 h of treatment. The highest reduction occurred in the rGO/Term treatment (log2RQ = −2.953 +/− 0.2990). After 24 h of treatment, no statistically significant changes were observed.

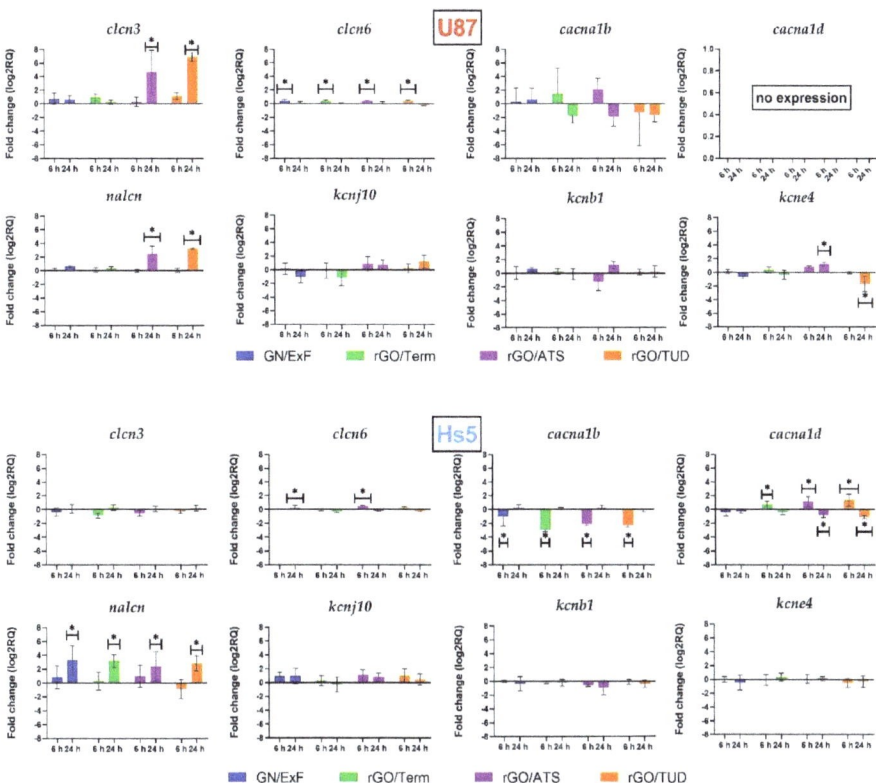

Figure 4. Analysis of the mRNA expression level of voltage-gated ion channels after treatment at 25 µg/mL concentration for 6 and 24 h in U87 and Hs5 cells. The results are calculated relative to the control values. Log2RQ (log2 relative quantitation) values for all genes are normalized to the RPL13A housekeeping gene. Statistical significance between the control and the treated cells is indicated by an asterisk and was assessed using Bonferroni's multiple comparisons test. Differences with p-value < 0.05 were considered significant. One asterisk (*) means p-value < 0.05. **Abbreviations**: rGO, reduced graphene oxide; GN, graphene; C, control group (untreated group); ExF, exfoliation; Term, thermal; ATS, ammonium thiosulphate; TUD, thiourea dioxide; *clcn3*, chloride voltage-gated channel 3; *clcn6*, chloride voltage-gated channel 6; *cacna1b*, calcium voltage-gated channel subunit alpha1 B; *cacna1d*, calcium voltage-gated channel subunit alpha1 D; *nalcn*, sodium leak channel, non-selective; *kcnj10*, potassium inwardly rectifying channel subfamily J member 10; *kcnb1*, potassium voltage-gated Channel subfamily B member 1; *kcne4*, potassium voltage-gated channel subfamily E regulatory subunit 4; rpl13a, ribosomal protein L13a; log2RQ, log2 relative quantitation.

2.3.3. Sodium Channels

In the Hs5 cell line, there was an increase in *nalcn* expression in all examined groups after 24 h of flake treatment. The expression changes were as follows: GN/ExF (log2RQ = 3.810 +/− 0.978), rGO/Term (log2RQ = 3.683 +/− 0.978), rGO/ATS (log2RQ = 2.904 +/− 0.978), and rGO/TUD (log2RQ = 3.338 +/− 0.978). After 6 h of treatment, no significant changes were observed. In the U87 cell line, an increase in *nalcn* expression was observed after 24 h of treatment with rGO/ATS (log2RQ = 2.472 +/− 0.978) and rGO/TUD (log2RQ = 3.200 +/− 0.978) flakes. There were no significant changes in *nalcn* gene expression after 6 h of treatment in the U87 cell line.

2.3.4. Potassium Channels

In our study, we analyzed the expression of two potassium ion channel genes (*kcnb1* and *kcne4*). For both Hs5 fibroblast cells and U87 tumor cells, no statistically significant changes in *kcnb1* gene expression were observed. In Hs5 cells, there was no statistically significant change in *kcne4* expression in any of the studied groups. In U87, there was an increase in *kcne4* expression after 24 h treatment by rGO/ATS flakes (log2RQ = 1.192 +/− 0.33452) and a decrease (log2RQ = −1.734 +/− 0.33452) after 24 h of rGO/TUD treatment. No statistically significant changes in the expression of the *kcnj10* gene were observed in either Hs5 fibroblast cells or U87 tumor cells.

2.4. Expression of Membrane Receptors

Extracellular receptors expression was analyzed using Human Receptor Antibody Array (Targets: 4-1BB (TNFRSF9), ALCAM (CD166), CD80 (B7-1), BCMA (TNFRSF17), CD14, CD30 (TNFRSF8), CD40 Ligand (TNFSF5), CEACAM-1, DR6 (TNFRSF21), Dtk, Endoglin (CD105), ErbB3, E-Selectin, Fas (TNFRSF6), Flt-3 Ligand, GITR (TNFRSF18), HVEM (TNFRSF14), ICAM-3 (CD50), IL-1 R4 (ST2), IL-1 R1, IL10 Rbeta, IL-17RA, IL-1 R gamma, IL-21R, LIMPII, Lipocalin-2 (NGAL), L-Selectin (CD62L), LYVE-1, MICA, MICB, NRG1-beta 1, PDGF R beta, PECAM-1 (CD31), RAGE, TIM-1 (KIM-1), TRAIL R3 (TNFRSF10C), Trappin-2, uPar, VCAM-1 (CD106), XEDAR) (Table S2). The results are shown in Figure 5 and Figure S1. Significant changes in the expression of endoglin, MICA, and uPar receptors were seen in Hs5 cells. An increase in the expression of the uPar receptor in the Hs5 cell line occurred in all treatment groups. On the contrary, in U87 tumor cells, the effect is the opposite. A reduction in uPar receptor expression in all the studied groups was noticed. In Hs5, there was an increase in CD105 receptor expression in the rGO/ATS and rGO/TUD groups. In U87, there was a decrease in the expression of the CD105 receptor in all groups.

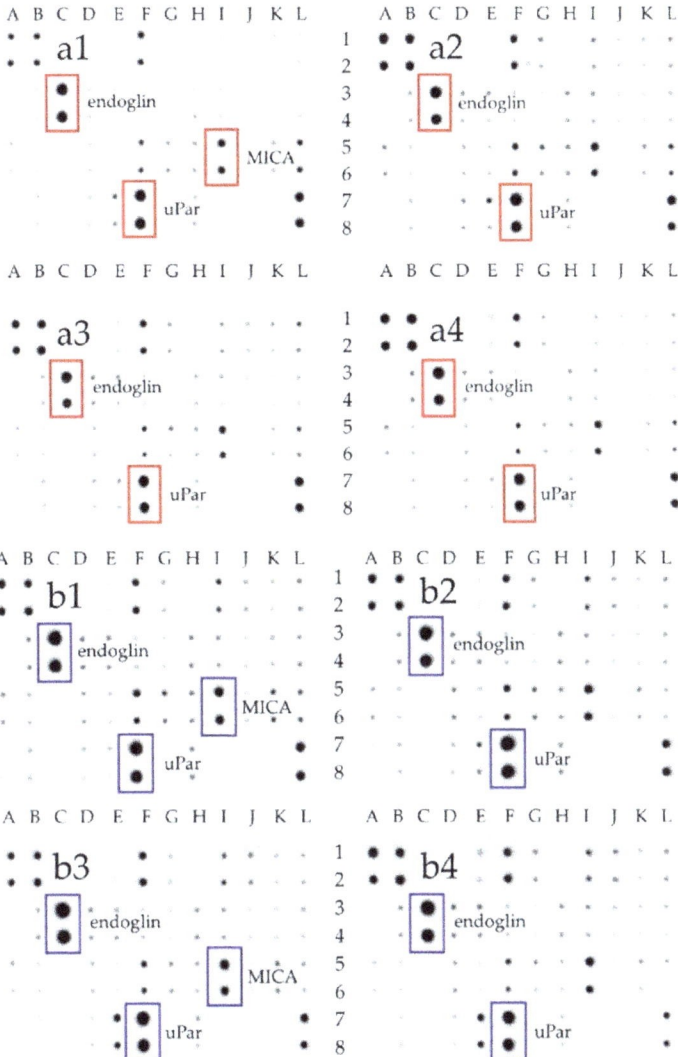

Figure 5. Antibody array analysis of human cell membrane receptor (original drafts) in U87 glioma cells (**a1–a4**) and Hs5 fibroblast cells (**b1–b4**), with or without 24 h treatment (C: control, **a1,b1**), with graphene flakes (GN/ExF: **a2,b2**) and two types of reduced graphene oxides (rGO/ATS: **a3,b3**; rGO/TUD: **a4,b4**). The full array map and uncropped images are available in the supplement material. Results were normalized and compared to a dots control sample. The dots with the location (C3, C4) indicate the expression of endoglin (CD105), (F7, F8) indicate uPar (CD87), and (I5, I6) indicate MICA. **Abbreviations**: rGO, reduced graphene oxide; GN, graphene; C, control group (untreated group); ExF, exfoliation; Term, thermal; ATS, ammonium thiosulphate; TUD, thiourea dioxide; CD105, endoglin; uPar, urokinase plasminogen activator surface receptor; MICA, major histocompatibility complex class I chain-related protein A.

3. Discussion

Previous studies have shown that reduced graphene oxides (rGOs), compared to graphene GN, does not exert identical effects on glioblastoma multiforme cancer cells. The effectiveness of the tested materials was examined in terms of viability, metabolic activity, cell cycle dynamics, and the level of apoptosis. The results indicate that GN and rGOs activate the mitochondria-dependent apoptotic pathway by reducing the potential of the mitochondrial membrane. This study has proved that rGOs have a stronger cytotoxic effect than GN. We used the MTT assay to assess whether graphene and its derivatives also affect the viability and metabolic activity of fibroblast Hs5 and glioblastoma U87 cells. The highest decrease in metabolic activity in glioblastoma U87 cells, at 8.69% ± 12.88%, was found in the group treated with rGO/TUD at a concentration of 100 µg/mL. Interestingly, in the group treated with rGO/ATS, the lowest viability, at 37.7% ± 12.55%, occurred at a concentration of 5 µg/mL rGO/ATS. In other groups treated with GN/ExF and rGO/Term, ~50% mortality was observed at concentrations ranging from 25 to 100 µg/mL. Reducing the oxygen content and increasing defects in the connections between carbons in rGOs compared to GO resulted in an increase in the number of delocalized electrons on the surface of the graphene flakes and oxygen groups, including hydroxyl, carboxyl, and epoxy [14]. It was hypothesized that the cytotoxicity of rGOs may mainly result from direct contact with the glioblastoma cell membrane and may lead to the disruption of signaling pathways in the plasma membrane or direct interaction with cell structures, which are sensitive to electrochemical potential (cell membrane, e.g., ion channels and extracellular receptors). Studying the interactions between graphene materials and cell membranes may reveal the underlying mechanisms of the cytotoxicity of these materials.

Using FTIR analysis, a detailed examination was performed to compare the presence of characteristic functional groups in both reduced graphene oxides (rGO/Term, rGO/ATS, rGO/TUD) and graphene (GN/ExF). This analysis confirmed the results obtained in a previous study [14]. The FTIR analysis presented in this paper shows the presence of hydroxyl groups (-OH) on the surface of the flakes, derived from the native aqueous solution in which the flake suspensions were prepared. Further, the presence of C=O connections may be associated with a greater affinity of the material for the attachment of hydroxyl (OH-) groups. Thus, the more C=O bonds that are hydrophilic in nature, the more OH- hydroxyl groups. The presence of carbonyl groups (C=O) and carboxyl groups (COO-) was also observed. The highest expression was observed in the rGO/TUD treatment. In a study by Loryuenyong et al., the presence of hydroxyl (-OH), carbonyl (C=O), and epoxy (C-O) groups was confirmed in GO and rGO. Moreover, the high intensity of the main peaks in GO confirms the presence of a large number of oxygen functional groups after the oxidation process [25]. Emiru and Ayele obtained a similar band arrangement of the FTIR spectrum of GO and rGO, as presented in this article. Among other things, they observed such functional groups as OH-, COH-, COOH-, and CO- [26]. It showed that with the use of chemical methods, the content of functional groups is reduced. The most effective method for GO reduction is the thermal process because the obtained rGO/Term had the highest rate of reduction [14].

It is known that graphene materials can modulate electron transfer in redox reactions. The speed of the redox reactions on the graphene surface depends on the migration of electrons across the reactive surface and the subsequent transfer along the surface. Pan et al., during an analysis of the resistance in the studied graphene materials, proved that the higher the degree of graphene folding, the more difficult it was to transfer electrons [27]. The same relationship between the degree of folding and resistance was observed in a previous study [14]. Thus, the more corrugated the surface of the carbon material (rGO/Term < rGO/ATS < rGO/TUD), the lower the resistance of the graphene flakes.

The obtained selectivity of graphene flakes towards cancer cells could be determined by the differences in cell membrane potential (Vm) between cancer and noncancerous cells. Cell surface charge is a key biophysical parameter that depends on the composition of the cytoplasmic membrane and the physiological state of cells. Cone's theory [28] proposes a general correlation between proliferation and Vm, as he showed the significant depolarization of Vm during the malignant transformation of normal cells [29,30]. Direct comparisons of the in vitro and in vivo Vm levels of normal and cancer cells showed that the cancer cells were more depolarized than their normal counterparts [17]. A cell depolarizes when Vm is relatively less negative (tumor cells, proliferating: 0 to −50 mV), while a hyperpolarized cell has more negative Vm (normal cells, nonproliferating; −50 to −90 mV) [31]. Glioblastoma cells express the potential of −14 mV of depolarized cell membrane, which determines the lower repulsive forces. Therefore, the flakes with reduced graphene (rGO/ATS and rGO/TUD in particular) possess a lower negative surface charge and can probably adhere to glioblastoma U87 cell membranes more easily.

When analyzing the results of the cell membrane potential (Vm) of U87 and Hs5, a decrease in cell membrane potential was observed over time after treatment with rGO/ATS and rGO/TUD flakes. However, the decrease in potential in fibroblast Hs5 cells was smaller than that in U87 cells: for rGO/ATS material, it was 258.4 RFU and 421 RFU, respectively, and for rGO/TUD material, 384.3 RFU and 561.7 RFU, respectively. Bondar et al. observed that the membrane potential in HeLa cells, in which apoptosis was thermally induced, shifted negatively by about 4.2 mV compared to control cells. This was probably the result of the redistribution of phosphatidylserine, containing a negatively charged carboxyl group, from the inner to the outer lipid layer of the cell membrane [16]. Therefore, the reduction of the U87 glioma cell membrane potential after treatment with rGO/ATS and rGO/TUD flakes may be partially secondary to the flakes due to the induction of apoptosis in cells.

Graphene flakes, apart from direct contact with the lipid double membrane, are in contact with channels, including ion channels. Ion channels are membrane proteins that open or close a plasmatic membrane, depending on a voltage gradient or ligand binding. They are essential for cell proliferation and play a key role in malignant glioma by influencing the shape and volume of glioblastoma cells, which, in turn, may influence the invasiveness and migration of tumors [32].

The effect of the graphene flakes on the expression of voltage-gated ion channels, participating in both electrical and chemical signaling pathways [33], was assessed. Wang et al. showed that between 18 genes of ion channels (voltage-gated and ligand-gated), the expression of 16 genes (*cacna1d*, *clcn6*, *glrb*, *gria2*, *grid1*, *kcnab1*, *kcnb1*, *kcnd2*, *kcnj10*, *kcnma1*, *kcnqn3*, *nalcn*, *p2rx7*, *scn1a*, and *vdac2*) was reduced compared to normal tissue [34]. Based on these studies, genes encoding voltage-dependent ion channel subunits (*cacna1d*, *clcn6*, *kcnab1*, *kcnb1*, *kcnj10*, and *nalcn*) were selected to determine the effect of graphene derivatives on channel gene expression. The following genes were also added for the analysis: *clcn3* [35], *cacna1b* [36], and *kcne4* [37].

The presented study shows that the expression of *nalcn* in U87 glioma tumor cells was significantly higher (log2RQ = 6.15) than in Hs5 fibroblasts. Ouwerkerk et al. reported that the concentration of Na^+ ions in malignant tumors increased in comparison to noncancer tissues [38]. Moreover, the expression of *nalcn* in astrocytes, the glial cells that glioblastoma is derived from, was markedly low [39]. The expression of *nalcn* in U87 glioma cells was significantly higher after treatment with rGO/ATS and rGO/TUD. In Hs5 cells, the same effect was observed in the case of each treatment with graphene derivative material. It was suggested that the increase of intracellular Na^+ occurs in the early phase of apoptosis [40,41]. Moreover, several studies also reported an elevation of cytoplasmic Na^+ in the late phase of apoptosis [42,43]. Thus, an increased expression of *nalcn*, induced by the graphene flakes, can provoke apoptosis in glioblastoma cells by stimulating sodium influx into the cells, which increases the cytoplasmic concentration of Na^+.

clcn6 is predominantly localized in the intracellular vesicles of the endoplasmic reticulum (late endosomes) and the cell membrane. Our study showed a significant increase of *clcn6* during the initial treatment of U87 glioblastoma cells in all tested groups. On the contrary, Hs5 cells revealed a significant increase of *clcn6* expression in the GN/ExF and rGO/ATS treatment groups. The observed overexpression leads to lysosomal acidification [44]. Neagoe et al. clearly proved that ClC-6 mediates the Cl^-/H^+ exchange, which affects its coupling, conductivity, and ion selectivity features [45]. Increased expression of the *clcn6* gene in U87 glioma cells leads to cytoplasmic alkalization. In fact, intracellular alkalization and extracellular acidification are commonly observed in malignant tumors. The altered activity of cell transporters, internal enzymes, and pH gradient in the cancer cell membrane plays a pivotal role in tumor progression and metastasis [46,47]. Acidification of the cytoplasm leads to the activation of apoptotic pathways in cancer cells [48]. Based on that, we conclude that graphene flakes (GN/ExF) and rGOs in U87 glioblastoma cells block Cl^-/H^+ transport. Consequently, it leads to the acidification of the cytoplasm and the activation of apoptosis. Simultaneously, in response to that blockage, an increase in *clcn6* gene expression can occur.

Four classes of potassium channels are distinguished: Kv channels (voltage-gated), KCa^{2+} channels (calcium-activated), Kir channels (inward-rectifier potassium channel), and K2P channels (two-pore channels) [49,50]. Based on these studies, selected markers of Kv and Kir classes were verified. In the channels of the Kv class, we analyzed expression *kcnb1* (Kv2.1) and *kcne4* channels, as well as *kcnj10*, which belongs to the Kir potassium channel class. Major changes were observed in the expression of the *kcne4* channel belonging to the Kv potassium channel class. This study confirmed a significant increase in the expression of the *kcne4* channel (log2RQ = 1.192) after 24 h treatment of U87 glioma cells with rGO/ATS. Meanwhile, the reduction in *kcne4* expression (log2RQ = -1.734) was observed after 24 h treatment by rGO/TUD. No changes in KCNE4 expression were noted in Hs5 fibroblasts. KCNE4 expression was characterized by a 2.9-fold increase in glioma compared to the healthy tissues [51]. The *kcne4* genes encode single proteins of the transmembrane domain with an extracellular *N*-terminus and an intracellular *C*-terminus. Therefore, the abovementioned proteins cannot form functional ion channels. They can function as the accessory subunits for various ion channels and regulate their biophysical and pharmacological properties in parallel [52–54].

The study confirmed no changes in *cacna1b* expression in U87 cells after treatment with different graphene materials. In Hs5 cells, the expression was decreased in all experimental groups in the initial treatment. The results confirmed that the expression of the *cacna1b* gene in the fibroblasts was significantly higher (log2RQ = 7.52) than in U87 cells. Wang et al. reported, based on brain and breast cancer studies, that *cacna1b* is expressed at a low level in tumor cells. Brain cancers, including glioblastoma, oligodendroglia, anaplastic astrocytoma, diffuse astrocytoma, and glioblastoma, show a significant reduction of *cacna1b* expression compared to the control groups [55].

In our studies, we did not observe the expression of the *cacna1d* gene encoding the subunit of calcium channels in U87 cells. Changes in *cacna1d* expression occurred only under rGO influence in Hs5 cells. At the initial 6 h treatment, an increase of *cacna1d* was observed after rGO/Term, rGO/ATS, and rGO/TUD treatment. However, 24 h of treatment decreased its expression in both rGO/ATS and rGO/TUD groups. The primary increase in *cacna1d* expression can induce a greater influx of Ca^{2+} ions into the cells. Therefore, after 24 h of treatment, we noticed the reduction in *cacna1d* expression, which equalized the earlier influx of Ca^{2+} ions.

Analysis of *clcn3* expression in U87 glioma cells showed an increase in expression after treatment with graphene flakes (GN/ExF) and reduced graphene oxide flakes (rGO/Term, rGO/ATS, rGO/TUD). In particular, significant changes were visible in the rGO/ATS and rGO/TUD groups. Recent studies have shown that *clcn3* is highly expressed in GBM, and it plays a significant role in cell survival, proliferation, and malignancy [56,57]. Sontheimer et al. showed that decreased expression of *clcn3* channels inhibits the migration of glioblastoma cells in vitro and in vivo [35]. An increase in *clcn3* expression can stimulate the invasiveness of U87 glioma cells. No parallel changes were observed in Hs5 fibroblasts after treatment with carbon flakes.

Additionally, using the protein matrix, which allows the determination of the expression of 40 different receptor proteins involved in different signaling pathways (Figure 5), we assessed the expression of selected extracellular receptors. As a result, it was noticed that the expression of the uPar protein was increased in U87 cells compared to Hs5 cells. Raghu et al. confirmed significantly higher uPar protein expression in U87 cells in comparison to normal HMEC cells [58]. Analysis of uPar expression in U87 glioma cells after treatment with graphene flakes (GN/ExF) and reduced graphene oxide (rGO/ATS, rGO/TUD) showed a significant reduction of its expression. It was mainly affected by the treatment with reduced graphene oxides. uPar is responsible for the degradation of extracellular matrix (ECM) components attached to the cell surface [59]. It contains three domains connected to the concave structure, which is the binding site for vitronectin. However, since the binding sites of vitronectin and uPa are distinct, uPar can bind simultaneously to both ligands, regulating proteolysis, adhesion, and cell signaling [60]. The association between uPa and uPar can cause the cleavage of the adjacent uPar molecules. The cleaved uPar does not support plasminogen, which is mediated by the activation of uPa on the cell surface [61]. Thus, the flakes of graphene and reduced graphene oxides act similarly to uPa by activating the cleavage of adjacent uPar receptors. They can also block the attachment sites of uPar and vitronectin and, therefore, inhibit uPar signaling [61,62].

Endoglin (CD105) is the other factor that had a changed expression after treatment with GN and rGO flakes. It is transmembrane homodimeric protein localized in the endothelial cells of blood vessels. It is a component of the transforming growth factor β (TGFβ) receptor signaling pathway. CD105 plays a pivotal role in angiogenesis and vasculogenesis processes, preventing apoptosis in hypoxic endothelial cells [63]. It was observed that CD105 is correlated with cancer prognosis (particularly in pediatric cases), but its role in high-grade gliomas remains unclear [64]. Our study showed a significantly higher endoglin expression in U87 glioma cells than in Hs5 fibroblasts. In other studies, a high expression of CD105 in neoplastic tissue, such as in meningiomas [65] or childhood brain tumors [64], was also confirmed. The presented study showed a decrease in CD105 expression in U87 glioma cells after treatment with GN and rGOs in all treatment groups. Muenzner et al. confirmed that the endoglin carboxy-terminal domain is required to inhibit cell detachment [66]. Therefore, treatment with GN and rGO, resulting in decreased endoglin expression, can stimulate cell adhesion and, consequently, leads to the reduction of the ability of cancer cell migration.

4. Materials and Methods

4.1. Production and Preparation of GN and rGO

Graphene and reduced graphene oxides were supplied by the Łukasiewicz Research Network—Institute of Microelectronics and Photonics, Warsaw. Direct graphite exfoliation using Capstone (a fluorinated surfactant) was used to obtain graphene flakes, designated as GN/ExF. Flakes of rGO were produced by reducing previously prepared graphene oxide (GO). GO was obtained by graphite oxidation and exfoliation, according to a modified Marcano method. Reduced graphene oxide, designated as rGO/ATS, was created by reducing GO with ammonium thiosulphate for 20 h at 95 °C. The molar ratio of the reducing agent to GO was 3:1, and the reduction was conducted at neutral pH. rGO/Term was created by reducing GO powder in an oven at 1000 °C for 1 h under a nitrogen atmosphere. Reduced rGO/TUD graphene oxide was prepared by reducing GO via exposure to thiourea dioxide at 85 °C for 1.5 h. The molar ratio of the reducing agent to GO was 5:1, and pH was set as 9. rGO/ATS and rGO/TUD were purified through pressure filtration on a membrane, and then dialysis was used. During these steps, a significant number of residue chemicals were removed from materials; however, there were still some sulfur ions [14].

4.2. Characterization of GN and rGOs

4.2.1. Fourier Transform IR (FT-IR) Spectrometer Analysis

The samples were processed in pellet form. Each pellet consisted of 0.007 g of graphene and 0.200 g of potassium bromide. Infrared absorption of the coatings in the spectral range of 400–4000 cm^{-1} was performed using a model iS50 Fourier transform IR (FT-IR) spectrometer (Thermo Fisher Scientific, Wilmington, DE, USA). Spectra were recorded with a resolution of 4 cm^{-1} using a high sensitivity MCT-B detector (mercury cadmium telluride). The measurements were performed in the transmitation mode. In each case, data from 120 scans were collected to construct a single spectrum.

4.2.2. Scanning Electron Microscopy of Flakes

The morphology of graphene and reduced graphene oxides was examined using scanning electron microscopy (SEM, Hitachi S-3000 N, Minato-ku, Tokyo, Japan). SEM images were taken in the AEE mode to scan the graphene with the table current.

4.3. Cell Cultures

Human glioblastoma U87 MG (ATCC® HTB-14™) and bone marrow stromal Hs5 (ATCC® CRL-11882™) cell lines were obtained from the American Type Culture Collection (Manassas, VA, USA) and maintained in Dulbecco's modified Eagle's medium (DMEM), supplemented with 10% fetal bovine serum (Life Technologies, Houston, TX, USA) and 1% antibiotic-antimycotic mixture containing penicillin and streptomycin (Life Technologies, Houston, TX, USA). Cultures were maintained at 37 °C under 5% CO_2 and 95% humidity in an INCOMED153 (Memmert GmbH & Co. KG, Schwabach, Germany).

4.4. Cell Membrane Potential Assay

A Cellular Membrane Potential Assay Kit (ab176764, Abcam, Cambridge, UK) was used to detect changes in membrane potential. Cells were plated in 96-well black plates. Approximately 3×10^4 cells were seeded onto each well. After incubation for 24 h, the culture medium was replaced with medium containing GN and rGO at a final concentration of 25 µg/mL. After 24 h of incubation with graphene and graphene derivatives, the growth medium was removed from each well and replaced with 100 µL of diluted (1:10) assay buffer, with the addition of 1.5 µL MP sensor dye loading. The plate was incubated for 30 min at RT in the dark, and membrane potential change was analyzed using an ELISA reader (Infinite M200, Tecan, Durham, NC, USA) by measuring fluorescence at Ex/Em = 530/570 nm.

4.5. Isolation of Total RNA and cDNA Synthesis

For the isolation of total RNA, U87 and Hs5 cells (2×10^5) were cultured on a six-well plate in three independent replicates. A water solution of GN and rGO flakes was added to the cultures at a concentration of 25 µg/mL and incubated two independent times for 6 h and 24 h. Cells were detached from the plates by trypsinization, centrifuged for 5 min at $400 \times g$, and washed twice with phosphate-buffered saline (PBS; Thermo Fisher Scientific, Wilmington, DE, USA). The cell pellet was resuspended in freshly prepared lysis buffer from a PureLink™ RNA Mini Kit (Thermo Fisher Scientific, Wilmington, DE, USA) containing 1% 2-mercaptoethanol and vortexed at high speed until the cell pellet was completely dispersed and the cells were lysed. The supernatant was transferred to new tubes, mixed with one volume of 70% ethanol, and then transferred to a spin cartridge. Further steps were performed according to the manufacturer's protocol. The isolated RNA was measured using a NanoDrop 2000 spectrophotometer (Thermo Fisher Scientific, Wilmington, DE, USA). A cDNA High Capacity Reverse Transcription Kit (Applied Biosystems, Foster City, CA, USA) was used. The procedure was performed with the following cycle conditions: 10 min at 25 °C, 120 min at 37 °C, and 5 min at 4 °C using a 2720 Thermal Cycler (Thermo Fisher Scientific, Wilmington, DE, USA). cDNA concentration was measured on a NanoDrop 2000 spectrophotometer and stored for further analysis at −80 °C.

4.6. Gene Expression

The reaction was carried out using 48-well plates and Power SYBR™ Green PCR Master Mix (Thermo Fisher Scientific, Wilmington, DE, USA); 100 ng of cDNA was used for each reaction. The following genes were examined: *clcn3*, *clcn6*, *cacna1b*, *cacna1d*, *nalcn*, *kcnj10*, *kcnb1*, and *kcne4*. Gene-specific primers (Table 2) were purchased from Genomed

(Warsaw, Poland), and *rpl13a* was used as the reference gene [67]. The reaction was performed using the Step One™ Real-Time PCR System (Thermo Fisher Scientific, Wilmington, DE, USA). Conditions of the reaction were set as specified by the manufacturer. Each sample was analyzed in duplicate. The ΔΔCt method was used to determine mRNA expression by real-time PCR: ΔΔCT = ΔCT test sample—ΔCT calibrator sample. RQ = $2^{-\Delta\Delta CT}$.

Table 2. Primers used to assess the expression of genes involved in voltage-dependant ion channels.

Genes	Forward Primers (5'-3')	Reverse Primers (5'-3')	References
clcn3	CTGTGCCGCCTCTAAGCC	ACTGTAGTTCGACTCGCTGAA	Primer blast
clcn6	ACCTGGAAGTTTTGGAGACCAT	TGAGTTGGGTGAAGAGTCGC	Primer blast
cacna1b	CCCTTGCTGTCAACATCTGGT	GGATGGGTGAGGAGTTGGC	Primer blast
cacna1d	ACTCGGGCTATCCAGAAGTAG	CTTGCCCAAAGAAAAGACTGC	Primer blast
nalcn	CGCCGTAGACTGTGGTTTTG	AATGACGCTGATGATGGCAC	Primer blast
kcnj10	TCAGAAGACGGGCGAAACAA	GCGAGCCTAAGCAAGACTCA	Primer blast
kcnb1	CCATTCTGCCATACTATGTCACC	AGCAAGCCCAACTCATTGTAG	[68]
kcne4	CACCGCTACCTGAAAACCCT	TTGATCGTGGCAGAGTGAGC	Primer blast
rpl13a	CATAGGAAGCTGGGAGCAAG	GCCCTCCAATCAGTCTTCTG	[67]

4.7. Human Receptor Antibody Array

For protein analysis, glioma cell line U87 and fibroblast cell line Hs5 were treated with graphene (GN) or reduced graphene oxides (rGOs) at a concentration of 25 μg/mL and incubated for 24 h. The cells were scraped off, centrifuged, and washed twice in PBS. Cells not treated with graphene flakes were used as a control. The cell pellet was resuspended in a diluted lysis buffer containing protease and phosphatase inhibitors (Sigma-Aldrich, St. Louis, MO, USA) according to the manufacturer's instructions. Frozen metal balls and TissueLyser (Qiagen, Hilden, Germany) were used for homogenization at 50 Hz for 10 min on a shaking frozen cartridge. The samples were then centrifuged (30 min; $14,000\times g$; 4 °C), and the supernatant was collected. Protein concentration was determined using a bicinchoninic acid kit (Sigma-Aldrich, St. Louis, MO, USA). Analysis of receptor cell membranes was performed using an antibody array (ab211065; Abcam, Cambridge, UK). The assay was performed in accordance with the manufacturer's instructions, using lysates containing 400 μg/mL of total protein per membrane. Membranes were visualized using the Azure Biosystem C400 (Azure, Dublin, CA, USA) [63]. The results shown in Figure S1 were obtained by analysis in ImageJ. Results were normalized and compared to a dot control sample.

4.8. Statistical Analysis

The data were analyzed using a two-way analysis of variance with GraphPad Prism 8.4.3 (GraphPad Software Inc., La Jolla, CA, USA). Differences between groups were tested using Bonferroni's multiple comparison test and Dunnett's multiple comparisons test. All mean values are presented using standard deviation or standard error. Differences at $p < 0.05$ were considered significant.

5. Conclusions

Different types of graphene derivatives may activate separate cell pathways. rGOs can affect cells as the result of contact with the glioblastoma cell membrane via its functional groups, which are presented on the surface of the examined flakes. This is the first study confirming that graphene flakes have a significant effect on the expression of voltage-dependent ion channel (VGIC) genes in U87 glioma cells. The mechanism of the effect of graphene-derived flakes on membrane proteins depends on the number of the specific structure of voltage-dependent ion channels and extracellular receptors. It was also demonstrated that major changes in the expression of voltage-dependent ion channel genes were observed in *clcn3*, *nalcn*, and *kcne4* after treatment with rGO/ATS and rGO/TUD flakes. Moreover, we present that the examined graphene forms or different types of graphene derivatives (GN and rGOs) can affect the expression of extracellular receptors (uPar and endoglin) in U87 cells, significantly reducing their expression. We showed that GN and rGOs decrease uPar expression by acting as inhibitors of neoplastic-promoted proteolysis. Therefore, the mobility of mesenchymal-type cells may be inhibited. Moreover, GN and rGOs decrease the expression of endoglin (CD105) and probably provide an increase in cell adhesion, consequently reducing cancer cell migration. In conclusion, we suggest that the presence of oxygen-containing functional groups of rGO, including their number and types, may be the most important feature of the examined flakes for their future medical application.

Supplementary Materials: The following are available online at https://www.mdpi.com/1422-0067/22/2/515/s1.

Author Contributions: Conceptualization, M.G. and J.S.; methodology, B.S.-C.; validation, K.D. and O.W.-P., formal analysis, M.S. (Malwina Sosnowska); investigation, J.S., J.J., D.N., S.J., and A.S.-G.; resources, M.W.; writing—original draft preparation, J.S.; writing—review and editing, J.S., M.G., B.S.-C., and M.S. (Maciej Szmidt); visualization, M.W.; supervision, M.G.; project administration, M.S. (Maciej Szmidt); funding acquisition, M.G. All authors have read and agreed to the published version of the manuscript.

Funding: This research was funded by The National Centre for Research and Development (grant number LIDER/144/L-6/14/NCBR/2015).

Institutional Review Board Statement: Not applicable.

Informed Consent Statement: Not applicable.

Data Availability Statement: The data that support the findings of this study are available from the corresponding author upon reasonable request.

Acknowledgments: The manuscript is a part of a Ph.D. thesis by Jaroslaw Szczepaniak.

Conflicts of Interest: The authors declare no conflict of interest.

References

1. Wen, P.Y.; Reardon, D.A. Neuro-oncology in 2015: Progress in glioma diagnosis, classification and treatment. *Nat. Rev. Neurol.* **2016**, *12*, 69–70. [CrossRef]
2. Zhang, Y.; Cruickshanks, N.; Yuan, F.; Wang, B.; Pahuski, M.; Wulfkuhle, J.; Gallagher, I.; Koeppel, A.F.; Hatef, S.; Papanicolas, C.; et al. Targetable T-type Calcium Channels Drive Glioblastoma. *Cancer Res.* **2017**, *77*, 3479–3490. [CrossRef] [PubMed]
3. Chen, J.-H.; Jang, C.; Xiao, S.; Ishigami, M.; Fuhrer, M.S. Intrinsic and extrinsic performance limits of graphene devices on SiO_2. *Nat. Nanotechnol.* **2008**, *3*, 206–209. [CrossRef] [PubMed]
4. Choi, W.; Lahiri, I.; Seelaboyina, R.; Kang, Y.S. Synthesis of Graphene and Its Applications: A Review. *Crit. Rev. Solid State Mater. Sci.* **2010**, *35*, 52–71. [CrossRef]
5. Geim, A.K.; Novoselov, K.S. The rise of graphene. *Nat. Mater.* **2007**, *6*, 183–191. [CrossRef] [PubMed]
6. Compton, O.C.; Nguyen, S.T. Graphene Oxide, Highly Reduced Graphene Oxide, and Graphene: Versatile Building Blocks for Carbon-Based Materials. *Small* **2010**, *6*, 711–723. [CrossRef]
7. Acik, M.; Mattevi, C.; Gong, C.; Lee, G.; Cho, K.; Chhowalla, M.; Chabal, Y.J. The Role of Intercalated Water in Multilayered Graphene Oxide. *ACS Nano* **2010**, *4*, 5861–5868. [CrossRef]
8. Gao, W. The chemistry of graphene oxide. In *Graphene Oxide: Reduction Recipes, Spectroscopy, and Applications*; Springer International Publishing: Cham, Switzerland, 2015; ISBN 9783319155005.
9. Grodzik, M.; Sawosz, E.; Wierzbicki, M.; Orlowski, P.; Hotowy, A.; Niemiec, T.; Szmidt, M.; Mitura, K.; Chwalibog, A. Nanoparticles of carbon allotropes inhibit glioblastoma multiforme angiogenesis in ovo. *Int. J. Nanomed.* **2011**, *6*, 3041–3048.
10. Wierzbicki, M.; Sawosz, E.; Grodzik, M.; Kutwin, M.; Jaworski, S.; Chwalibog, A. Comparison of anti-angiogenic properties of pristine carbon nanoparticles. *Nanoscale Res. Lett.* **2013**, *8*, 195. [CrossRef]
11. Chwalibog, A.; Jaworski, S.; Sawosz, E.; Grodzik, M.; Winnicka, A.; Prasek, M.; Wierzbicki, M. In vitro evaluation of the effects of graphene platelets on glioblastoma multiforme cells. *Int. J. Nanomed.* **2013**, *8*, 413–420. [CrossRef]
12. Jaworski, S.; Sawosz, E.; Kutwin, M.; Wierzbicki, M.; Hinzmann, M.; Grodzik, M.; Winnicka, A.; Lipińska, L.; Wlodyga, K.; Chwalibog, A. In vitro and in vivo effects of graphene oxide and reduced graphene oxide on glioblastoma. *Int. J. Nanomed.* **2015**, *10*, 1585–1596. [CrossRef]

13. Jaworski, S.; Strojny, B.; Sawosz, E.; Wierzbicki, M.; Grodzik, M.; Kutwin, M.; Daniluk, K.; Chwalibog, A. Degradation of Mitochondria and Oxidative Stress as the Main Mechanism of Toxicity of Pristine Graphene on U87 Glioblastoma Cells and Tumors and HS-5 Cells. *Int. J. Mol. Sci.* **2019**, *20*, 650. [CrossRef]
14. Szczepaniak, J.; Strojny, B.; Sawosz, E.; Jaworski, S.; Jagiello, J.; Winkowska, M.; Szmidt, M.; Wierzbicki, M.; Sosnowska, M.; Balaban, J.; et al. Effects of Reduced Graphene Oxides on Apoptosis and Cell Cycle of Glioblastoma Multiforme. *Int. J. Mol. Sci.* **2018**, *19*, 3939. [CrossRef] [PubMed]
15. Martelli, C.; King, A.; Simon, T.; Giamas, G. Graphene-Induced Transdifferentiation of Cancer Stem Cells as a Therapeutic Strategy against Glioblastoma. *ACS Biomater. Sci. Eng.* **2020**, *6*, 3258–3269. [CrossRef]
16. Bondar, O.V.; Saifullina, D.V.; Shakhmaeva, I.I.; Mavlyutova, I.I.; Abdullin, T.I. Monitoring of the Zeta Potential of Human Cells upon Reduction in Their Viability and Interaction with Polymers. *Acta Nat.* **2012**, *4*, 78–81. [CrossRef]
17. Marmo, A.A.; Morris, D.M.; Schwalke, M.A.; Iliev, I.G.; Rogers, S. Electrical Potential Measurements in Human Breast Cancer and Benign Lesions. *Tumor Biol.* **1994**, *15*, 147–152. [CrossRef]
18. Fiorillo, M.; Verre, A.F.; Iliut, M.; Peiris-Pagés, M.; Ozsvari, B.; Gandara, R.; Cappello, A.R.; Sotgia, F.; Vijayaraghavan, A.; Lisanti, M.P. Graphene oxide selectively targets cancer stem cells, across multiple tumor types: Implications for non-toxic cancer treatment, via "differentiation-based nano-therapy". *Oncotarget* **2015**, *6*, 3553–3562. [CrossRef]
19. Pollak, J.; Rai, K.G.; Funk, C.C.; Arora, S.; Lee, E.; Zhu, J.; Price, N.D.; Paddison, P.J.; Ramirez, J.-M.; Rostomily, R.C. Ion channel expression patterns in glioblastoma stem cells with functional and therapeutic implications for malignancy. *PLoS ONE* **2017**, *12*, e0172884. [CrossRef]
20. Simon, O.J.; Müntefering, T.; Grauer, O.; Meuth, S.G. The role of ion channels in malignant brain tumors. *J. Neuro-Oncol.* **2015**, *125*, 225–235. [CrossRef]
21. Parsons, D.W.; Jones, S.; Zhang, X.; Lin, J.C.-H.; Leary, R.J.; Angenendt, P.; Mankoo, P.; Carter, H.; Siu, I.-M.; Gallia, G.L.; et al. An Integrated Genomic Analysis of Human Glioblastoma Multiforme. *Science* **2008**, *321*, 1807–1812. [CrossRef]
22. Joshi, A.D.; Parsons, D.W.; Velculescu, V.E.; Riggins, G.J. Sodium ion channel mutations in glioblastoma patients correlate with shorter survival. *Mol. Cancer* **2011**, *10*, 17. [CrossRef] [PubMed]
23. Cuddapah, V.A.; Robel, S.; Watkins, S.; Sontheimer, H. A neurocentric perspective on glioma invasion. *Nat. Rev. Neurosci.* **2014**, *15*, 455–465. [CrossRef] [PubMed]
24. Weaver, A.K.; Liu, X.; Sontheimer, H. Role for calcium-activated potassium channels (BK) in growth control of human malignant glioma cells. *J. Neurosci. Res.* **2004**, *78*, 224–234. [CrossRef] [PubMed]
25. Loryuenyong, V.; Totepvimarn, K.; Eimburanapravat, P.; Boonchompoo, W.; Buasri, A. Preparation and Characterization of Reduced Graphene Oxide Sheets via Water-Based Exfoliation and Reduction Methods. *Adv. Mater. Sci. Eng.* **2013**, *2013*, 923403. [CrossRef]
26. Emiru, T.F.; Ayele, D.W. Controlled synthesis, characterization and reduction of graphene oxide: A convenient method for large scale production. *Egypt. J. Basic Appl. Sci.* **2017**, *4*, 74–79. [CrossRef]
27. Pan, M.; Zhang, Y.; Shan, C.; Zhang, X.; Gao, G.; Pan, B. Flat Graphene-Enhanced Electron Transfer Involved in Redox Reactions. *Environ. Sci. Technol.* **2017**, *51*, 8597–8605. [CrossRef]
28. Cone, C.D. Unified theory on the basic mechanism of normal mitotic control and oncogenesis. *J. Theor. Biol.* **1971**, *30*, 151–181. [CrossRef]
29. Tokuoka, S.; Morioka, H. The membrane potential of the human cancer and related cells. I. *Gan* **1957**, *48*, 353–354. [CrossRef]
30. Johnstone, B.M. Micro-Electrode Penetration of Ascites Tumour Cells. *Nature* **1959**, *183*, 411. [CrossRef]
31. Yang, M.; Brackenbury, W.J. Membrane potential and cancer progression. *Front. Physiol.* **2013**, *4*, 185. [CrossRef]
32. Molenaar, R.J. Ion Channels in Glioblastoma. *ISRN Neurol.* **2011**, *2011*, 590249. [CrossRef] [PubMed]
33. Hille, B. *Ion Channels of Excitable Membranes*, 3rd ed.; Sinauer Associates, Inc.: Sunderland, MA, USA, 2001; ISBN 0878933212.
34. Wang, R.; Gurguis, C.I.; Gu, W.; Ko, E.A.; Lim, J.; Bang, H.; Zhou, T.; Ko, J.-H. Ion channel gene expression predicts survival in glioma patients. *Sci. Rep.* **2015**, *5*, 11593. [CrossRef] [PubMed]
35. Sontheimer, H. An Unexpected Role for Ion Channels in Brain Tumor Metastasis. *Exp. Biol. Med.* **2008**, *233*, 779–791. [CrossRef] [PubMed]
36. Phan, N.N.; Wang, C.-Y.; Chen, C.-F.; Sun, Z.; Lai, M.-D.; Lin, Y.-C. Voltage-gated calcium channels: Novel targets for cancer therapy. *Oncol. Lett.* **2017**, *14*, 2059–2074. [CrossRef] [PubMed]
37. Solé, L.; Roura-Ferrer, M.; Pérez-Verdaguer, M.; Oliveras, A.; Calvo, M.; Fernández-Fernández, J.M.; Felipe, A. KCNE4 suppresses Kv1.3 currents by modulating trafficking, surface expression and channel gating. *J. Cell Sci.* **2009**, *122*, 3738–3748. [CrossRef]
38. Ouwerkerk, R.; Jacobs, M.A.; Macura, K.J.; Wolff, A.C.; Stearns, V.; Mezban, S.D.; Khouri, N.F.; Bluemke, D.A.; Bottomley, P.A. Elevated tissue sodium concentration in malignant breast lesions detected with non-invasive 23Na MRI. *Breast Cancer Res. Treat.* **2007**, *106*, 151–160. [CrossRef]
39. Cahoy, J.D.; Emery, B.; Kaushal, A.; Foo, L.C.; Zamanian, J.L.; Christopherson, K.S.; Xing, Y.; Lubischer, J.L.; Krieg, P.A.; Krupenko, S.A.; et al. A Transcriptome Database for Astrocytes, Neurons, and Oligodendrocytes: A New Resource for Understanding Brain Development and Function. *J. Neurosci.* **2008**, *28*, 264–278. [CrossRef]
40. Fernández-Segura, E.; Cañizares, F.J.; Cubero, M.A.; Warley, A.; Campos, A. Changes in Elemental Content During Apoptotic Cell Death Studied by Electron Probe X-Ray Microanalysis. *Exp. Cell Res.* **1999**, *253*, 454–462. [CrossRef]

41. Skepper, J.N.; Karydis, I.; Garnett, M.R.; Hegyi, L.; Hardwick, S.J.; Warley, A.; Mitchinson, M.J.; Cary, N.R.B. Changes in elemental concentrations are associated with early stages of apoptosis in human monocyte-macrophages exposed to oxidized low-density lipoprotein: An X-ray microanalytical study. *J. Pathol.* **1999**, *188*, 100–106. [CrossRef]
42. Arrebola, F.; Fernández-Segura, E.; Campos, A.; Crespo, P.V.; Skepper, J.N.; Warley, A. Changes in intracellular electrolyte concentrations during apoptosis induced by UV irradiation of human myeloblastic cells. *Am. J. Physiol. Cell Physiol.* **2006**, *290*, C638–C649. [CrossRef]
43. Arrebola, F.; Zabiti, S.; Cañizares, F.J.; Cubero, M.A.; Crespo, P.V.; Fernández-Segura, E. Changes in intracellular sodium, chlorine, and potassium concentrations in staurosporine-induced apoptosis. *J. Cell. Physiol.* **2005**, *204*, 500–507. [CrossRef] [PubMed]
44. Poët, M.; Kornak, U.; Schweizer, M.; Zdebik, A.A.; Scheel, O.; Hoelter, S.; Wurst, W.; Schmitt, A.; Fuhrmann, J.C.; Planells-Cases, R.; et al. Lysosomal storage disease upon disruption of the neuronal chloride transport protein ClC-6. *Proc. Natl. Acad. Sci. USA* **2006**, *103*, 13854–13859. [CrossRef]
45. Neagoe, I.; Stauber, T.; Fidzinski, P.; Bergsdorf, E.-Y.; Jentsch, T.J. The Late Endosomal ClC-6 Mediates Proton/Chloride Countertransport in Heterologous Plasma Membrane Expression. *J. Biol. Chem.* **2010**, *285*, 21689–21697. [CrossRef] [PubMed]
46. Webb, B.A.; Chimenti, M.S.; Jacobson, M.P.; Barber, D.L. Dysregulated pH: A perfect storm for cancer progression. *Nat. Rev. Cancer* **2011**, *11*, 671–677. [CrossRef] [PubMed]
47. Damaghi, M.; Wojtkowiak, J.W.; Gillies, R.J. pH sensing and regulation in cancer. *Front. Physiol.* **2013**, *4*, 370. [CrossRef] [PubMed]
48. Matsuyama, S.; Reed, J.C. Mitochondria-dependent apoptosis and cellular pH regulation. *Cell Death Differ.* **2000**, *7*, 1155–1165. [CrossRef]
49. Liu, J.; Qu, C.; Han, C.; Chen, M.-M.; An, L.-J.; Zou, W. Potassium channels and their role in glioma: A mini review. *Mol. Membr. Biol.* **2019**, *35*, 76–85. [CrossRef]
50. Du, J.; Haak, L.L.; Phillips-Tansey, E.; Russell, J.T.; McBain, C.J. Frequency-dependent regulation of rat hippocampal somato-dendritic excitability by the K$^+$ channel subunit Kv2.1. *J. Physiol.* **2000**, *522*, 19–31. [CrossRef]
51. Biasiotta, A.; D'Arcangelo, D.; Passarelli, F.; Nicodemi, E.M.; Facchiano, A. Ion channels expression and function are strongly modified in solid tumors and vascular malformations. *J. Transl. Med.* **2016**, *14*, 285. [CrossRef]
52. McCrossan, Z.A.; Abbott, G.W. The MinK-related peptides. *Neuropharmacology* **2004**, *47*, 787–821. [CrossRef]
53. Li, Y.; Sung, Y.U.; McDonald, T.V. Voltage-Gated Potassium Channels: Regulation by Accessory Subunits. *Neuroscientist* **2006**, *12*, 199–210. [CrossRef]
54. Kanda, V.A.; Abbott, G.W. KCNE Regulation of K$^+$ Channel Trafficking—A Sisyphean Task? *Front. Physiol.* **2012**, *3*, 231. [CrossRef] [PubMed]
55. Wang, C.-Y.; Lai, M.-D.; Phan, N.N.; Sun, Z.; Lin, Y.-C. Meta-Analysis of Public Microarray Datasets Reveals Voltage-Gated Calcium Gene Signatures in Clinical Cancer Patients. *PLoS ONE* **2015**, *10*, e0125766. [CrossRef] [PubMed]
56. Hong, S.; Bi, M.; Wang, L.; Kang, Z.; Ling, L.; Zhao, C. CLC-3 channels in cancer (Review). *Oncol. Rep.* **2015**, *33*, 507–514. [CrossRef] [PubMed]
57. Lui, V.C.H.; Lung, S.S.S.; Pu, J.K.S.; Hung, K.N.; Leung, G.K.K. Invasion of human glioma cells is regulated by multiple chloride channels including ClC-3. *Anticancer Res.* **2010**, *30*, 4515–4524.
58. Montuori, N.; Cosimato, V.; Rinaldi, L.; Rea, V.E.A.; Alfano, D.; Ragno, P. uPAR regulates pericellular proteolysis through a mechanism involving integrins and fMLF-receptors. *Thromb. Haemost.* **2013**, *109*, 309–318. [CrossRef]
59. Raghu, H.; Lakka, S.S.; Gondi, C.S.; Mohanam, S.; Dinh, D.H.; Gujrati, M.; Rao, J.S. Suppression of uPA and uPAR Attenuates Angiogenin Mediated Angiogenesis in Endothelial and Glioblastoma Cell Lines. *PLoS ONE* **2010**, *5*, e12458. [CrossRef]
60. Huai, Q.; Zhou, A.; Lin, L.; Mazar, A.P.; Parry, G.C.; Callahan, J.; Shaw, D.E.; Furie, B.; Furie, B.C.; Huang, M. Crystal structures of two human vitronectin, urokinase and urokinase receptor complexes. *Nat. Struct. Mol. Biol.* **2008**, *15*, 422–423. [CrossRef]
61. Høyer-Hansen, G.; Rønne, E.; Solberg, H.; Behrendt, N.; Ploug, M.; Lund, L.R.; Ellis, V.; Danø, K. Urokinase plasminogen activator cleaves its cell surface receptor releasing the ligand-binding domain. *J. Biol. Chem.* **1992**, *267*, 18224–18229.
62. Høyer-Hansen, G.; Behrendt, N.; Ploug, M.; Danø, K.; Preissner, K.T. The intact urokinase receptor is required for efficient vitronectin binding: Receptor cleavage prevents ligand interaction. *FEBS Lett.* **1997**, *420*, 79–85. [CrossRef]
63. Smith, S.J.; Tilly, H.; Ward, J.H.; MacArthur, D.C.; Lowe, J.; Coyle, B.; Grundy, R.G. CD105 (Endoglin) exerts prognostic effects via its role in the microvascular niche of paediatric high grade glioma. *Acta Neuropathol.* **2012**, *124*, 99–110. [CrossRef] [PubMed]
64. Birlik, B.; Canda, S.; Ozer, E. Tumour vascularity is of prognostic significance in adult, but not paediatric astrocytomas. *Neuropathol. Appl. Neurobiol.* **2006**, *32*, 532–538. [CrossRef] [PubMed]
65. Barresi, V.; Cerasoli, S.; Vitarelli, E.; Tuccari, G. Density of microvessels positive for CD105 (endoglin) is related to prognosis in meningiomas. *Acta Neuropathol.* **2007**, *114*, 147–156. [CrossRef] [PubMed]
66. Muenzner, P.; Rohde, M.; Kneitz, S.; Hauck, C.R. CEACAM engagement by human pathogens enhances cell adhesion and counteracts bacteria-induced detachment of epithelial cells. *J. Cell Biol.* **2005**, *170*, 825–836. [CrossRef] [PubMed]
67. Aithal, M.G.S.; Rajeswari, N. Validation of Housekeeping Genes for Gene Expression Analysis in Glioblastoma Using Quantitative Real-Time Polymerase Chain Reaction. *Brain Tumor Res. Treat.* **2015**, *3*, 24–29. [CrossRef] [PubMed]
68. Wang, H.-Y.; Wang, W.; Liu, Y.-W.; Li, M.-Y.; Liang, T.-Y.; Li, J.-Y.; Hu, H.-M.; Lu, Y.; Yao, C.; Ye, Y.-Y.; et al. Role of KCNB1 in the prognosis of gliomas and autophagy modulation. *Sci. Rep.* **2017**, *7*, 14. [CrossRef]

Article

Towards Performant Design of Carbon-Based Nanomotors for Hydrogen Separation through Molecular Dynamics Simulations

Sebastian Muraru [1,2] and Mariana Ionita [1,2,*]

[1] Faculty of Medical Engineering, University Politehnica of Bucharest, GhPolizu 1-7, 011061 Bucharest, Romania; sebmuraru@gmail.com

[2] Advanced Polymer Materials Group, University Politehnica of Bucharest, GhPolizu 1-7, 011061 Bucharest, Romania

* Correspondence: mariana.ionita@polimi.it

Received: 10 November 2020; Accepted: 14 December 2020; Published: 16 December 2020

Abstract: Clean energy technologies represent a hot topic for research communities worldwide. Hydrogen fuel, a prized alternative to fossil fuels, displays weaknesses such as the poisoning by impurities of the precious metal catalyst which controls the reaction involved in its production. Thus, separating H_2 out of the other gases, meaning CH_4, CO, CO_2, N_2, and H_2O is essential. We present a rotating partially double-walled carbon nanotube membrane design for hydrogen separation and evaluate its performance using molecular dynamics simulations by imposing three discrete angular velocities. We provide a nano-perspective of the gas behaviors inside the membrane and extract key insights from the filtration process, pore placement, flux, and permeance of the membrane. We display a very high selectivity case ($\omega = 180°\ ps^{-1}$) and show that the outcome of Molecular Dynamics (MD) simulations can be both intuitive and counter-intuitive when increasing the ω parameter ($\omega = 270°\ ps^{-1}$; $\omega = 360°\ ps^{-1}$). Thus, in the highly selective, $\omega = 180°\ ps^{-1}$, only H_2 molecules and 1–2 H_2O molecules pass into the filtrate area. In the $\omega = 270°\ ps^{-1}$, H_2, CO, CH_4, N_2, and H_2O molecules were observed to pass, while, perhaps counter-intuitively, in the third case, with the highest imposed angular velocity of $360°\ ps^{-1}$ only CH_4 and H_2 molecules were able to pass through the pores leading to the filtrate area.

Keywords: hydrogen separation; rotating carbon nanotube membrane; molecular dynamics

1. Introduction

Predominantly, hydrogen meant to be used as fuel is obtained through steam reforming gas, with the following chemical reactions involved [$CH_4 + H_2O \rightarrow CO + 3H_2$; $CO + H_2O \rightarrow CO_2 + H_2$]. Any impurities, such as CH_4, CO_2, CO, N_2, and H_2O molecules can significantly reduce the performance of a hydrogen fuel cell by poisoning its precious metal catalyst driving forward the electrochemical reactions [1,2]. Thus, separating the H_2 molecules out of the mixture of gases is crucial for keeping fuel cells efficient. Mature separation technologies such as pressure swing absorption and cryogenic distillation have lost ground to newer solutions, e.g., separation membranes due to their affordability and easy operation [2–5]. Many of these have been designed using different metals, silica, zeolite, various polymers, and carbon, yet without coming upon an ideal solution [1,3–7].

Carbon, as one of the most common and versatile elements in the universe, has continuously attracted significant interest within the scientific community over the span of several decades. Constant discovery of new allotropes, each displaying unique properties, allowed for carbon to be used for an extraordinary array of purposes and applications [8,9]. In particular, synthetic allotropes such as the one-dimensional carbon nanotubes [10] and the two-dimensional graphene [11] have ignited enormous interest within the materials science community. Despite being impermeable to water and gas molecules, techniques such as ion bombardment or e-beam lithography [3,12] can be used to create tailored nano-pores and permeabilize these structures.

Molecular dynamics (MD) simulations represent a computational technique useful in acquiring a nano, atom-level perspective of all molecules making up a defined system. Complete trajectories are built step-by-step for each atom present in a simulation box. Although it is used ultimately to complement experimental research, it may also be used to uncover key insights entirely on its own. This is possible because a computational researcher can explore scenarios that may prove too costly, time-consuming, or simply inaccessible yet to experimental research. Here, we make use of MD simulations to explore certain useful parameters during hydrogen separation using a theoretical framework. Several MD-only studies have investigated the molecular mechanisms useful in describing the separation potential of carbon-based structures, either for gas separation [12–18] or water desalination [19,20].

The recently observed rotational motion of double-walled carbon nanotubes exposed to an electric field [21] opened the door to more complex MD simulations of carbon-based nanomotors, investigating areas such as behavior in water, water transport, and desalination [21–24]. To our knowledge, there are no existing MD studies on carbon-based nanomotors evaluating gas separation. In this study, we explored the impact of an imposed angular velocity on the performance of H_2 separation of a rotating partially double-walled carbon nanotube design out of a mixture of gases containing H_2, CH_4, CO, CO_2, N_2, and H_2O molecules. The gases, placed inside the rotating carbon nanotube could either exit through the nanopores present along the length of the nanotube or through one of its ends which was left open. Performance was determined by analyzing the filtrated molecules, pore placement, and exploring parameters such as flux and permeance. In a practical setting, the angular velocity imposed on the rotating carbon nanotube can be generated by the presence of an electric field, either directly or through a gear-based mechanism, for which we have provided a segment on our design, leaving it pore-free. We found a favorable case of imposed angular velocity leading to impressive selectivity ($\omega = 180°$ ps^{-1}) and showed that the outcome of MD simulations can be both intuitive and counter-intuitive when increasing the ω parameter ($\omega = 270°$ ps^{-1}; $\omega = 360°$ ps^{-1}). The design allows for the gas mixture to be exposed to a high surface area; each gas molecule always traveling towards a pore. H_2 molecules were able to quickly exit the rotating nanotube by the 10ns mark and no gas molecules were able to cross back into the membrane leading to overall increased efficiency. In addition, flux values were comparable to graphene pores corresponding to 10-16 removed atoms. Overall, we have shown that the angular velocity plays a significant role in hydrogen separation through a rotating nanotube and discussed key parameters involved in the separation process.

2. Results and Discussion

2.1. Choosing Pore Shape and Size

In this study, the pores' shapes and sizes, shown in Figure 1A,B, were chosen based on our previous investigation on gas separation membranes [18]. The larger pore, corresponding to the removal of 14-atoms out of a graphene sheet, was proven to allow all molecules of the described gas mixture to pass through, while the smaller pore, corresponding to the removal of a 6-atom graphene ring, was shown to allow only the passage of H_2 and H_2O molecules.

An important aspect to mention when conducting investigations similar to ours, obvious when looking at Figure 2A–C,G–I,M–O, is that some deformation of the carbon nanotube occurred during each of the simulations due to the imposed rotational movements. This aspect can be observed easiest by looking at the no gas area. The same area is specifically shown in Figure 3. These structural deformations can lead to changes that occur in the shape and size of the pores situated on the rotating carbon nanotube and thus different gas separation behaviors. In addition, looking at Figure 2D–F,J–L,P–R, no gas molecules ever cross back into the membrane. The small fluctuations seen in the case of H_2 molecules are due to the calculation method (the H_2 number on the graphs always goes back up quickly).

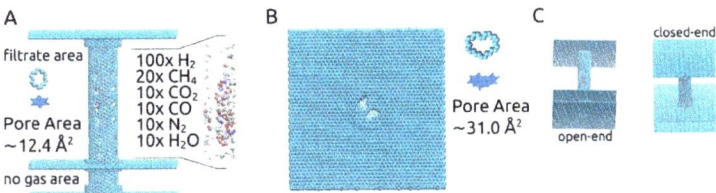

Figure 1. (**A**) Outline of the rotating carbon nanotube membrane, together with gas mixture and small pore details; (**B**) View of the open-end together with the larger pore details; (**C**) Perspectives on the membrane showing the open-end and the closed-end.

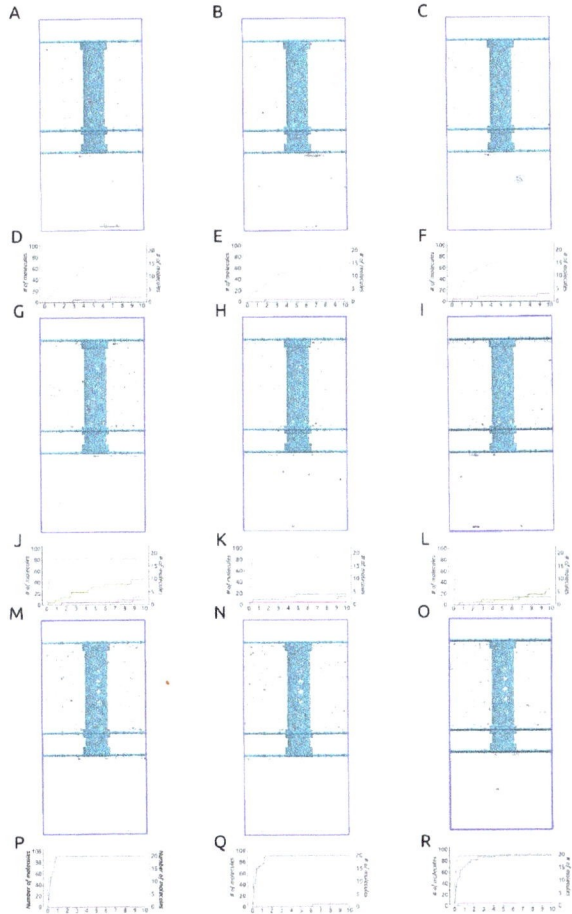

Figure 2. (**A–C**). ($\omega = 180°$ ps^{-1}), (**G–I**). ($\omega = 270°$ ps^{-1}), (**M–O**). ($\omega = 360°$ ps^{-1}) Final frame of the MD simulations (10 ns); (**D–F**). ($\omega = 180°$ ps^{-1}), (**J–L**). ($\omega = 270°$ ps^{-1}), (**P–R**). ($\omega = 360°$ ps^{-1}) Evolution of the number of molecules passing into the filtrate area during simulations time.

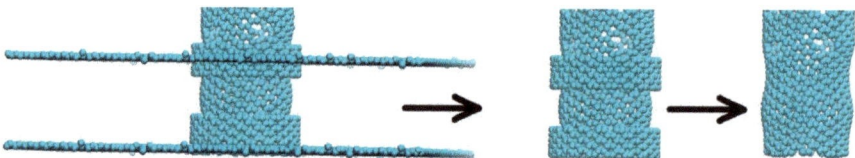

Figure 3. Example of the structural deformation occurring in the rotating carbon nanotube due to imposed angular velocity.

Effective pore areas A_p were determined using a Monte Carlo hit-and-miss procedure [14,16,18], considering the effective carbon atom radius $R_{eff} = R_{m,c}/\sqrt{2}$ and $R_{m,c}$ equal to 0.17 nm. The approximate results are shown numerically in Figure 1A,B, meaning ~12.4Å^2 for the smaller pores located along the carbon nanotube and ~31.0Å^2 for the larger pores placed in the graphene layer at one of its ends. Blue-filled shapes for each pore type showing their effective pore areas are also shown in Figure 1A,B (images not to scale). The two values were calculated considering the pores to be sculpted in a planar 2D graphene sheet. This approach fits very well in the case of the larger pore. For the smaller pore, however, determining the area in such a manner is less accurate due to being sculpted in a carbon nanotube with a chiral vector determined by the indices (12, 16). Nevertheless, it still provides a useful approximate value. Using the same method for calculating the smaller pore's area, now when sculpted in the chiral nanotube, we ended up with the following approximate values: 12.26 Å^2 at 0ns before the start of the simulations, 12.51Å^2 for $\omega = 180°$ ps^{-1}, 13.02Å^2 for $\omega = 270°$ ps^{-1} and 14.70Å^2 for $\omega = 360°$ ps^{-1}.

Thus, given the imposed angular velocities to the carbon nanotube, changes in the shape and size of the smaller pores took place. The shape and size of the pores, together with the angular velocity of the nanotube are the main parameters behind the distinct results obtained here for each ω case. This aspect helps explain the results obtained in this paper. The influence of an imposed angular velocity on the geometry of a carbon nanotube and pores sculpted on its surface can be investigated in future studies.

The rationale behind keeping an "open" end to the membrane, represented by the graphene layer with the large pores placed as pictured in Figure 1A,B, was to allow for the gas mixture to be fed into the carbon nanotube. This obviously led to some H_2 molecules to be lost as they were able to exit the membrane through the pores situated at the end of the tube instead of crossing through the smaller pores into the filtrate area. However, we deemed this "open" end setup more realistic and practical. Given our previous results [18], which confirmed that all gas molecules present in the gas mixture used here are able to pass through the 31.0 Å^2 pore, we focused the current investigation on the separation process that would occur once the gas mixture is already loaded into the rotating carbon nanotube membrane and the effect of the imposed angular velocity.

2.2. Simulation Results

Curiously, given the setups described, all three ω cases yielded significantly different results, consistent throughout all three runs, as can be observed in Figure 2. We must mention that in all nine simulations, all H_2 molecules had either crossed through the smaller pores or the larger pores by the 10 ns mark. Thus, all of them had left the rotating carbon nanotube by the time-point at which we stopped our simulations. On top of that, given the three tested values for the imposed angular velocity, at least 60% of the H_2 molecules were able to cross into the filtrate area within the simulated time in all given setups.

2.3. Angular Velocity $\omega = 180°$ ps^{-1}

In the case of $\omega = 180°$ ps^{-1}, consistent with our previous investigations [18], only H_2 and H_2O molecules were able to pass through the smaller pores. The process of H_2 molecules passing through the smaller pores and entering the filtrate area took place in about 7–8 ns, with ~20 H_2 molecules passing within the first nanosecond and some other 40 to 50 molecules within the next 6–7 ns. In regards to H_2O molecules, only 1–2 molecules were able to pass into the filtrate area, while most of them (6–7) were still found inside the carbon nanotube at the end of the 10 ns. The reason for this is that although the smaller pores should theoretically allow them to pass through, these are found to cluster together inside the nanotube due to hydrogen bonding. In all our simulations this prevents most of them from leaving the rotating membrane, given the pore sizes used. No other gas molecules were able to pass through the smaller pores at this angular velocity.

Taking a look at the process of separation occurring throughout our simulations, we observed that some of the gas molecules inside the nanotube position themselves as if blocking the smaller pores and remain positioned so, despite the rotating motion. This aspect is shown in Supplementary Video S1 and in Figure 4A with CH_4, CO, H_2O, N_2, and H_2 molecules blocking some of the pores. We also show, in Figure 4B, that by the end of the simulation, at 10ns, pores were still blocked by the larger molecules at a time at which H_2 molecules were no longer found inside the rotating nanotube. We think that due to the imposed rotational motion and the size and shape of the chosen smaller pores, the smaller molecules such as H_2 and H_2O are able to eventually exit through the pores they are blocking. Thus, with an imposed ω of 180° ps^{-1}, some degree of fouling will occur, leaving fewer pores open for H_2 molecules to cross and leading more to pass through the open end of the membrane in exchange for the high selectivity offered. Between 59 and 73 out of 100 H_2 molecules were able to pass into the filtrate area throughout the three simulations with an imposed angular velocity of 180° ps^{-1}. Interestingly, no CO_2 molecules were found to block any of the smaller pores throughout all simulations but were instead able to pass through the larger pores as can be seen in Figure 2I.

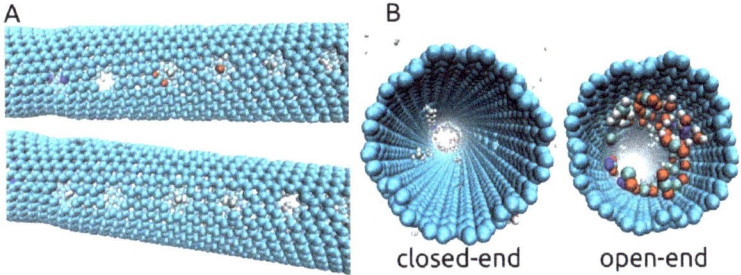

Figure 4. (**A**). Cases of N_2, CH_4, CO, and H_2 gas molecules shown to block some of the pores situated along the rotating central carbon nanotube; (**B**). Closed-end and open-end views at the end of the simulations (ω = 180° ps^{-1}) showing the agglomeration of the molecules at one of the ends and the CH_4 molecules still blocking some of the pores.

At ω = 180° ps^{-1}, as it can be observed in Supplementary Video S1, some H_2 adsorption to the wall of the rotating carbon nanotube took place, which allowed the gas molecules to travel with the nanotube in its rotating motion. However, the phenomenon lasted for a short time (a few ps) before the adsorbed molecules were disturbed. We think both the imposed high angular velocity and the placement of the pores along the membrane, with each pore crossing disturbing the nearby gas molecules, help prevent adsorption from taking place and thus allow all H_2 molecules to leave the nanotube by the end of the simulation.

Another important aspect to discuss in the presented rotating carbon nanotube membrane design regards the placement of the pores leading to the filtrate area along the tube. In our case, as shown in Figure 1, we have placed the 24 smaller pores in the central part of the nanotube, leaving both ends with no pores to the filtrate area. As shown in Figure 5, at the end of all simulations (10 ns), unless found blocking one of the smaller pores, all molecules that remained in the rotating nanotube can be seen gathered at one of its ends. Without smaller pores nearby, the agglomeration of molecules becomes somewhat "stagnant" despite the rotational movement of the tube (see Supplementary Video S2).

This aspect may be beneficial to the flux of H_2 to the filtrate area as the smaller pores are then less likely to be blocked.

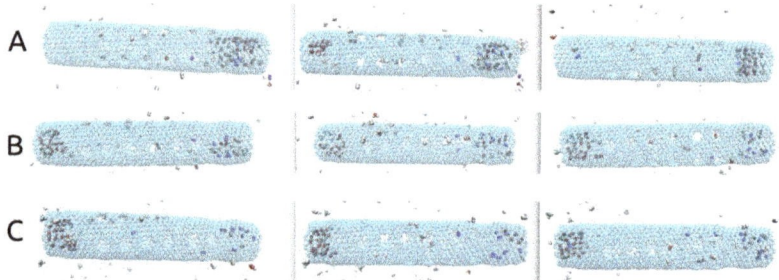

Figure 5. (**A**) ($\omega = 180°$ ps^{-1}), (**B**) ($\omega = 270°$ ps^{-1}), (**C**) ($\omega = 360°$ ps^{-1}) Representations of the rotating central carbon nanotube at 10ns simulation time showing the gas molecules gathered at one of its ends.

Additionally, inspecting Figure 5, all H_2 molecules exited by the 10ns mark, and only CH_4, CO, CO_2, N_2, and H_2O molecules (clustered) can be seen inside. To provide an explanation for this we looked at the movement of H_2 molecules.

As shown in Supplementary Video S2, the molecules in the mixture with a smaller mass, meaning H_2, CH_4, and H_2O were able to move slightly quicker when inside the rotating nanotube. On top of that, due to their small volume, which allowed them to "squeeze" through the larger molecules and their tendency to adsorb to the wall of the nanotube, H_2 molecules had a high number of collisions. Together with the chirality of the nanotube, we think these aspects lead to the exiting of all H_2 molecules within the 10 ns timeframe as they are more likely to travel along the nanotube and exit either towards the filtrate area or through the "open" end. The movement of H_2 molecules within rotating carbon nanotubes of different chiralities could be investigated in future studies to determine whether chirality can drive the gas molecules due to their tendency to adsorb to the wall of the nanotube.

2.4. Angular Velocity $\omega = 270°$ ps^{-1}

For the higher imposed angular velocities, significantly different outcomes were observed compared to the $\omega = 180°$ ps^{-1} case. The outcomes were consistent throughout all three repetitions corresponding to a certain ω value. In the case of $\omega = 270°$ ps^{-1}, H_2, CO, CH_4, N_2, and H_2O molecules were observed to pass through the smaller pores, as shown in Figure 2G–L. Almost all H_2 molecules that passed into the filtrate area (~80–85) did so in the first half nanosecond. Again, no H_2 molecules were found inside the tube at the end of the 10 ns. Most molecules that passed through the smaller pores did so after blocking the pore for a while and exited either due to a collision with another gas molecule or due to the rotational motion of the nanotube, which allowed for slight changes in the orientation of the molecules. At this imposed angular velocity, despite that more H_2 molecules reached the filtrate area, there was far less selectivity leading to a poor separation performance.

The movement of the H_2 molecules inside the tube was slightly different compared to the $\omega = 180°$ ps^{-1} case, please see Supplementary Video S3. Due to the higher imposed angular velocity, H_2 molecules that were not blocking a pore were far less able to adsorb to the wall of the tube and travel in a synchronized manner with its rotational motion (as observed when $\omega = 180°$ ps^{-1}). Thus, due to the collisions with the wall, some H_2 molecules traveling in a rotational motion nearby, did so both in the sense of the rotation and in the anti-sense, but slower when compared to the rotating wall. Essentially, this should have led to a higher number of collisions taking place inside the tube, which, alongside with the enlargement of the pores due to the larger angular velocity, enabled most molecules to exit into the filtrate area. CH_4 molecules were also observed to pass through the pores after blocking it for a while and following a collision. No CO_2 molecules were able to pass through the smaller pores. The results obtained for $\omega = 270°$ ps^{-1} case are somewhat intuitive in that the higher angular velocity led to the expansion of the smaller pores and plummeted its selectivity while allowing more H_2 molecules to pass into the filtrate area. From these simulations one can also see the following insight: as most remaining gas molecules at the 10ns mark are either found blocking one of the smaller pores or situated at one of the ends of the tube, the placement of the smaller pores along the center of the rotating carbon nanotube membrane could limit the passage of larger gas molecules into the filtrate area, thus improving the performance of the membrane.

2.5. Angular Velocity $\omega = 360°$ ps^{-1}

In the final case of $\omega = 360°$ ps^{-1}, as shown in Figure 2M–R, 89 to 94 H_2 molecules were able to pass through the smaller pores in less than 0.5 ns. The H_2 molecules that passed through the larger pores did so within 0.3 ns since the start of the simulations. Curiously, the only other gases which passed into the filtrate area at the end of the simulated time were CH_4 and H_2O molecules. Consistent throughout all three simulations with $\omega = 360°$ ps^{-1}, all 20 CH_4 molecules passed through the smaller pores and, similarly to the previously mentioned cases, $\omega = 180°$ ps^{-1} and $\omega = 270°$ ps^{-1}, only one H_2O molecule managed to do so, due to the clustering of the water molecules inside the tube. The fact that these results were consistent throughout all three repetitions shows, counter-intuitively, that using higher angular velocities values can lead to unique and unexpected insights. The fact that almost 90–95% of H_2 molecules placed initially in the tube, together with 100% of the CH_4 molecules and only 10% of the water molecules were able to pass into the filtrate area could be useful in future studies working on improving the current design by making use of a gradual separative process. In such a case, the second step would involve only the separation of H_2 and CH_4 molecules, as, curiously, no other gas molecules were able to pass into the filtrate area.

Observing the movement of the molecules inside the rotating carbon nanotube at $\omega = 360°$ ps^{-1} shows the molecules with a smaller mass, H_2, CH_4, and H_2O moving significantly faster than the larger CO, N_2, and CO_2 molecules (see Supplementary Video S4). On top of that, due to collisions with the walls of the tube, the same smaller mass molecules rotate around their own center of mass much quicker than the larger ones, with CH_4 being a special case and rotating faster than all linear molecules due to its geometry. Further on, given the quick rotation of the tube, no gas molecules were seen blocking any of the smaller pores or being able to adsorb to the rotating walls of the membrane. Thus, we think that when a gas molecule reaches the vicinity of a pore, due to the high angular velocity of the nanotube, the right orientation has to be found quickly in order to exit, which did not happen in the case of the larger linear molecules with a higher mass, but did take place in the case of the non-linear molecules with a smaller mass, meaning CH_4 and H_2O molecules. That is, gas molecules had a very short time to match an orientation that would have allowed them to cross the pore. After nearing a pore and colliding with the rotating wall, a small mass non-linear molecule was more likely to quickly rotate around its mass center and change its orientation quickly, which increased its chances of crossing. However, despite its linear structure, H_2 molecules crossed the smaller pores easily due to their very small mass and volume as these aspects allowed them to change their orientation quickly and match that necessary to exit through the pore. More details, solely in regards to pore crossing can be reached in further investigations through ab initio molecular dynamics (AIMD) techniques.

2.6. Flux and Permeance

To further characterize the rotating carbon nanotube design for gas separation, we calculated the flux of H_2 molecules through the smaller pores. Thus, in order to determine the total flux we counted the number of crossings from both inside and outside the tube and then made use of the formula below:

$$Flux = \frac{Crossing\ s_{H2}/N_A}{A_{innersurface} \times Time} \qquad (1)$$

where N_A represents Avogadro's number, $A_{inner\_surface}$ represents the inner surface of the tube and time corresponds to the time point of the last H_2 crossing during the simulation. In addition, given that the majority of H_2 molecules exit in a time window significantly shorter than the duration of the whole simulation, we have calculated the flux within the time frame in which 80% of H_2 molecules had left the rotating carbon nanotube membrane. The inner surface of the tube was determined to be 59.854 nm^2 using the formula $2\pi rh$. The radius of the tube, r, was determined using the formula:

$$Radius = \frac{a}{2\pi} \times \sqrt{(m \times m + m \times n + n \times n)} \qquad (2)$$

where a = 0.246 nm and m and n represent the chiral indices of the nanotube. The results are shown in Figure 6A. Given the calculated total flux values, permeance values for our membrane design were estimated and shown both in mol/m^2sPa and gas permeation units (GPUs) in Figure 6B.

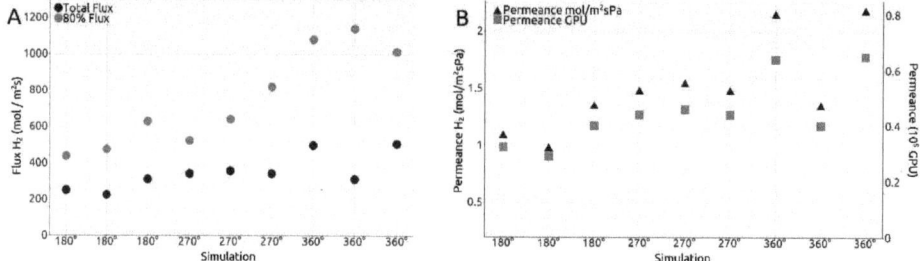

Figure 6. (**A**). Calculated total flux and 80% flux values; (**B**). Calculated permeance values in mol m^{-2} s^{-1} Pa^{-1} and Gas Permeation Units (GPU).

Dividing the flux values presented in Figure 6A to the number of smaller pores present along the inner surface of the carbon nanotube, we obtain the intervals shown in Table 1:

Table 1. Resulting flux intervals after dividing flux values to the number of smaller pores.

Angular Velocity-ω (deg/ps)	Total Flux (mol/m²s)	80% Flux (mol/m²s)
180	9.0–13.0	18.0–26.0
270	14.0–15.0	21.5–34.0
360	13.0–21.0	42.0–47.5

Despite the very different setups, comparing the total flux values obtained here (shown in Table 1) after division to the number of relevant pores, with the values presented in our previous research for a different design with the same gas separation purpose, the results are not significantly different from the interval 4.5–14.0 mol/m²s observed in [18] on the third filtration layer. The 80% flux intervals presented in Table 1 are higher, yet that is unsurprising, but these importantly highlight the manner in which the H_2 molecules cross the smaller pores in the current design, with most of them managing to exit the rotating nanotube in the first 5.0 ns or 0.5 ns depending on the ω case, yet much quicker than the last 20%.

Previous results were exceptional in terms of selectivity and permeability due to the gradual separation through multiple graphene layers each with its own specific pore size [18]. In the current design, we observed different behaviors, yet the exit of all H_2 molecules took place in a very short time and no H_2 molecules were left inside the nanotube at the end of the simulated 10 ns. Thus, nanopores used for filtration can be customized for certain molecules not only by selecting their size and 2D shape but also by using an imposed angular velocity and the forces involved in the consequent rotational motion, thus changing its geometry. Therefore, using a rotating carbon nanotube membrane allowed the gas molecules to be exposed to most of the surface area of the membrane, uncovering interesting distinct behaviors, dependent on the imposed angular velocity. Nevertheless, alongside the advantage of the quick exiting of all H_2 molecules within the membrane, a disadvantage of the current design is the loss of some of them due to the "open" end. Future studies may investigate different scenarios for improving the efficiency of the membrane. The easiest scenario to imagine being blocking the open end with a full graphene layer with no pores.

Comparing to Sun's results [16], the total flux values obtained in Figure 6A are comparable to a graphene pore size of 10 to 12 atoms, while the 80% flux values are comparable to a 12-atom graphene pore for $\omega = 180°$ ps^{-1}, 14-atom graphene pore for 270° ps^{-1} and 16-atom graphene pore for 360° ps^{-1}.

Naively comparing the flux values obtained for each ω case to each other, a clear trend can be seen showing that a higher imposed angular velocity leads to generally higher flux. However, an outlier can be seen in Figure 6A, with one total flux value in the $\omega = 360°$ ps^{-1} significantly smaller than the other two. This was also one of the reasons we decided to calculate the 80% flux values. The corresponding value to the outlier for the 80% flux is higher than the two other similar simulations, which shows that most of the H_2 molecules in the outlier simulation crossed into the filtrate area quicker and left the other 20% needing more time to cross due to, potentially, fewer collisions between molecules. Thus, with this explanation, we deem the outlier valid. One must take into account, however, that the heights of the total flux and the 80% flux values are not perfectly proportional due to events such as pore blocking, exit of other gas molecules, etc.

Similar to flux calculations, we estimated permeance values for the rotating carbon nanotube membrane, as shown in Figure 6B. Similar conclusions can be taken out of the permeance values as with the total flux calculations. We find all our simulations to indicate a value within the following intervals: 0.0097 to 0.0215 mol^{-3}/m^2sPa and 0.30 to 0.65 $\times 10^5$ GPU, and thus in the vicinity of those calculated for our previous design, with values situated between 0.005 to 0.03 mol^{-3}/m^2sPa and 0.20 to 0.85 $\times 10^5$ GPU, results already highlighted as superior to state-of-the-art solutions [18].

3. Materials and Methods

3.1. Membrane Prototype

The structure of the membrane is shown in Figure 1A,B: a central carbon nanotube with indices (12, 16) and 10 nm in length is held in place by two 10 nm × 10 nm graphene sheets placed at its ends. At one end the graphene sheet is intact ("closed-end"), while at the other, two pores corresponding to the removal of 14-atoms were made ("open-end"). The central carbon nanotube is partially doubled by three short carbon nanotubes with the indices (19, 19) and 0.5 nm in length. The radius difference corresponds to the 0.334 nm distance met in graphite sheets, situated closely, at 0.3358 nm. Two of the short carbon nanotubes are placed at the central nanotube's ends and one is placed 2 nm away from its open end, which is then inserted in an additional 10 nm × 10 nm graphene layer, dividing the volume outside the central carbon nanotube into the "filtrate area" and "no gas area". The nanotubes and graphene sheets were generated using VMD [25]. To make the design more applicable to experimental settings, the "no gas area" represents a 2 nm-long pore-free segment for moving the central carbon nanotube through nano gear-based mechanisms. Such may be a very complex undertaking for molecular dynamics studies due to the need of obtaining a constant rotating motion and constant velocity and thus remains to be investigated in future studies. Along the center of the carbon nanotube, four columns of six smaller pores, corresponding to the removal of six atoms making up a hexagonal ring, are present. A gas mixture of 100× H_2, 20× CH_4, 10× CO_2, 10× CO, 10× N_2, and 10× H_2O molecules are placed randomly inside the central carbon nanotube, which is free to rotate around its own axis.

3.2. Simulation Details and Force Field Parameters

All simulations were run using the OPLS-AA force field, together with GROMACS 2018.3. The rotational motion of the central carbon nanotube was achieved using the Enforced Rotation module and iso-pf potential. Three different angular velocities were imposed, with ω = 180° ps^{-1}, 270° ps^{-1}, and 360° ps^{-1}. For each case, three repetitions were run using the rotational force constant 500 kJ mol^{-1} nm^{-2}. All simulations were run for 20,000,000 steps, using a step size of 0.5 fs and summing up to 10 ns simulated time. Coordinates for visualizations were saved every 5 fs. Simulation box size was 10 nm × 10 nm × 20 nm and contained ~15,000 atoms. Periodic boundary conditions were active in the X and Y directions. On the Z-axis, the top and bottom of the simulation box contained two walls made out of Kr atoms (Lenard Jones (LJ) 10–4) with a number density of 5 nm^{-2}. In order to confer a repulsive character, interactions with the Kr atoms were cut at 0.1 nm. VdW and Coulomb cutoff was set at 1.0 nm. The water model used was SPC/E. The Verlet cut-off scheme was used, together with the V-rescale thermostat at 300 K.

Most parameters for the gas molecules were taken from previous studies using the OPLS-AA forcefield [12]. H_2 and N_2 molecules were modeled as three-site models, with one virtual mass-less atom in the center of the molecules, while CO_2 molecules were built as a five-site model, thus using two virtual atoms, as shown in Lemkul's tutorial [26]. CH_4 and CO molecules were built without the aid of virtual atoms and thus contained five, respectively two atoms.

Carbon atoms making up the setup of the membrane were considered uncharged LJ spheres with cross-section 0.34 nm, potential wall depth 0.36 kJ mol^{-1}, C-C bond length 1.42 Å, C-C-C bending angle 120°, and C-C-C-C planar angles maintained by harmonic potentials with springs constants of 322.55 kcal mol^{-1} Å^{-2}, 53.35 kcal mol^{-1} rad^{-2} and 3.15 kcal mol^{-1} [27].

For data analysis we made use of Python libraries such as NumPy, bokeh [28] and MDAnalysis [29,30].

4. Conclusions

We have presented a rotational partial double-walled carbon nanotube hydrogen separation membrane design relying on the recently observed rotational motion of double-walled carbon nanotubes when exposed to an electric field [21] and investigated its gas separation performance under three different imposed angular velocities (180° ps^{-1}, 270° ps^{-1}, and 360° ps^{-1}), while observing the movement of the gases inside the membrane. We have shown that the outcomes of the simulations are dependent on the structural changes occurring in the rotating membrane and affecting the size and shape of the smaller pores leading to the filtrate area. We have presented a case of very high selectivity for the current setup (ω = 180° ps^{-1}) and highlighted the advantages of the design such as the fast exit of the H_2 molecules from within the nanotube. We have provided explanations for its interesting strong points such as the exit of all H_2 molecules within simulated time irrespective of the behavior of the other gases, a phenomenon consistent through all our simulations. Given the significantly different simulation outcomes, we deem that in an experimental setting, tightly controlling the angular velocity would be paramount. Given the imposed angular velocity cases presented, we have shown that the outcome of MD simulations can be both intuitive and counter-intuitive: compared to the 180° ps^{-1} case, an increase in the flux of H_2 molecules in the 270° ps^{-1} case came with a plummeting of the selectivity and allowed many other gas molecules to pass into the filtrate area; however, a further increase of the imposed angular velocity to 360° ps^{-1} lead to the exit through the smaller pores of only H_2 and CH_4 molecules in spite of the increased pore size, making for a very interesting insight useful in gradual separation designs. Compared to other designs, in the presented prototype, the gas mixture was exposed to a large surface area with each gas molecule always traveling towards a pore. In addition, no gas molecules were able to cross back into the membrane.

We have provided flux and permeance estimates and found them comparable to our previous designs and superior to existing experimental solutions [18]. On top of that, our 80% flux values were found comparable to 12-atoms, 14-atoms, and 16-atoms graphene pores [16]. Thus, we have shown that the angular velocity plays a significant role in hydrogen separation through a rotating nanotube.

We deem further experimentation involving MD investigations on novel gas separation membrane designs will lead to valuable insights for the hydrogen fuel industry.

Supplementary Materials: The following are available online at http://www.mdpi.com/1422-0067/21/24/9588/s1.

Author Contributions: Conceptualization, M.I. and S.M.; methodology, S.M.; software, S.M.; validation, S.M.; formal analysis, S.M; investigation, M.I and S.M.; resources, M.I and S.M.; data curation, S.M.; writing—original draft preparation, S.M.; writing—review and editing, S.M. and M.I.; visualization, S.M.; supervision, M.I.; project administration, M.I.; funding acquisition, M.I. All authors have read and agreed to the published version of the manuscript.

Funding: This work was supported by a grant of the Executive Agency for Higher Education, Research, Development and innovation funding (UEFISCDI), project number PN-III-P1-1.1-TE-2016-24-2, contract TE 122/2018. Sebastian Muraru acknowledges funding support from a grant of the Ministry of Research and Innovation, Operational Program Competitiveness Axis 1—Section E, Program co-financed from European Regional Development Fund under the project number 154/25.11.2016, P_37_221/2015, "A novel graphene biosensor testing osteogenic potency; capturing best stem cell performance for regenerative medicine" (GRABTOP).

Conflicts of Interest: The authors declare no conflict of interest.

Abbreviations

MD   Molecular Dynamics

References

1. Bernardo, G.; Araújo, T.; Lopes, T.D.S.; Sousa, J.; Mendes, A. Recent advances in membrane technologies for hydrogen purification. *Int. J. Hydrogen Energy* **2020**, *45*, 7313–7338. [CrossRef]
2. Zhu, L.; Xuea, Q.; Li, X.; Jin, Y.; Zheng, H.; Wu, T.; Guo, Q. Theoretical Prediction of Hydrogen Separation Performance of Two-Dimensional Carbon Network of Fused Pentagon. *ACS Appl. Mater. Interfaces* **2015**, *7*, 28502–28507. [CrossRef]
3. Bernardo, P.; Clarizia, G. 30 years of membrane technology for gas separation. *Chem. Eng. Trans.* **2013**, *32*, 1999–2004.
4. Ockwig, N.W.; Nenoff, T.M. Membranes for Hydrogen Separation. *Chem. Rev.* **2007**, *107*, 4078–4110. [CrossRef] [PubMed]
5. Saraswathi, M.S.S.A.; Nagendran, A.; Rana, D. Tailored polymer nanocomposite membranes based on carbon, metal oxide and silicon nanomaterials: A review. *J. Mater. Chem. A* **2019**, *7*, 8723–8745. [CrossRef]
6. Li, P.; Wang, Z.; Qiao, Z.; Liu, Y.; Cao, X.; Li, W.; Wang, J.; Wang, S. Recent developments in membranes for efficient hydrogen purification. *J. Membr. Sci.* **2015**, *495*, 130–168. [CrossRef]
7. Liang, C.Z.; Chung, T.-S.; Lai, J.-Y. A review of polymeric composite membranes for gas separation and energy production. *Prog. Polym. Sci.* **2019**, *97*, 101141. [CrossRef]
8. Tiwari, S.K.; Kumar, V.; Huczko, A.; Oraon, R.; De Adhikari, A.; Nayak, G.C. Magical Allotropes of Carbon: Prospects and Applications. *Crit. Rev. Solid State Mater. Sci.* **2016**, *41*, 257–317. [CrossRef]
9. Hirsch, A. The era of carbon allotropes. *Nat. Mater.* **2010**, *9*, 868–871. [CrossRef]
10. Monthioux, M.; Kuznetsov, V.L. Who should be given the credit for the discovery of carbon nanotubes? *Carbon* **2006**, *44*, 1621–1623. [CrossRef]
11. Novoselov, K.S.; Geim, A.K.; Morozov, S.V.; Jiang, D.; Zhang, Y.; Dubonos, S.V.; Grigorieva, I.V.; Firsov, A.A. Electric Field Effect in Atomically Thin Carbon Films. *Science* **2004**, *306*, 666–669. [CrossRef] [PubMed]
12. Jiao, S.; Xu, Z. Selective Gas Diffusion in Graphene Oxides Membranes: A Molecular Dynamics Simulations Study. *ACS Appl. Mater. Interfaces* **2015**, *7*, 9052–9059. [CrossRef] [PubMed]
13. Ye, H.; Li, D.; Ye, X.; Zheng, Y.; Zhang, Z.; Zhang, H.; Chen, Z. An adjustable permeation membrane up to the separation for multicomponent gas mixture. *Sci. Rep.* **2019**, *9*, 7380. [CrossRef]
14. Sun, C.; Bai, B. Fast mass transport across two-dimensional graphene nanopores: Nonlinear pressure-dependent gas permeation flux. *Chem. Eng. Sci.* **2017**, *165*, 186–191. [CrossRef]
15. Shan, M.; Xuea, Q.; Jing, N.; Ling, C.; Zhang, T.; Yan, Z.; Zheng, J. Influence of chemical functionalization on the CO_2/N_2 separation performance of porous graphene membranes. *Nanoscale* **2012**, *4*, 5477–5482. [CrossRef] [PubMed]

16. Sun, C.; Boutilier, M.S.H.; Au, H.; Poesio, P.; Bai, B.; Karnik, R.; Hadjiconstantinou, N.G. Mechanisms of Molecular Permeation through Nanoporous Graphene Membranes. *Langmuir* **2014**, *30*, 675–682. [CrossRef]
17. Wesołowski, R.P.; Terzyk, A.P. Pillared graphene as a gas separation membrane. *Phys. Chem. Chem. Phys.* **2011**, *13*, 17027–17029. [CrossRef]
18. Muraru, S.; Ionita, M. Super Carbonaceous Graphene-based structure as a gas separation membrane: A Non-Equilibrium Molecular Dynamics Investigation. *Compos. Part B Eng.* **2020**, *196*, 108140. [CrossRef]
19. Boretti, A.; Al-Zubaidy, S.; Vaclavikova, M.; Al-Abri, M.; Castelletto, S.; Mikhalovsky, S. Outlook for graphene-based desalination membranes. *NPJ Clean Water* **2018**, *1*, 5. [CrossRef]
20. Roy, K.; Mukherjee, A.; Raju, M.N.; Chakraborty, S.; Shen, B.; Li, M.; Du, D.; Peng, Y.; Lu, F.; Cruzatty, L.C.G. Outlook on the bottleneck of carbon nanotube in desalination and membrane-based water treatment—A review. *J. Environ. Chem. Eng.* **2020**, *8*, 103572. [CrossRef]
21. Bailey, S.W.D.; Amanatidis, I.; Lambert, C.J. Carbon Nanotube Electron Windmills: A Novel Design for Nanomotors. *Phys. Rev. Lett.* **2008**, *100*, 256802. [CrossRef]
22. Feng, J.-W.; Ding, H.-M.; Ren, C.-L.; Ma, Y.-Q. Pumping of water by rotating chiral carbon nanotube. *Nanoscale* **2014**, *6*, 13606–13612. [CrossRef] [PubMed]
23. Tu, Q.; Yang, Q.; Wang, H.; Li, S. Rotating carbon nanotube membrane filter for water desalination. *Sci. Rep.* **2016**, *6*, 26183. [CrossRef] [PubMed]
24. Rahman, M.; Chowdhury, M.M.; Alam, K. Rotating-Electric-Field-Induced Carbon-Nanotube-Based Nanomotor in Water: A Molecular Dynamics Study. *Small* **2017**, *13*, 1603978. [CrossRef] [PubMed]
25. Humphrey, W.; Dalke, A.; Schulten, K. VMD: Visual molecular dynamics. *J. Mol. Graph.* **1996**, *14*, 33–38. [CrossRef]
26. Lemkul, J.A. From Proteins to Perturbed Hamiltonians: A Suite of Tutorials for the GROMACS-2018 Molecular Simulation Package [Article v1.0]. *Living J. Comput. Mol. Sci.* **2019**, *1*, 5068. [CrossRef]
27. Chen, J.; Wang, X.; Dai, C.; Chen, S.; Tu, Y. Adsorption of GA module onto graphene and graphene oxide: A molecular dynamics simulation study. *Phys. E Low Dimens. Syst. Nanostruct.* **2014**, *62*, 59–63. [CrossRef]
28. Bokeh Development Team. Bokeh: Python Library for Interactive Visualization. 2020. Available online: https://bokeh.org (accessed on 10 October 2020).
29. Gowers, R.J.; Linke, M.; Barnoud, J.; Reddy, T.J.E.; Melo, M.N.; Seyler, S.L.; Domanski, J.; Dotson, D.L.; Buchoux, S.; Kenney, I.M.; et al. MDAnalysis: A Python Package for the Rapid Analysis of Molecular Dynamics Simulations. In Proceedings of the 15th Python in Science Conference (SciPy), Austin, TX, USA, 11–17 July 2016; pp. 98–105. [CrossRef]
30. Michaud-Agrawal, N.; Denning, E.J.; Woolf, T.B.; Beckstein, O. MDAnalysis: A toolkit for the analysis of molecular dynamics simulations. *J. Comput. Chem.* **2011**, *32*, 2319–2327. [CrossRef]

Publisher's Note: MDPI stays neutral with regard to jurisdictional claims in published maps and institutional affiliations.

© 2020 by the authors. Licensee MDPI, Basel, Switzerland. This article is an open access article distributed under the terms and conditions of the Creative Commons Attribution (CC BY) license (http://creativecommons.org/licenses/by/4.0/).

Article

Analysis of Biomechanical Parameters of Muscle Soleus Contraction and Blood Biochemical Parameters in Rat with Chronic Glyphosate Intoxication and Therapeutic Use of C_{60} Fullerene

Dmytro Nozdrenko [1], Olga Abramchuk [2], Svitlana Prylutska [1,3], Oksana Vygovska [4], Vasil Soroca [1], Kateryna Bogutska [1], Sergii Khrapatyi [5], Yuriy Prylutskyy [1], Peter Scharff [6] and Uwe Ritter [6,*]

1. Department of Biophysics and Medical Informatic, Taras Shevchenko National University of Kyiv, 01601 Kyiv, Ukraine; ddd@univ.kiev.ua (D.N.); psvit_1977@ukr.net (S.P.); vmsoroka@gmail.com (V.S.); bogutska_ki@knu.ua (K.B.); prylut@ukr.net (Y.P.)
2. Lesya Ukrainka Volyn National University, 43025 Lutsk, Ukraine; Abramchuk.Olga@vnu.edu.ua
3. National University of Life and Environmental Science of Ukraine, 03041 Kyiv, Ukraine
4. Bogomolets National Medical University of Kyiv, 01601 Kyiv, Ukraine; ovvigovskaya@gmail.com
5. Interregional Academy of Personnel Management, 03039 Kyiv, Ukraine; khrapatiysv@gmail.com
6. Institute of Chemistry and Biotechnology, Technical University of Ilmenau, 98693 Ilmenau, Germany; peter.scharff@tu-ilmenau.de
* Correspondence: uwe.ritter@tu-ilmenau.de

Abstract: The widespread use of glyphosate as a herbicide in agriculture can lead to the presence of its residues and metabolites in food for human consumption and thus pose a threat to human health. It has been found that glyphosate reduces energy metabolism in the brain, its amount increases in white muscle fibers. At the same time, the effect of chronic use of glyphosate on the dynamic properties of skeletal muscles remains practically unexplored. The selected biomechanical parameters (the integrated power of muscle contraction, the time of reaching the muscle contraction force its maximum value and the reduction of the force response by 50% and 25% of the initial values during stimulation) of muscle soleus contraction in rats, as well as blood biochemical parameters (the levels of creatinine, creatine phosphokinase, lactate, lactate dehydrogenase, thiobarbituric acid reactive substances, hydrogen peroxide, reduced glutathione and catalase) were analyzed after chronic glyphosate intoxication (oral administration at a dose of 10 µg/kg of animal weight) for 30 days. Water-soluble C_{60} fullerene, as a poweful antioxidant, was used as a therapeutic nanoagent throughout the entire period of intoxication with the above herbicide (oral administration at doses of 0.5 or 1 mg/kg). The data obtained show that the introduction of C_{60} fullerene at a dose of 0.5 mg/kg reduces the degree of pathological changes by 40–45%. Increasing the dose of C_{60} fullerene to 1 mg/kg increases the therapeutic effect by 55–65%, normalizing the studied biomechanical and biochemical parameters. Thus, C_{60} fullerenes can be effective nanotherapeutics in the treatment of glyphosate-based herbicide poisoning.

Keywords: glyphosate; C_{60} fullerene; muscle soleus of rat; biomechanical parameters; blood biochemical parameters

1. Introduction

Glyphosate (N-(phosphonomethyl) glycine) is a non-selective herbicide most commonly used for weed control. Among herbicides, it ranks first in the world in production. Many agricultural crops are genetically engineered to tolerate glyphosate. This significantly increases the effectiveness of weed control in these crops. The effect of glyphosate on a plant is due to the fact that it inhibits the components of the enzyme system of the shikimate pathway of biosynthesis of benzoic aromatic compounds [1]. Animals do not have such an enzyme system and therefore this herbicide is considered to be relatively harmless to them. The half-lethal dose (LD_{50}) of glyphosate is >5 kg/kg of rat body weight with a single administration [1].

In recent years, there has been a growing worldwide concern about the possible direct and indirect health effects of the widespread use of glyphosate. In 2015, the World Health Organization reclassified glyphosate as likely carcinogenic to humans [2]. There is considerable controversy regarding its carcinogenicity and toxicity, with very different opinions of the scientists and regulatory bodies involved in the glyphosate study. One of the key aspects of this controversy is the extent of pathological changes in laboratory animals that are caused by glyphosate [3]. The most convincing data indicate that glyphosate causes hemangiosarcomas, tumors and malignant lymphomas, renal and liver adenomas, nervous and miotic disorders of a wide spectrum of severity [3].

Glyphosate also causes numerous morphological, physiological and biochemical disorders in the cells and organs of animals, including mammals. It worsens the condition of the gastrointestinal tract: a violation of its contractile function is observed already at a concentration of 3 mg/L; the violation of motility continues after the removal of glyphosate from the incubation solution [4]. The use of glyphosate as a herbicide in agriculture can lead to the presence of its residues and metabolites (aminomethylphosphonic acid) in food for human consumption and thus pose a threat to human health. The authors [5] found that glyphosate reduces energy metabolism in the brain, its amount increases in white muscle fibers.

At the same time, the effect of chronic use of glyphosate on the dynamic properties of skeletal muscles remains practically unexplored. It was shown that the activity of acetylcholinesterase (AChE) did not change in the muscles and brain of animals exposed to glyphosate during the first 96 h. On the contrary, the expression of this enzyme in muscle tissue changed [6]. The consequence of these pathological processes is disorders in the dynamics of contraction of the muscular system of varying severity. The results [7] show that glyphosate intoxication increases energy expenditure to maintain homeostasis. In particular, there was a decrease in the level of glycogen and triglycerides in all organs and an increase in lipid peroxidation (LPO).

The mechanisms for the toxicity of glyphosate-based drugs are complex. It is difficult to separate the toxicity of glyphosate from the toxicity of the drug as a whole, or to determine the contribution of surfactants to the overall toxicity. As a result, the treatment of poisoning occurs for a long time, symptomatic and ineffective [8].

Studies of the effect of a glyphosate-based herbicide on AChE enzyme activity and oxidative stress at concentrations of 0.5–10.0 mg/L for 96 h followed by an equal recovery period indicate the presence of LPO and AChE inhibition. The results also showed an increased level of thiobarbituric acid reactive substances (TBARS) at all tested herbicide concentrations, which remained elevated even after a recovery period [9]. According to the authors, the triggering mechanisms of the onset of these pathological cascades are associated precisely with the formation of a large number of free radicals. They initiate LPO, cause direct inhibition of mitochondrial enzymes of the respiratory chain and their ATPase activity, inactivation of glyceraldehyde 3-phosphate dehydrogenase and membrane sodium channels.

The ability of C_{60} fullerenes to inactivate free radicals was described back in 1991 [10]. One C_{60} molecule simultaneously captures 34 methyl radicals, effectively inactivates the superoxide anion radical and hydroxyl radicals in vitro system, protecting cell membranes from oxidation [11]. It is assumed that biocompatible and water-soluble C_{60} fullerenes [12] can be considered as powerful scavengers of free radicals during the development of ischemia and fatigue processes in skeletal muscle [13,14]. In our previous works on in vivo models, it was shown that the usage of safe doses of water-soluble C_{60} fullerene at the initiation of various pathologies leads to significant positive therapeutic effects, in particular, during acute liver injury, colorectal cancer, obesity, acute cholangitis and hemiparkinsonism [15–20].

Based on the above data, the purpose of this work was to estimate the therapeutic effect of water-soluble C_{60} fullerene, as a powerful antioxidant, on the development of muscle pathologies in rat skeletal muscle caused by chronic glyphosate intoxication.

2. Results and Discussion

2.1. AFM Analysis

It is known that the size of C_{60} fullerene particles in aqueous solution strongly correlates with their specific biological properties and toxicity. So, the antibacterial activity of C_{60} fullerene is connected with its ability to undergo aggregation [21]; the macrophage apoptosis induced by aqueous C_{60} fullerene aggregates changes the mitochondrial membrane potential [22]; the respiratory toxicity and immunotoxicity of C_{60} fullerenes in mice and rats after nose inhalation strictly depends on their nano- and micro-size [23]; depending on the size C_{60} fullerenes can inhibit BK_{Ca} but not K_V channels in pulmonary artery smooth muscle cells [24], penetrate through plasma membrane inside the cell [25] or be adsorbed on the surface of the membrane [26]. Therefore, the size effect of C_{60} fullerene particles in aqueous solution is considered now to be very important.

The atomic force microscopy (AFM) study of C_{60} fullerene films deposited from an aqueous solution revealed a high degree of molecules dispersion in solution. It turned out that C_{60} fullerene aqueous solution ($C_{60}FAS$) contains both single C_{60} fullerene (see the objects with a height of ~0.7 nm in Figure 1a) and its labile nanoaggregates (objects with a height of 1.4–60 nm in Figure 1b). The majority of C_{60} molecules were located chaotically and separately along the surface, or in the form of bulk clusters. Thus, $C_{60}FAS$ is a polydisperse colloid nanofluid. This result is in a good agreement with our previous probe microscopic data [27,28].

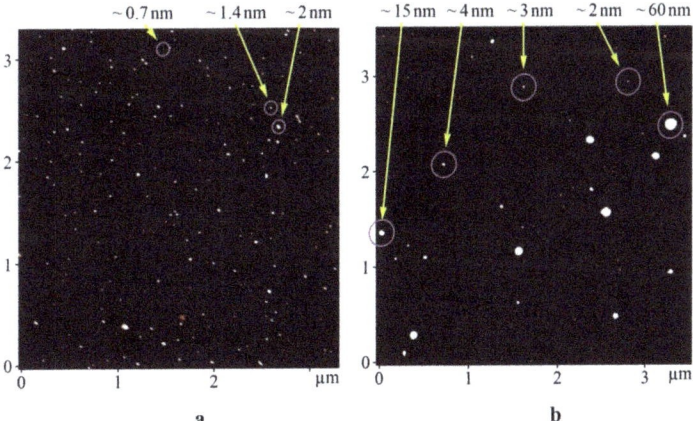

Figure 1. AFM images (tapping mode) of C_{60} fullerene nanoparticles on the mica surface (concentration 0.15 mg/mL) (**a**) objects with a height of ~0.7 nm (**b**) objects with a height of 1.4–60 nm.

In addition, the stability of the used $C_{60}FAS$ was evaluated by the zeta potential measurement. This value was shown to be -30.3 mV at room temperature. Such a high (by absolute value) zeta potential for the $C_{60}FAS$ indicates its high stability (low tendency for nanoparticle aggregation over time) and suitability for further biological research.

2.2. Biomechanical Analysis

In the process of analysis of the force curves obtained during stimulation of muscle soleus by 5 s pools for 1500 s after chronic intoxication of animals with glyphosate for 30 days, serious disorders in muscle dynamics are visible (Figure 2). The integrated power of muscle contraction during the whole period of stimulation decreased to $41 \pm 3\%$ of the control values (Figure 3). A significant reduction in the force response ended in complete muscle rigidity after 1200 s. However, in animals treated with C_{60}FAS, this parameter was 57 ± 2% and 68 ± 4% at doses of C_{60} fullerene 0.5 and 1 mg/kg, respectively. It should be noted that in this case, the muscle responded with a contractile response throughout the stimulation period, not falling below 30% of the limit (Figures 2 and 3). The time of reduction of the force response by 50% and 25% from the initial values increased from 103 ± 11 s and 790 ± 17 s after glyphosate poisoning to 760 ± 8 s and 1213 ± 14 s and 940 ± 21 s and 1820 ± 24 s after therapeutic use of C_{60}FAS at doses of 0.5 and 1 mg/kg, respectively. The maximum and minimum recorded forces of muscle contraction throughout stimulation were 0.81 ± 0.10 N and 0.30 ± 0.05 N after glyphosate poisoning, 1.65 ± 0.20 N and 1.72 ± 0.20 N and 2.67 ± 0.30 N and 2.93 ± 0.30 N after therapeutic use of C_{60}FAS at doses of 0.5 and 1 mg/kg, respectively (Figures 2 and 3).

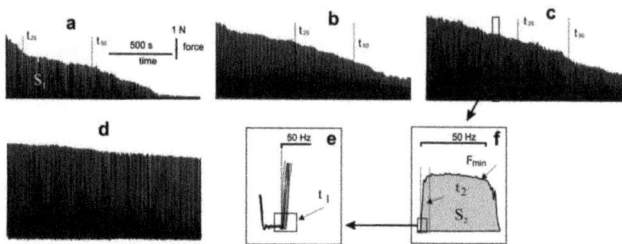

Figure 2. Curves of the generation of the contraction force of muscle soleus rat after chronic intoxication with glyphosate for 30 days: (**a**), (**b**), (**c**) and (**d**)—the curves of muscle contraction for 1500 s with the administration to the animals of glyphosate, glyphosate and C_{60}FAS at doses of 0.5 and 1 mg/kg, respectively, and with the administration to the animals of distilled water (control group); (**e**) mechanograms of single contractions; (**f**) an example of calculating the time of the onset of a muscle response. S_1 is the integrated power of muscle contraction throughout the entire period of stimulation; S_2 is the integrated power in a single contraction; F_{min} is the minimum value of force generation in a single contraction; t_{50} and t_{25} are the time of decreasing the maximum force response to 50% and 25% of the initial amplitude of muscle force; t_1 and t_2 are the time of the onset of the muscle response and the force reaching its maximum value in a single contraction.

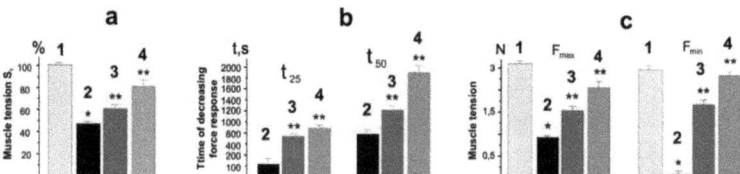

Figure 3. Parameters of contractile activity of muscle soleus rat after chronic intoxication with glyphosate for 30 days: (**a**) integrated power of muscle contraction throughout the entire period of stimulation (S_1), presented as a percentage of control values; (**b**) time of decreasing the force response by 50% (t_{50}) and 25% (t_{25}) from the initial values; (**c**) maximum (F_{max}) and minimum (F_{min}) fixed forces of muscle contraction throughout the entire duration of stimulation. 1—control group (native muscle); 2—the glyphosate group; 3—the glyphosate+C_{60} fullerene (0.5 mg/kg) group; 4—the glyphosate + C_{60} fullerene (1 mg/kg) group; * $p < 0.05$ relative to the control group; ** $p < 0.05$ relative to the glyphosate group.

A decrease in the strength activity of a muscle during glyphosate poisoning can be explained by a violation of energy metabolism. So, in a study [29], the authors found that a high concentration of the herbicide led to a significant decrease in the energy reserve in the muscles, showing an unfavorable sublethal effect on energy metabolism and, consequently, on the dynamic properties of the muscular system in general. The recorded significant positive therapeutic effect of C_{60} fullerene may be associated exclusively with its antioxidant properties, which reduce the degree of damage to cell membranes. To confirm this, we analyzed the biomechanical parameters of single muscle contractions (Figure 4).

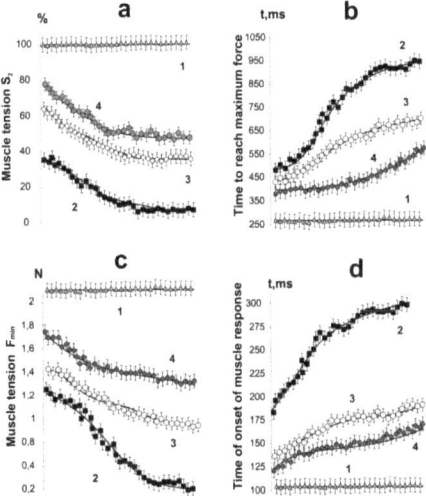

Figure 4. Parameters of single contractions of muscle soleus rat after chronic intoxication with glyphosate for 30 days, caused by 5 s stimulation with a frequency of 50 Hz: (**a**) integrated muscle power (S_2), calculated from the total area of the force curves as a percentage of the control values; (**b**) time to reach the maximum force response; (**c**) minimum (F_{min}) fixed force of muscle contraction; (**d**) time of onset of muscle response to stimulation. 1—control group (native muscle); 2—the glyphosate group; 3—the glyphosate+C_{60} fullerene (0.5 mg/kg) group; 4—the glyphosate+C_{60} fullerene (1 mg/kg) group.

The dynamics of the contractile component is determined by the sensitive mechanisms of interaction of motor neuron pools with actin and myosin myofilaments. The influence of pathological factors on these processes leads either to a complete dysfunction of this process, or to its desynchronization. As a result, the whole muscle, as a dynamic system, is unable to adequately implement the pools of neural activity coming from the central nervous system (CNS). The nature and level of such dysfunctions is directly related to the level of development of pathological processes in both muscle and nervous tissue. The results of studies [30] show that the effect of the glyphosate-based herbicide affects the CNS of rats, possibly altering the neurotransmitter systems that regulate locomotor activity.

The experiment made it possible to trace the therapeutic effect of C_{60}FAS on different regions of the generation of the force response of the rat muscle after chronic intoxication with glyphosate (Figure 4). A change in the time the force reaches its maximum level is one of the most important parameters of the kinetics of skeletal muscle contraction. This component of muscle dynamics is especially important in controlling hand contraction in humans. Pathological processes occurring in the nervous or muscle tissue lead to its increase, which complicates, and in some cases completely blocks the possibility of accurate positioning of the joint with the damaged muscle [31]. After taking glyphosate for 30 days, this parameter increased significantly. It should be noted that this increase was progressive with growing in the number of contractions: from 470 ± 27 ms with the first contraction to 954 ± 33 ms with the last contraction (in control, 250 ± 11 ms). These values changed significantly after the therapeutic use of C_{60}FAS: with a dose of C_{60} fullerene of 0.5 mg/kg, this time was 430 ± 22 ms and 650 ± 29 ms, respectively, and with a dose of 1 mg/kg—367 ± 19 ms and 543 ± 24 ms, respectively (Figure 4). Thus, the protective effect of C_{60}FAS was more than 30% in the first and more than 65% in the second cases.

To understand the features of muscle dynamics during the development of a pathological process, it is important to analyze the rate of processing of stimulation pools emanating from the CNS into the mechanical component of contraction and the possibility of modifying the kinetics of contraction under the influence of pathological changes. A change in the time of the onset of muscle response after nerve stimulation is one of the most important parameters of the kinetics of skeletal muscle contraction. Analysis of the data obtained showed a significant increase in this parameter after glyphosate poisoning from 175 ± 22 ms with the first contraction to 298 ± 27 ms with the last compared with the control—102 ± 8 ms. C_{60}FAS therapy at a dose of C_{60} fullerene 0.5 mg/kg reduced this time to 132 ± 19 ms and 184 ± 17 ms during the first and last muscle contractions, respectively, and with a dose of C_{60} fullerene 1 mg/kg—to 120 ± 20 ms and 165 ± 14 ms, respectively. It should be noted that significant decrease in this indicator under the action of both doses of C_{60} fullerene (Figure 4): the protective effect of C_{60}FAS was more than 65% in the first case and more than 75% in the second case.

A change in the level of minimum force of muscle contraction generation is an indicator of significant changes caused by pathological processes in the myocyte. This indicator is not associated with neuropathic damage and its analysis gives an idea of violations of the force generation system within the muscle fiber. When performing fairly simple single-joint movements, this marker is the main indicator of muscle dysfunction, the phenomenological analysis of which makes it possible to establish the presence of causal relationships between the levels of decrease in the biomechanical activity of muscles and the development of the pathological process [32]. With a constant level of the minimum force of more than 2 N in the control, its drop with the use of glyphosate ranged from 1.3 ± 0.1 N to zero. C_{60}FAS therapy increased the level of the minimum force of muscle contraction to $(1.4–0.9) \pm 0.1$ N at a dose of C_{60} fullerene 0.5 mg/kg and up to $(1.7–1.4) \pm 0.1$ N at a dose of C_{60} fullerene 1 mg/kg, respectively. In this case, the protective effect of C_{60}FAS was more than 50% in the first case and more than 75% in the second case.

All these changes ultimately lead to a change in the overall strength activity of the muscle, which can be quantified by the value of the integrated power. A change in this parameter can be associated with disorder in both neural component and muscular component of the studied pathology [33]. With chronic use of glyphosate, the integrated power decreased from $41 \pm 3\%$ to zero. C_{60}FAS therapy brought this level to $63 \pm 3\%$—$47 \pm 5\%$ at a dose of C_{60} fullerene 0.5 mg/kg and $80 \pm 6\%$—$54 \pm 2\%$ at a dose of C_{60} fullerene of 1 mg/kg, respectively. The protective effect of C_{60}FAS was more than 50% in the first case and more than 60% in the second case.

2.3. Biochemical Analysis

Analysis of biochemical markers of rat blood, in particular creatinine, creatine phosphokinase (CPK), lactate (LA) and lactate dehydrogenase (LDH), makes it possible to assess the physiological changes occurring in skeletal muscle and the effect of a therapeutic drug on pathological processes in it. Studies have shown that the levels of the selected markers have a pronounced tendency to increase in the blood of rats intoxicated with glyphosate and decrease during C_{60}FAS therapy (Figure 5).

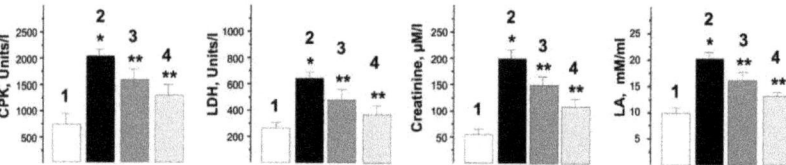

Figure 5. Biochemical parameters of rat blood (CPK, LDH, creatinine and LA) after chronic glyphosate intoxication for 30 days. 1—control group (native muscle); 2—the glyphosate group; 3—the glyphosate+C_{60} fullerene (0.5 mg/kg) group; 4—the glyphosate+C_{60} fullerene (1 mg/kg) group; * $p < 0.05$ relative to the control group; ** $p < 0.05$ relative to the glyphosate group.

The change in the concentration of CPK, an enzyme from the energy supply system of musculoskeletal cells, from 756 ± 26 U/L in the norm to 1950 ± 33 U/L after glyphosate intoxication, in our opinion, may be the result of destruction of myocyte walls caused by the influence of the herbicide, with partial release of intramyocytic enzymes into the extracellular space. With the use of C_{60}FAS, the CPK level decreased by $23.2 \pm 3\%$ and $31.7 \pm 2\%$ at doses of C_{60} fullerene 0.5 and 1 mg/kg, respectively (Figure 5).

Analysis of changes in the level of LDH made it possible to assess the overall health of the injured muscle. The increase in the LDH level after administration of glyphosate increased from 254 ± 13 U/L (normal) to 659 ± 26 U/L and is evidence of the development of significant dysfunctions of the neuromuscular drug and, as a consequence, the development of fatigue processes. After therapeutic use of C_{60}FAS, the LDH level decreased by $27 \pm 3\%$ and $31 \pm 2\%$ at doses of C_{60} fullerene 0.5 and 1 mg/kg, respectively.

The change in creatinine level from 50 ± 2 μM/L in the control to 196 ± 4 μM/L with chronic intake of glyphosate confirms previously obtained data that increased serum creatinine level is an important factor for predicting the severity of glyphosate poisoning [34]. C_{60}FAS therapy led to a significant decrease in its levels to 157 ± 3 μM/L and 112 ± 4 μM/L at doses of C_{60} fullerene 0.5 and 1 mg/kg, respectively. In our opinion, the decrease in the creatinine fraction in this case is caused by the antioxidant properties of C_{60} fullerene, its ability to reduce inflammatory reactions and protect the membranes of skeletal muscle cells from nonspecific free radical destruction by efficient absorption of free radicals [35].

Contraction of skeletal muscles leads to the accumulation of LA and H^+ ions and, accordingly, to acidification of the intra- and extracellular media, which reduces the production of ATP and suppresses the activity of Na^+, K^+-ATPase. This leads to a delay in the generation of action potentials and reduces muscle activity. Pathological processes in the myocyte increase this imbalance towards acidification of the medium and, thus, the LA level is an important marker for assessing the degree of muscle activity. Analysis of the LA level showed its increase from 10 ± 1 mM/mL (normal) to 19 ± 2 mM/mL after using glyphosate. The use of C_{60}FAS therapy reduced its level to 16 ± 2 mM/mL and 14 ± 1 mM/mL at doses of C_{60} fullerene 0.5 and 1 mg/kg, respectively.

Glyphosate is an endocrine disruptor in chronic ingestion, exhibiting high cytotoxicity. The previously obtained results [36] show that it affects survival due to deregulation of the cell cycle and metabolic changes that can alter mitochondrial oxygen consumption, increase free radical levels, damage DNA, cause hypoxia, accumulation of mutations and, ultimately, cell death. It was also shown that after exposure to the herbicide for 8 days at a concentration of 0.95 mg/L, there was an increase in the amount of TBARS in muscle and brain tissues. An increase in reduced glutathione (GSH) level also indicated a compensatory response of the body against toxic conditions. Oxidative stress that arose during the period of exposure to the herbicide was probably caused by increased LPO [30]. Thus, a change in the level of endogenous antioxidants is an important marker that determines the degree of physiological disorders in muscle cells during glyphosate intoxication.

Figure 6 shows the results of measurements of indicators of pro- and antioxidant balance in the blood of experimental rats. The data obtained indicate increased levels of peroxidation and oxidative stress as well as endogenous antioxidants with the use of the herbicide. The increase in these biochemical markers compared to control values was $218 \pm 19\%$, $251 \pm 14\%$, $280 \pm 19\%$ and $250 \pm 24\%$ for TBARS, hydrogen peroxide (H_2O_2), GSH and catalase (CAT) activity, respectively.

The level of these markers decreased significantly after therapeutic use of C_{60}FAS. So, the TBARS level decreased to $170 \pm 11\%$ and $120 \pm 8\%$ of the control values, H_2O_2—$160 \pm 14\%$ and $114 \pm 11\%$, GSH—$150 \pm 12\%$ and $128 \pm 9\%$, CAT activity—$140 \pm 14\%$ and $119 \pm 10\%$ at doses of C_{60} fullerene 0.5 and 1 mg/kg, respectively.

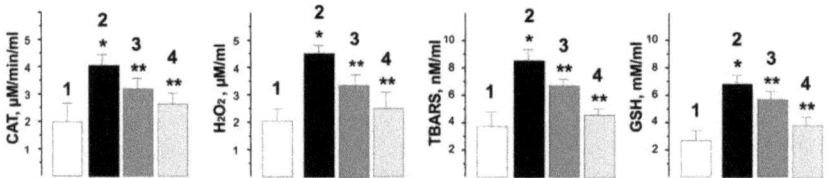

Figure 6. Indicators of pro- and antioxidant balance (CAT, H_2O_2, TBARS and GSH) in the blood of rats after chronic intoxication with glyphosate for 30 days. 1—control group (native muscle); 2—the glyphosate group; 3—the glyphosate+C_{60} fullerene (0.5 mg/kg) group; 4—the glyphosate+C_{60} fullerene (1 mg/kg) group; * $p < 0.05$ relative to the control group; ** $p < 0.05$ relative to the glyphosate group.

Summarizing, the proposed therapy with the use of low doses of water-soluble C_{60} fullerenes, possessing membranotropic [25,37] and powerful antioxidant properties [38], leads to positive biomechanical and biochemical changes in the character of contractile processes in the skeletal muscles of rats with chronic glyphosate intoxication.

3. Materials and Methods

To obtain C_{60}FAS (maximum concentration 0.15 mg/mL), a method based on the transfer of these carbon molecules from toluene to water followed by sonication was used [27,39]. The prepared C_{60}FAS was stored at a temperature of +4 °C for 12 months.

The AFM (Solver Pro M system, NT-MDT, Moscow, Russia) was performed to determine the size of C_{60} fullerene particles in the prepared aqueous solution. A drop of investigated solution was transferred on the atomic-smooth substrate to deposit layers. Measurements were carried out after complete evaporation of the solvent. For AFM study, a freshly broken surface of mica (SPI supplies, V-1 grade) was used as a substrate. Measurements were carried out in a semicontact (tapping) mode with AFM probes of the RTPESPA150 (Bruker, 6 N/m, 150 kHz) type.

The zeta potential was measured to assess the stability of the prepared C_{60}FAS using the Zetasizer Nano-ZS90 technique (Malvern, Worcestershire, UK).

The experiments were performed on male Wistar rats aged 3 months weighing 170 ± 5 g. The study protocol was approved by the bioethics committee of Taras Shevchenko National University of Kyiv in accordance with the rules of the European Convention for the Protection of Vertebrate Animals Used for Experimental and Other Scientific Purposes and the norms of biomedical ethics in accordance with the Law Of Ukraine №3446—IV 21.02.2006, Kyiv, on the Protection of Animals from Cruelty during medical and biological research.

In total, 40 rats divided into four groups (10 animals each) were used in the study. Glyphosate was administered daily at a dose of 10 μg/kg of animal weight orally using a metal catheter for 30 days ($n = 10$). The animals of the control group ($n = 10$) were injected with an equivalent volume of distilled water for 30 days. C_{60}FAS was administered at doses of 0.5 ($n = 10$) and 1 mg/kg of animal weight ($n = 10$) immediately after administration of the herbicide for 30 days. Measurements of the studied parameters (see below) in all groups were performed on the 31st day after the start of the experiment.

It should be noted that the use of selected doses of C_{60}FAS are based on previous experimentally established data, which showed a high protective effect of water-soluble C_{60} fullerenes [13,14,19]. Additionally, it should be noted that the doses of C_{60} fullerene used in our experiments are significantly lower than the LD_{50} value, which was 600 mg/kg body weight when administered orally to rats [40] and 721 mg/kg when administered intraperitoneally to mice [25].

Anesthesia of animals was performed by intraperitoneal administration of nembutal (40 mg/kg). Preparation of the experiment included the cannulation (*a. carotis communis sinistra*) for the therapeutic administration of the drug and pressure measurement, tracheotomy and laminectomy at lumbar spinal cord level. Muscle soleus of rat was released from the surrounding tissues. Its tendon was cut across in distal part, which was connected to the force sensors. For modulated stimulation of efferents, the ventral roots were cut at the points of their exit from the spinal cord. Stimulation of efferents was performed by electrical pulses lasting 2 ms, generated by the generator, through platinum electrodes. The control of the external load on the muscle was performed using a system of mechanical stimulators. Perturbation of the load was carried out by a linear electromagnetic motor [41].

The choice of muscle soleus for this study is due to the fact that this muscle contains the maximum number of slow fibers, which is important for accurate and high-quality fixation of fast-acting processes, occurring in the anterior front of the tetanus, in pathology.

To induce muscle contraction, a stimulation signal with a frequency of 50 Hz and a duration of 5 s was used without a relaxation period. The total duration of stimulation was 1500 s. The current strength, at which the muscle began to contract, was considered a threshold, and further stimulation was performed with a current strength of 1.3–1.4 thresholds.

To record the force of skeletal muscle contraction, we used the original strain gauge that consists of force and length sensors, a synchronous pulse generator and a thermal control system [13].

In the process of analyzing the obtained results, the following parameter was used: the integrated power of muscle contraction (calculated area under the force curve), which is an indicator of the overall performance of the muscle with the applied stimulation pools. The development of muscle contractile activity was assessed by calculating the time of the decrease in the force response by 50% and 25% of the initial values during stimulation. We also analyzed the time to reach the maximum value of the muscle contraction force and the delay in the onset of the muscle response.

The level of enzymes content in the blood of experimental animals (creatinine, CPK, LA, LDH, TBARS, H_2O_2, GSH and CAT), as marker of muscle injury [42], was determined using clinical diagnostic equipment—a haemoanalyzer [13].

Statistical processing of results was performed by methods of variation statistics using software Original 9.4. We conducted at least six repetitions for each measurement. Data are expressed as the means ± SEM for each group. The differences among experimental groups were detected by one-way ANOVA followed by Bonferroni's multiple comparison test. Values of $p < 0.05$ were considered significant.

4. Conclusions

The obtained results indicate that the therapeutic administration of water-soluble C_{60} fullerenes at a dose of 0.5 mg/kg reduces the degree of pathological changes in rats caused by chronic glyphosate intoxication by 40–45%. Increasing the dose of water-soluble C_{60} fullerenes to 1 mg/kg increases the therapeutic effect by 55–65%, normalizing the studied biomechanical and biochemical parameters. Considering the fact that poisoning with glyphosate compounds has a lethality of up to 20% and there is currently no antidote to them, and the basis for the treatment of systemic toxicity is deactivation and aggressive supportive therapy [34], the proposed C_{60} fullerene therapy of this type of intoxication opens up new prospects for clinical trials.

Author Contributions: Biomechanical analysis, D.N., O.A. and V.S.; biochemical analysis, S.P., O.V. and K.B.; sample preparation, U.R. and P.S.; characterization of the samples, S.K. and Y.P.; coordination the research work, analysis of the data and preparing of the manuscript, Y.P. All authors have read and agreed to the published version of the manuscript.

Funding: This research received no external funding.

Institutional Review Board Statement: The study protocol was approved by the bioethics committee of Taras Shevchenko National University of Kyiv in accordance with the rules of the European Convention for the Protection of Vertebrate Animals Used for Experimental and Other Scientific Purposes and the norms of biomedical ethics in accordance with the Law Of Ukraine №3446—IV 21.02.2006, Kyiv, on the Protection of Animals from Cruelty during medical and biological research.

Informed Consent Statement: Not applicable.

Data Availability Statement: Not applicable.

Acknowledgments: This research was supported by the Ministry of Education and Science of Ukraine. We acknowledge support for the publication costs by the Open Access Publication Fund of the Technische Universität Ilmenau.

Conflicts of Interest: Authors declare that they have no conflict of interest.

References

1. Benbrook, C.M. Trends in glyphosate herbicide use in the United States and globally. *Environ. Sci. Eur.* **2016**, *28*, 1–15. [CrossRef] [PubMed]
2. Van Bruggen, A.H.C.; He, M.M.; Shin, K.; Mai, V.; Jeong, K.C.; Finckh, M.R.; Morris, J.G., Jr. Environmental and health effects of the herbicide glyphosate. *Sci. Total Environ.* **2018**, *616–617*, 255–268. [CrossRef] [PubMed]
3. Portier, C.J. A comprehensive analysis of the animal carcinogenicity data for glyphosate from chronic exposure rodent carcinogenicity studies. *Environ. Health* **2020**, *19*, 18. [CrossRef] [PubMed]
4. Chłopecka, M.; Mendel, M.; Dzieka, N.; Wojciech, K. Glyphosate affects the spontaneous motoric activity of intestine at very low doses—In vitro study. *Pestic. Biochem. Physiol.* **2014**, *113*, 25–30. [CrossRef] [PubMed]
5. Zhidenko, A.A.; Bibchuk, E.V.; Barbukho, E.V. Effect of glyphosate on the energy exchange in carp organs. *Ukr. Kyi Biokhimichnyi Zhurnal* **2013**, *85*, 22–30. [CrossRef] [PubMed]
6. Lopes, F.M.; Caldas, S.S.; Primel, E.G.; da Rosa, C.E. Glyphosate adversely affects Danio rerio males: Acetylcholinesterase modulation and oxidative stress. *Zebrafish* **2017**, *14*, 97–105. [CrossRef] [PubMed]
7. Dornelles, M.F.; Oliveira, G.T. Toxicity of atrazine, glyphosate, and quinclorac in bullfrog tadpoles exposed to concentrations below legal limits. *Environ. Sci. Pollut. Res. Int.* **2016**, *23*, 1610–1620. [CrossRef] [PubMed]
8. Bradberry, S.M.; Proudfoot, A.T.; Vale, J.A. Glyphosate poisoning. *Toxicol. Rev.* **2004**, *23*, 159–167. [CrossRef]
9. Cattaneo, R.; Clasen, B.; Loro, V.L.; de Menezes, C.C.; Pretto, A.; Baldisserotto, B.; Santi, A.; de Avila, L.A. Toxicological responses of Cyprinus carpio exposed to a commercial formulation containing glyphosate. *Bull. Environ. Contam. Toxicol.* **2011**, *87*, 597–602. [CrossRef]
10. Krusic, P.J.; Wasserman, E.; Keizer, P.N.; Morton, J.R.; Preston, K.F. Radical reactions of C_{60}. *Science* **1991**, *254*, 1183–1185. [CrossRef]
11. Ferreira, C.A.; Ni, D.; Rosenkrans, Z.T.; Cai, W. Scavenging of reactive oxygen and nitrogen species with nanomaterials. *Nano Res.* **2018**, *11*, 4955–4984. [CrossRef]
12. Halenova, T.; Raksha, N.; Savchuk, O.; Ostapchenko, L.; Prylutskyy, Y.; Ritter, U.; Scharff, P. Evaluation of the biocompatibility of water-soluble pristine C_{60} fullerenes in rabbit. *BioNanoScience* **2020**, *10*, 721–730. [CrossRef]
13. Nozdrenko, D.M.; Zavodovsky, D.O.; Matvienko, T.Y.; Zay, S.Y.; Bogutska, K.I.; Prylutskyy, Y.I.; Ritter, U.; Scharff, P. C_{60} fullerene as promising therapeutic agent for the prevention and correction of functioning skeletal muscle at ischemic injury. *Nanoscale Res. Lett.* **2017**, *12*, 1–9. [CrossRef]
14. Vereshchaka, I.V.; Bulgakova, N.V.; Maznychenko, A.V.; Gonchar, O.O.; Prylutskyy, Yu.I.; Ritter, U.; Moska, W.; Tomiak, T.; Nozdrenko, D.M.; Mishchenko, I.V.; et al. C_{60} fullerenes diminish the muscle fatigue in rats comparable to N-acetylcysteine or β-alanine. *Front. Physiol.* **2018**, *9*, 517. [CrossRef]
15. Halenova, T.I.; Vareniuk, I.M.; Roslova, N.M.; Dzerzhynsky, M.E.; Savchuk, O.M.; Ostapchenko, L.I.; Prylutskyy, Yu.I.; Ritter, U.; Scharff, P. Hepatoprotective effect of orally applied water-soluble pristine C_{60} fullerene against CCl4-induced acute liver injury in rats. *RSC Adv.* **2016**, *6*, 100046–100055. [CrossRef]
16. Lynchak, O.V.; Prylutskyy, Yu.I.; Rybalchenko, V.K.; Kyzyma, O.A.; Soloviov, D.; Kostjukov, V.V.; Evstigneev, M.P.; Ritter, U.; Scharff, P. Comparative analysis of the antineoplastic activity of C_{60} fullerene with 5-fluorouracil and pyrrole derivative in vivo. *Nanoscale Res. Lett.* **2017**, *12*, 1–6. [CrossRef]
17. Halenova, T.I.; Raksha, N.G.; Vovk, T.B.; Savchuk, O.M.; Ostapchenko, L.I.; Prylutskyy, Yu.I.; Kyzyma, O.A.; Ritter, U.; Scharff, P. Effect of C_{60} fullerene nanoparticles on the diet-induced obesity in rats. *Int. J. Obes.* **2018**, *42*, 1987–1998. [CrossRef]
18. Kuznietsova, H.M.; Lynchak, O.V.; Dziubenko, N.V.; Osetskyi, V.L.; Ogloblya, O.V.; Prylutskyy, Yu.I.; Rybalchenko, V.K.; Ritter, U.; Scharff, P. Water-soluble C_{60} fullerenes reduce manifestations of acute cholangitis in rats. *Appl. Nanosci.* **2019**, *9*, 601–608. [CrossRef]
19. Maznychenko, A.V.; Mankivska, O.P.; Sokolowska, I.V.; Kopyak, B.S.; Tomiak, T.; Bulgakova, N.V.; Gonchar, O.O.; Prylutskyy, Y.I.; Ritter, U.; Mishchenko, I.V.; et al. C_{60} fullerenes increase the intensity of rotational movements in non-anesthetized hemiparkinsonic rats. *Acta Neurobiol. Exp.* **2020**, *80*, 32–37.
20. Kuznietsova, H.; Dziubenko, N.; Herheliuk, T.; Prylutskyy, Yu.; Tauscher, E.; Ritter, U.; Scharff, P. Water-soluble pristine C_{60} fullerene inhibits liver alterations associated with hepatocellular carcinoma in rat. *Pharmaceutics* **2020**, *12*, 794. [CrossRef]
21. Lyon, D.Y.; Adams, L.K.; Falkner, J.C.; Alvarezt, P.J. Antibacterial activity of fullerene water suspensions: Effects of preparation method and particle size. *Environ. Sci. Technol.* **2006**, *40*, 4360–4366. [CrossRef]

22. Zhang, B.; Bian, W.; Pal, A.; He, Y. Macrophage apoptosis induced by aqueous C_{60} aggregates changing the mitochondrial membrane potential. *Environ. Toxicol. Pharmacol.* **2015**, *39*, 237–246. [CrossRef]
23. Sayers, B.C.; Germolec, D.R.; Walker, N.J.; Shipkowski, K.A.; Stout, M.D.; Cesta, M.F.; Roycroft, J.H.; White, K.L.; Baker, G.L.; Dill, J.A.; et al. Respiratory toxicity and immunotoxicity evaluations of microparticle and nanoparticle C_{60} fullerene aggregates in mice and rats following nose-only inhalation for 13 weeks. *Nanotoxicology* **2016**, *10*, 1458–1468. [CrossRef]
24. Melnyk, M.I.; Ivanova, I.V.; Dryn, D.O.; Prylutskyy, Yu.I.; Hurmach, V.V.; Platonov, M.; Al Kury, L.T.; Ritter, U.; Soloviev, A.I.; Zholos, A.V. C_{60} fullerenes selectively inhibit BKCa but not Kv channels in pulmonary artery smooth muscle cells. *Nanomed. Nanotechnol. Biol. Med.* **2019**, *19*, 1–11. [CrossRef] [PubMed]
25. Prylutska, S.V.; Grebinyk, A.G.; Lynchak, O.V.; Byelinska, I.V.; Cherepanov, V.V.; Tauscher, E.; Matyshevska, O.P.; Prylutskyy, Yu.I.; Rybalchenko, V.K.; Ritter, U.; et al. In vitro and in vivo toxicity of pristine C_{60} fullerene aqueous colloid solution. *Fuller. Nanotub. Carbon Nanostructures* **2019**, *27*, 715–728. [CrossRef]
26. Schuetze, C.; Ritter, U.; Scharff, P.; Bychko, A.; Prylutska, S.; Rybalchenko, V.; Prylutskyy, Yu. Interaction of N-fluorescein-5-isothiocyanate pyrrolidine-C_{60} compound with a model bimolecular lipid membrane. *Mater. Sci. Eng. C* **2011**, *31*, 1148–1150. [CrossRef]
27. Ritter, U.; Prylutskyy, Yu.I.; Evstigneev, M.P.; Davidenko, N.A.; Cherepanov, V.V.; Senenko, A.I.; Marchenko, O.A.; Naumovets, A.G. Structural features of highly stable reproducible C_{60} fullerene aqueous colloid solution probed by various techniques. *Fuller. Nanotub. Carbon Nanostructures* **2015**, *23*, 530–534. [CrossRef]
28. Skamrova, G.B.; Laponogov, I.; Buchelnikov, A.S.; Shckorbatov, Y.G.; Prylutska, S.V.; Ritter, U.; Prylutskyy, Y.I.; Evstigneev, M.P. Interceptor effect of C_{60} fullerene on the in vitro action of aromatic drug molecules. *Eur. Biophys. J.* **2014**, *43*, 265–276. [CrossRef]
29. Menéndez-Helman, R.J.; Miranda, A.L.; Dos Santos Afonso, M.; Salibián, A. Subcellular energy balance of Odontesthes bonariensis exposed to a glyphosate-based herbicide. *Ecotoxicol. Environ. Saf.* **2015**, *114*, 157–163. [CrossRef]
30. Gallegos, C.E.; Bartos, M.; Bras, C.; Gumilar, F.; Antonelli, M.C.; Minetti, A. Exposure to a glyphosate-based herbicide during pregnancy and lactation induces neurobehavioral alterations in rat offspring. *Neurotoxicology* **2016**, *53*, 20–28. [CrossRef]
31. Nozdrenko, D.M.; Miroshnychenko, M.S.; Soroca, V.M.; Korchinska, L.V.; Zavodovskiy, D.O. The effect of chlorpyrifos upon ATPase activity of sarcoplasmic reticulum and biomechanics of skeletal muscle contraction. *Ukr. Biochem. J.* **2016**, *88*, 82–88. [CrossRef]
32. Nozdrenko, D.M.; Abramchuk, O.M.; Soroca, V.M.; Miroshnichenko, N.S. Aluminum chloride effect on Ca^{2+},Mg^{2+}-ATPase activity and dynamic parameters of skeletal muscle contraction. *Ukr. Biochem. J.* **2015**, *87*, 38–45. [CrossRef]
33. Nozdrenko, D.M.; Korchinska, L.V.; Soroca, V.M. Activity of $Ca(2+),Mg(2+)$-ATPase of sarcoplasmic reticulum and contraction strength of the frog skeletal muscles under the effect of organophosphorus insecticides. *Ukr. Biochem. J.* **2015**, *87*, 63–69. [CrossRef]
34. Thakur, D.S.; Khot, R.; Joshi, P.P.; Pandharipande, M.; Nagpure, K. Glyphosate poisoning with acute pulmonary edema. *Toxicol. Int.* **2014**, *21*, 328–330. [CrossRef]
35. Gonchar, O.O.; Maznychenko, A.V.; Bulgakova, N.V.; Vereshchaka, I.V.; Tomiak, T.; Ritter, U.; Prylutskyy, Yu.I.; Mankovska, I.M.; Kostyukov, A.I. C_{60} fullerene prevents restraint stress-induced oxidative disorders in rat tissues: Possible involvement of the Nrf2/ARE-antioxidant pathway. *Oxid. Med. Cell. Longev.* **2018**, *2018*, 2518676. [CrossRef]
36. Stur, E.; Aristizabal-Pachon, A.F.; Peronni, K.C.; Agostini, L.P.; Waigel, S.; Chariker, J.; Miller, D.M.; Thomas, S.D.; Rezzoug, F.; DeTogni, R.S.; et al. Glyphosate-based herbicides at low doses affect canonical pathways in estrogen positive and negative breast cancer cell lines. *PLoS ONE* **2019**, *14*, e0219610. [CrossRef]
37. Russ, K.A.; Elvati, P.; Parsonage, T.L.; Dews, A.; Jarvis, J.A.; Ray, M.; Schneider, B.; Smith, P.J.S.; Williamson, P.T.F.; Violi, A.; et al. C_{60} fullerene localization and membrane interactions in RAW 264.7 immortalized mouse macrophages. *Nanoscale* **2016**, *8*, 4134–4144. [CrossRef]
38. Eswaran, S.V. Water soluble nanocarbon materials: A panacea for all? *Curr. Sci.* **2018**, *114*, 1846–1850. [CrossRef]
39. Scharff, P.; Carta-Abelmann, L.; Siegmund, C.; Matyshevska, O.P.; Prylutska, S.V.; Koval, T.V.; Golub, A.A.; Yashchuk, V.M.; Kushnir, K.M.; Prylutskyy, Yu.I. Effect of X-ray and UV irradiation of the C_{60} fullerene aqueous solution on biological samples. *Carbon* **2004**, *42*, 1199–1201. [CrossRef]
40. Gharbi, N.; Pressac, M.; Hadchouel, M.; Szwarc, H.; Wilson, S.R.; Moussa, F. Fullerene is a powerful antioxidant in vivo with no acute or subacute toxicity. *Nano Lett.* **2005**, *5*, 2578–2585. [CrossRef]
41. Nozdrenko, D.N.; Berehovyi, S.M.; Nikitina, N.S.; Stepanova, L.I.; Beregova, T.V.; Ostapchenko, L.I. The influence of complex drug cocarnit on the nerve conduction velocity in nerve tibialis of rats with diabetic polyneuropathy. *Biomed. Res.* **2018**, *29*, 3629.
42. Brancaccio, P.; Lippi, G.; Maffulli, N. Biochemical markers of muscular damage. *Clin. Chem. Lab. Med.* **2010**, *48*, 757–767. [CrossRef]

MDPI
St. Alban-Anlage 66
4052 Basel
Switzerland
Tel. +41 61 683 77 34
Fax +41 61 302 89 18
www.mdpi.com

International Journal of Molecular Sciences Editorial Office
E-mail: ijms@mdpi.com
www.mdpi.com/journal/ijms